T0317490

Functional Safety of Machinery

To Laura, Luca and Francesco

Live as if you were to die tomorrow. Learn as if you were to live forever
(Mahatma Gandhi)

Functional Safety of Machinery

How to Apply ISO 13849-1 and IEC 62061

Marco Tacchini
Technical Director at GT Engineering
Poncarale, Brescia
Italy

Published by John Wiley & Sons, Inc., Hoboken, New Jersey.
Published simultaneously in Canada.

Library of Congress Cataloging-in-Publication Data applied for:
Hardback ISBN: 9781119789048

Cover Design: Wiley
Cover Image: © Marco Tacchini

Set in 9.5/12.5pt STIXTwoText by Straive, Pondicherry, India

Contents

Preface

This book is about Risk Reduction in Machinery, achieved through **Functional Safety**.

Three International standards deal with the subject, **ISO 13849-1, IEC 62061,** and ISO 13849-2: The first two were subject to important revisions. The **second edition** of IEC 62061 was published in 2021, while the **fourth edition** of ISO 13849-1 was published in 2023. Therefore, it is a good opportunity to understand their latest approach.

Most machineries have risks associated with their use. In order to achieve an acceptable level of safety, risk reduction measures have to be adopted. In a forging press, glycol water can be used instead of oil, to eliminate the risk of fire. Fixed guards that protect from dangerous movement inside the machine can be used to protect operators. Those measures have nothing to do with **Functional safety** since no control system is involved in the risk reduction achieved by the fixed guards.

In other cases, there may be the need to install an interlocking device that detects a dangerous situation: for example, a person enters a dangerous area of the machine. In that case, when the guard is opened, all dangerous movements have to be stopped. In other situations, a high-pressure switch is needed to detect a dangerous status: in that case, a valve has to be closed, for example, to bring the process **to a safe state.** All what is just described belongs to the domain of Functional Safety; in other words, **a Safety function has just been defined and a Safety Control System** has to be designed to perform that specific safety function.

Safety is defined as the freedom from unacceptable risk of physical injury or of damage to people's health, either directly or indirectly, as a result of damage to property or to the environment. **Functional safety** is part of the overall risk reduction process that relies upon a system or equipment operating correctly in response to its inputs. For example, the activation of a level switch in a tank, which causes a valve to close and prevents a flammable liquid from entering the tank, can be an example of risk reduction through Functional Safety.

Both above-mentioned standards deal with Functional safety requirements intended to reduce the risk of injury or damage to the health of persons in the immediate vicinity of the machine and those directly involved in the use of the machine.

Every time we rely upon a control system to reduce a risk, the following question arises: What is the probability that the safety control system will fail and therefore the risk reduction measure does not work as expected? In other words, what is the likelihood a compressor does not stop in case of high pressure in a tank, or a dangerous movement does not stop in a Robot cell when a guard door is opened? This, once more, is the domain of Functional safety.

In the past fifteen years, Engineers were confronted with two standards: ISO 13849-1 and IEC 62061. Efforts were put in place to combine the two documents in one. With the publication of the new editions of both standards, they are now aligned as ever before.

This book is an effort to show that, despite two standards still exist, **there is a quite similar approach to Functional Safety of Machinery**. Some differences still remain, and they will be detailed. The book explains the mathematical basis of the two standards and how to put them into practice.

This book is written for machinery manufacturers and it expresses the ideas of the author only. This book tries to explain what the international standards prescribe, in a plain and simple language that gives priority to clearness and simplicity than to a correct formalism. These International standards remain the only reference from this point of view. Please address comments or suggestions to marco.tacchini@gt-engineering.it.

About the Structure of this Book

Despite the existence of three standards, the book looks at Functional safety of Machinery as **one approach.** This was possible thanks to the two new editions of IEC 62061 [12] and ISO 13849-1 [13]. Please consider that parts 1 and 2 of ISO 13849 belong together, although only part 1 is harmonized. Therefore, in the book, we will refer to ISO 13849-1 and 2 as one standard.

In **Chapter 1**, the basics of Reliability Engineering are summarised. The concept of **Random** and **Systematic failure** is detailed, because Functional safety is not only about Reliability Data and Formulas (the random part of failures) but also about a correct design, engineering, production, and maintenance of a Safety System. If the electrical control panel is not correctly designed (systematic failure), all the assumptions and calculations about random failures become meaningless. The $R(t)$, **Reliability function** is the starting point; $F(t) = 1 - R(t)$ is the **unreliability function.** PFH_D and PFD_{avg}, the key parameters, are based upon the unreliability function $F(t)$. The key concepts needed for a correct understanding of those two parameters are presented: the Failure Rates, the Mean Time to Failure (MTTF), the importance of a constant failure rate, the Weibull distribution, and the Markov modeling. Eventually, **safety systems working in high and low-demand mode of operation** are presented. **If you are not so interested in the mathematical background of functional safety, you can start directly with the next chapter**.

Chapter 2 contains a brief history of Functional Safety. EN 954-1 was the first European machinery safety standard dealing with Safety-related Control Systems. Soon later, IEC 61508 was published as the Functional Safety reference Standard. IEC 61511-1, for the Process Industry, came few years later, and IEC 62061 for the Machinery Sector was published in 2005. High and low-demand mode of operation concepts are introduced. What is a Safety Control System is explained as well; where it begins and it ends, since there is still confusion, for example, if a pneumatic cylinder is part of a Safety Control System or not.

Finally, **in Chapter 3**, the parameters used in both IEC 62061 and ISO 13849-1 are described. IEC 62061 introduced, in the new edition, the important Failure Rate λ_{NE}, or No Effect Failure, that does not contribute to the SFF of the subsystem. DC, SFF, $MTTF_D$, and CCF are explained in detail. The importance of HFT (**Hardware fault Tolerance**) is highlighted and, linked to that, the Route 1_H and the Proof Test concepts are presented.

In Chapter 4: both standards are introduced and it is clarified that **they should be used after a risk assessment** on the machine is done. The standards play a role when the risk reduction has to be done by a control system. With the two new editions, the importance of **subsystems** is now valid for both standards: all the considerations on Categories and Architectures have to be done at subsystem level and not for the whole Safety Function. The misunderstanding was mainly in the previous editions of ISO 13849-1, due to the heritage of EN 954-1. The chapter continues with the introduction and explanation of the PFH and SIL acronyms, and their relationship. Finally, the concept of Required SIL (sometimes indicated as SIL_r in the book) and PL_r is introduced and the methodology how to determine them, recommended by each of the two new international standards, is described. Examples on how to avoid Systematic Failures are presented with the concepts of Basic and Well-tried Safety Principles, used in both Standards. We also clarify that the Reset is often a Safety Function. Finally, the book details some technical aspects (like the Direct Opening Action), not detailed in the two new standards, but that are important to understand them correctly.

In Chapter 5, how to design and evaluate a safety function is presented in detail. The concept of Subsystems and Architectural Constraints is put into practice. More and more Safety Functions have software inside; therefore, it is important to understand the difference between **Limited and Full Variability Languages** and what the machine manufacturer has to do when using the former or the latter. Finally, we shed some light on **how to treat low-demand mode of operation in machinery.**

In **Chapter 6**, the five Categories of ISO 13849-1 are presented, with several insights into Category 2, probably the most difficult one to understand and use. Similarly, **Chapter 7** deals with the four Architectures of IEC 62061. **Chapter 8** describes the validation activity, which consists in making sure what was designed and engineered as a Safety Control System (IEC 62061) or a Safety Related Part of the Control System (ISO 13849-1), is implemented correctly.

The book ends with some final considerations in **Chapter 9**. Besides the frequency of demand, which distinguishes between a high or a low demand mode safety system, there is another important difference between the two approaches. In machinery, there is a much higher interaction between the Equipment Under Control and the people, between the machine and the operator. The majority of accidents in machinery do not happen while the machine or the manufacturing system is running, but during set-up, maintenance, and handling of production disturbances. Functional Safety practitioners should keep in mind that the final aim of this whole exercise should be to reduce the number of accidents. That is why, spending time and efforts in detailed calculations according to the two new standards discussed in this book makes sense only if there is a solid and thorough risk assessment behind it.

Acknowledgments

This book would not have seen the light without the mindset we have in GT Engineering, the consulting practice on Machinery and Process Safety I have the honour to coach. The team curiosity, openness and thirst for knowledge was fundamental in writing this book. At the time the book was written, here are the colleagues I thank for being part of the team: Claudia Bruno, Alessandro Castelli, Ezio Compagnoni, Andrea Federici, Matteo Guglielmina, Laura Terenghi, Matteo Zilioli and Guido Zotti.

I am also grateful to some experts I had the pleasure to meet in both the Technical Committee **ISO TC 199** Working Group 8 (ISO 13849-1) and in the Technical Committee **TC 44/MT 62061**. I thank them for their openness, patience and technical feedbacks to the questions I asked them over time.

I would like to thank **David Felinski**, President of B11 Standards Inc. for his support in the writing of chapter on Functional Safety in the USA. I also would like to thank **Patrick Gehlen**, International Standardization Manager at SIEMENS, for the many informal discussions we had on the subject of Functional Safety and for giving good feedbacks on some parts of the book.

I also would like to thank our customers for giving us the opportunity to put into practice what is written in this book. We learn from them as much as they learn from us. They keep us "feet on the ground" when dealing with Risk Assessment and Functional Safety of Machinery and Process plants.

A special recognition to **Loredana Cristaldi**, Full professor at Politecnico di Milano, Scuola di Ingegneria Industriale e dell'Informazione, for inviting us every year to discuss the principles of Functional Safety with her students and for the scientific publications done together and **Giuseppe Tomasoni**, Associate professor at the Department of Mechanical and Industrial Engineering, Brescia University, for inviting us every year to discuss the principles of Functional Safety with his students.

I would not have written this book without the specific support of some colleagues in GT Engineering: **Matteo Zilioli** did all the drawings, gave good comments on the book and prepared some of the examples. **Claudia Bruno** contributed with asking good questions and providing some of the answers; she prepared several examples you will find in the book and wrote most of chapter one. The whole Team did the final review of the text, providing excellent feedbacks and improving its readability.

Last but not least, I am grateful to my wife Laura, for having motivated me during the writing of the book, despite the many weekends and evenings spent on it.

About the Author

Marco Tacchini is Technical Director of GT Engineering, a Consulting Company (www.gt-engineering.it) based in Italy and specialized in CE Marking, Risk Assessment and Risk Reduction of Machinery, according to European Directive 2006/42/EC.

He is a member of several international Technical Committees dealing with Functional Safety, among which ISO/TC 199 Working Group 8 for **ISO 13849-1 and 2,** IEC/TC44 Maintenance Team 62061 for **IEC 62061.** He is also a member of TC 65/SC 65A/**MT 61508-1-2,** TC 65/SC 65A/**MT 61511,** and TC44/**PT 63394**.

Before You Start Reading this Book

When you start a journey, you may wonder what lies ahead, what difficulties you may find, and if you will reach the end of it. Since I would like your journey to be successful, **I will give you a few tips before you start**.

Figure 0.1 EUC, the process control system and safety instrumented system.

1. **What is Functional Safety?**

You need the domain of Functional Safety every time you decide to use an Automation System to reduce **the risk associated with a Machinery or a Process**. The risk is normally reduced by removing all the energies: those can be electrical (a motor that drives a dangerous movement), pneumatic, or hydraulic but also given by process fluids like methane gas for a burner or a pump that increases the pressure in a tank. Every time you decide that, in order to eliminate the risk, you need a pressure sensor that, in case of a high dangerous value, triggers the closure of a valve, that is when Functional Safety plays the key role. The issue is that one of the elements of the so-called **Safety Instrumented System** can fail.

2. **Why components fail?**

Components fail because of two reasons:

- They fail because they are not properly designed, manufactured, installed, used, or subject to correct maintenance. If we take the example of car tyres, if we use a car with the tires badly inflated, they are likely to fail faster than normal. These are **Systematic Failures:** they are failures due to mistakes in the design, manufacturing, installation or maintenance of the component. Systematic Failures are difficult to estimate and can only be reduced by making sure the whole process, from the component design up through the usage and maintenance of the product, is done properly.

  That is the reason for the importance of concepts like Systematic Capability or Systematic Safety Integrity of components, or of Safety-related Control Systems.

  Both ISO 13849-1 and IEC 62061 define good engineering practices to be followed in order to reduce the probability of Systematic Failures: they are called **Basic** and **Well-tried safety principles**. Moreover, both standards require a **Functional Safety Plan**. You may refer to Annex I in IEC 62061 or annex G in ISO 13849-1.

- Despite the whole process (from design to maintenance) is done according to correct rules and procedures, during their lifetime, components experience **Random Failures:** those are the failures that can be statistically estimated.

3. Why do you need special components to take care of the safety of a process or a machinery?

Any component can fail, regardless if it is suitable to be used in a Safety system or not.

Therefore, any **Process Control system**, for example the one that keeps the temperature in a Heat Treatment furnace under control, can fail, and the temperature may increase until it generates a dangerous situation.

If you are not familiar with functional safety, you may think that the occurrence of the event is so unlikely that it can be disregarded and nothing more is needed to be able to declare my furnace safe.

That is not the way Functional Safety reasons. Yes, the event has a low probability to happen, **but it can happen**!

In order to be able to CE mark the furnace, you need to install a Safety system **made with components having a known probability of failure**. That allows you to calculate the Reliability of your additional **Safety Layer**. Its Reliability has to be the higher, the higher is the risk linked to, in our example, the high temperature.

The probability of failure of a component can be given using parameters explained in this book:

- The failure rate, λ.
- The B_{10}.
- The Mean time to failure, MTTF
- The PFD_{avg}
- The PFH_D

4. Why is there a distinction between High and Low demand mode of operation?

This is one of the key concepts to understand if you want to be able to get to the end of this journey.

To reach a low probability of failure of a Safety System, the following should be done:

1. To choose components that have a **low probability of failure**, and
2. To **regularly test** if each component is still working, before a dangerous situation happens; in other words, before a demand is placed upon the Safety System. A demand can be, for example, a high dangerous pressure.

Both aspects are influenced by how often the Safety System is used. Consider again a new car that is kept in a garage and used once every five years, compared with one that is used daily. If you want to make sure the former works when you turn on the key, you would need to do regular checking, for example to switch on the engine every three months and verify if the mechanics is still in good shape. If the car were a **Safety System**, it would be defined as **working in low-demand mode**.

On the other hand, if you use the car every day, most of the checking is done “automatically” while you drive it. You may hear a strange noise that indicates the gearbox is faulty. This car would be a **Safety System working in high-demand mode**.

If you think for a moment to these examples, you understand that, depending upon the usage (high or low demand mode), the car manufacturer should design some components in a different way; think for example to the battery system.

You now understand why a pressure switch or a contactor or a valve used in high or in low-demand mode:

- may have **different failure rates**
- **require different types of testing**. If they work in high demand mode, most of the testing can be done in an automatic way (that is called **Functional Testing,** and it is achieved thanks to what is called the **Diagnostic Coverage**), while if it works in low demand mode, besides functional testing, it also requires off-line testing, called **Proof Test**.

5. Why are there so many standards dealing with Functional Safety?

Because there are different industries involved and each industry has tailored the principles stated in IEC 61508 series to its specific situation. In this book, we deal with two Industries:

- **The process Industry:** they are behind the IEC 61511 series of standards. They see functional safety mainly in low-demand mode.
- **The Machinery Industry:** it is the ISO 13849-1 standard and the IEC 62061 one. They see functional safety mainly in High-Demand mode.

Chapter 2 is written with the aim of clarifying this aspect.

6. Why are there so many formulas behind Functional Safety Theory?

Because the standards are written by engineers who love formulas, who love to introduce formulas as soon as they see an opportunity and who think formulas are the only thing they need to design a safe system. The issue is that you may decide to do bungee jumping safely by using the finite element analysis to design the elastic cord (Random Failures), but if you forget to attach the cord before jumping (Systematic Failure), the result will be a failure, even if all calculations were correct.

Before you dive into the formulas, you need to understand the key parameters used. You need to become familiar with the concepts of Failure Rate, Diagnostic Coverage, Safe Failure Fraction, and many others. They are all presented and explained in **Chapter 3**. You will find the formulas used in ISO 13849-1 and IEC 62061 in **Chapters 6** and **7,** respectively.

However, please be aware that standards are not written to explain an approach or a methodology; standards are written to be clear about the required safety aspects. A standard states what needs to be done and not necessarily why it was decided to do that way. That is the reason **Chapter 1** was written. It gives you the mathematical background needed to understand where the formulas are coming from and what they really mean. **If you struggle to understand it**, it is not a problem; you will still be able to run all the number crunching required by the two new standards, without any problem.

7. Why the validation part in ISO 13849-2 has been included in ISO 13849-1?

The normative part of ISO 13849-2 is now included in part 1 of ISO 13849. Part 2 remains valid for the informative annexes only. That is the meaning of the sentence:

> ***[ISO 13849-1] Introduction.*** *[...]*
> *The requirements of Clause 10 of ISO 13849-1 supersede the requirements of ISO 13849-2:2012 (excluding the informative annexes).*

That was done to give the Validation process the same importance as the rest of the **iterative process for the design of the safety-related control system**.

8. Why are there two standards dealing with Functional Safety of Machinery?

Probably because, when IEC 61508 series was published and consequently IEC 62061 was designed, there was the willingness to keep the basic approach of Categories of EN 954-1. IEC 61508 series was designed to make sure electronics could be used in Safety Systems successfully. In machinery, sensors and final elements were mainly electromechanical, and the concept of the categories was considered suitable. For that reason, EN 954-1 evolved into ISO 13849-1.

From that moment, manufacturers were confused why two standards were available and which one was the most suitable to assess their application. Some years ago, a Joint Working Group of ISO and IEC was set up with the aim to "merge" both standards as ISO/IEC 17305, but for various reasons they did not complete their assignment.

The main reason why this book was written is to show that the new editions of ISO 13849-1 and IEC 62061 are very much aligned in the general approach. They still use different terminology, like for example ISO uses the term **Safety Related Part of the Control System** (SRP/CS), while IEC now uses the term **Safety-related Control System** (SCS), but they mean the same concept. IEC uses PFH while ISO uses PFH_D, but they mean exactly the same thing: the unreliability of a safety system.

1

The Basics of Reliability Engineering

1.1 The Birth of Reliability Engineering

> ***[EN 764-7] [20] 3 Terms and definitions***
> ***3.14 Reliability.*** *Ability of a system or component to perform a required function under specified conditions and for a given period of time without failing.*

The first Reliability models appeared during World War I, and they were used in connection with **airplane performances**: the Reliability was measured as the number of **accidents per hour** of flight time.

In the 1930s, the Reliability concepts and statistical methods were used for **quality control** of industrial products and, in the 1940s, to analyze the missile system during World War II. At that time, Robert Lusser, a mathematician, established the so-called "Product probability law of series components."

Lusser discovered that the **Reliability of a system** is equal to the product of the reliabilities of the individual components which make up the system. If the system has many components, the system Reliability may therefore be rather low, even though the individual components have high Reliability values.

After the war, the interest in the United States was concentrated on intercontinental ballistic missiles and space research; this led to the creation of an association for engineers working with Reliability. The first journal on the subject, "IEEE Transactions on Reliability" came out in 1963, and several textbooks on the subject were published in that decade. The famous military standard MIL-STD-781 was created at that time. Around that period, also the much-used predecessor to military handbook 217 was published by RCA, Radio Corporation of America, and was used for the **prediction of failure rates of electronic components**.

In the following years, more pragmatic approaches were developed and used in the consumer industries. Reliability tools and tasks became more closely tied to the **engineering design process**.

Today, the study of Reliability engineering permits not only the **evaluation of the conformity** of a device over time but also **to compare different design solutions** with the same **functional characteristics**. It can also identify, inside an apparatus, subsystems or **critical elements** that could cause a failure or malfunction of the apparatus itself, needing corrective actions. For this reason, Reliability has an important role in modern design and constitutes a competitive element, even in the light of stricter **safety requirements**.

Functional Safety of Machinery: How to Apply ISO 13849-1 and IEC 62061, First Edition. Marco Tacchini.

1.1.1 Safety Critical Systems

A part of the Reliability studies deals with **Safety Critical Systems.** Those are systems whose failure could result in the loss of lives or significant damage to properties or to the environment [58].

In the 1970s, the design principles of safety-critical systems, both in Machinery and in the process Industry, were the following:

- **Single-channel system** (no redundancy). This architecture would be regarded as a basic design having minimum safety performance.
- **Dual-channel system** (redundancy) applicable to sensors, for example pressure switches, logic units, and final elements, like contactors and valves.
- **2 out of 3 voting systems (2oo3).** Those systems were used originally in the petrochemical industry: they give a good level of both Reliability and of Availability. **Reliability** measures the ability of a system to function correctly, whereas **Availability** measures how often the system is available for use, even though it may not be functioning correctly. For example, a server may run forever and so have ideal Availability, but may be unreliable, with frequent data corruption.
- All systems were using the concept of **Fail Safe**: a failure in any part of the system would lead to a safe state of the process or the machinery under control.

In the 1990s, a part of the Reliability of Critical System studies became known as **Functional Safety** and focussed on Electrical, Electronic, and Programmable Electronic (E/E/PE) systems. The reference standard became the IEC 61508 series.

1.2 Basic Definitions and Concepts of Reliability

According to IEC 60050-191, **Reliability is the ability of the item to remain functional**, to perform a required function, the item's task, **under given conditions for a given time interval**. The concept of "**performing a required function**" is complementary to that of a "**failure.**"

A numerical statement of Reliability must be specified by the definition of the required function, the operating conditions, and the mission duration.

Both the required function and the operating conditions can be time dependent, and this is the reason why it's important to define a **mission profile** related to the Reliability of the item's life. If the Mission Time is considered as a parameter of time, the Reliability function is then defined by the **time-dependent function** $R(t)$.

$R(t)$ is the probability that no failure, at item level, will occur in the interval $(0, t]$.

Reliability is based upon mathematical models, and it can be estimated thanks to the **observation** of items during their lifetime. That is done thanks to **measurements and statistical parameters** such as failure rate (λ), mean time to failure (MTTF), and mean time between failures (MTBF), which are presented in the following paragraphs.

1.3 Faults and Failures

One of the first concepts that needs to be clearly understood, for someone approaching the field of Functional Safety, is **the difference between a Fault and a Failure**.

1.3.1 Definitions

Hereafter are the definitions taken from IEC 61508-4 [8]:

> ***[IEC 61508-4] 3.6 Fault, failure and error***
> ***3.6.1 Fault****. Abnormal condition that may cause a reduction in, or loss of, the capability of a functional unit to perform a required function.*

In other words, a Fault is the situation where a system cannot perform anymore its required function. Figure 1.1 shows the two statuses where a control system can be: an "OK" state, where it works correctly and a "FAULT" state.

Bottom line, it is important not to confuse the concept of failure (event) with the concept of fault (associated with a particular state of a system).

When the system has a failure, it may stop working properly, and therefore it may move to a FAULT state. Here is the definition of Failure:

> ***[IEC 61508-4] 3.6 Fault, failure and error***
> ***3.6.4 Failure****. Termination of the ability of a functional unit to provide a required function or operation of a functional unit in any way other than as required.*

Reliability theory classifies failures in various ways [51], among which are **Primary Failure** (not due to other failures), **Secondary Failure**, **Early life Failure**, **Random Failure**, and **Wear out Failure**.

This can also be classified in: **Total failure** (when variations in the characteristics of the element are such to completely compromise its function) or **Partial failure** (when the variations of one or more characteristics of the element do not impede its complete functioning).

1.3.2 Random and Systematic Failures

In Functional safety, Failures are classified as either **random** (in hardware) or **systematic** (in hardware or software).

> ***[IEC 61508-4] 3.6 Fault, failure and error***
> ***3.6.5 Random Hardware Failure****. Failure, occurring at a random time, which results from one or more of the possible degradation mechanisms in the hardware.*

> ***[IEC 61508-4] 3.6 Fault, failure and error***
> ***3.6.6 Systematic Failure****. Failure, related in a deterministic way to a certain cause, which can only be eliminated by a modification of the design or of the manufacturing process, operational procedures, documentation or other relevant factors.*

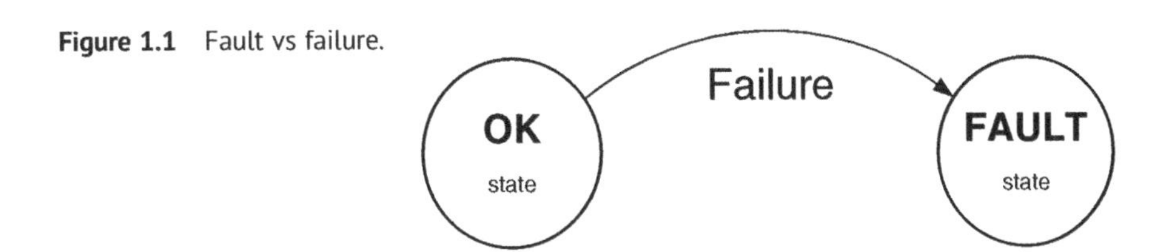

Figure 1.1 Fault vs failure.

Random Failures are therefore normally attributed to hardware. They are failures, occurring at a random time, which result in one or more of degradation of the component capability to perform its scope. There are many degradation mechanisms occurring at different rates, in different components and, since manufacturing tolerances cause components to fail due to these mechanisms after different times in operation, failures of total equipment, comprising many components, occur **at predictable rates but at unpredictable** (i.e. random) **times**. Based upon historical data, Random Failures can be characterized by a parameter called **Failure Rate λ**. In other words, a random hardware failure involves only the equipment; random failures can occur suddenly without warning or be the outcome of slow deterioration over time. These failures can be characterized by a single reliability parameter, the device failure rate, which can be controlled and managed using an asset integrity program.

Systematic failures can only be eliminated by a modification of the design or of the manufacturing process, operational procedures, or other relevant factors. Examples of causes of systematic failures include human error in the design, manufacture, installation, and operation of the hardware. If, **for example**, a product is not used correctly or it is used in the wrong environment, the risk of systematic failures exists. A systematic failure involves both the equipment and a human error; systematic failures exist from the time that human errors were made and continue to exist until they are corrected. A systematic failure can be eliminated after being detected, while random hardware failures cannot.

Failures are therefore either random or systematic; the latter can be hidden in the hardware or in the software program. A **major difference between random hardware failures and systematic failures** is that system failure rates (or other appropriate measures), arising from random hardware failures, can be predicted with reasonable accuracy, while systematic failures, by their very nature, cannot be accurately predicted. That means system failure rates arising from random hardware failures can be quantified with reasonable accuracy, while those arising from systematic failures cannot be statistically quantified because the events leading to them cannot easily be predicted. In other words, the reliability parameters of random hardware failures can be estimated from field feedbacks, while it is very difficult to do the same for systematic failures: a qualitative approach is preferred for systematic failures.

In high demand mode functional safety standards, IEC 62061 and ISO 13849-1, the issue of the presence of Systematic Failures is addressed by requiring a systematic approach to product development, engineering, manufacturing, and maintenance, the so-called **Management of Functional Safety**, and the application of the so-called **Basic and Well-tried Safety Principles** (§ 4.13).

When analyzing the Reliability of a Safety System, **all causes of failure,** both systematic and random, **that lead to an unsafe state, should be included**. Some of these failures, in particular the Random hardware failures, can be quantified using the **average frequency of failure in dangerous mode** or the **probability** of a safety-related protection system **failing to operate on demand**.

1.3.2.1 How Random is a Random Failure?

It cannot be stressed enough **the importance of a correct design, installation, and maintenance,** to avoid systematic failures: it often happens that **many failures that we consider to be random are preventable, to a large extent**. Please consider that failures occurring at a random time, which results from one or more degradation mechanisms in the hardware, may be treated as random to the extent that the failures cannot reasonably be prevented.

Only purely random failures can be characterized by a failure rate, including the surrogated failure rate, explained further on in the book. Purely random failures are sudden and complete failures that occur without warning. They are impossible to forecast by examining the item. Constant in time, random failures are usually limited to electronic devices; for electromechanical components, a constant **surrogated failure rate** is used (also referred to as **substitute failure rate**). Hardware failures that are related to deterioration mechanisms may be characterized by failure rates as if they were random, but failure rates depend heavily on the operating environment and on the effectiveness of preventive maintenance. Failure rates will vary by at least an order of magnitude between different applications and different maintenance organizations.

1.4 Probability Elements Beyond Reliability Concepts

We defined $R(t)$ as the probability that no failure, at item level, occurs in the interval $(0, t]$, where t is a random variable.

A **random variable** is a function able to assign a real number to each outcome in the sample space of a random experiment. It can be **continuous or discrete**. Continuous when it has an interval, finite or infinite, of real numbers for its range; discrete when it has a finite range of values.

The probability that a **random variable *X*** assumes a **well-defined value *x*** is described mathematically by the **probability distribution**, which can be continuous or discrete with respect to the random variable considered.

To better understand the Reliability concepts, a brief summary of the **axioms of probability by Kolmogorov (1933)** is reported:

Let (Ω, A) be a measurable space of events. Any event E is a subset of Ω. Assume that the set of all events is represented by a particular family of sets over Ω denoted A.

A probability measure is a real-value function mapping $\mathbb{P}: A \rightarrow \mathbb{R}$ satisfying:

1. *for any event, $E \in A$, $\mathbb{P}(E) \geq 0$*
2. $\mathbb{P}(\Omega) = 1$
3. *for any countably sequence of events $(E_i)_{i \geq 1}$ that are mutually exclusive ($E_i \cap E_j = 0$ if $i \neq j$),*

$$\mathbb{P}\left(\bigcup_{i=1}^{\infty} E_i\right) = \sum_{i=1}^{\infty} \mathbb{P}(E_i)$$

1.4.1 The Discrete Probability Distribution

Discrete probability distribution is usually indicated with $p(x)$, and it can be seen as the relationship between the possible values for the variable and the probability of each value. In this case, the probability distribution can be drawn as shown in Figure 1.2.

This distribution has the following property:

$$\sum_i p(x_i) = 1 = \mathbb{P}(\Omega)$$

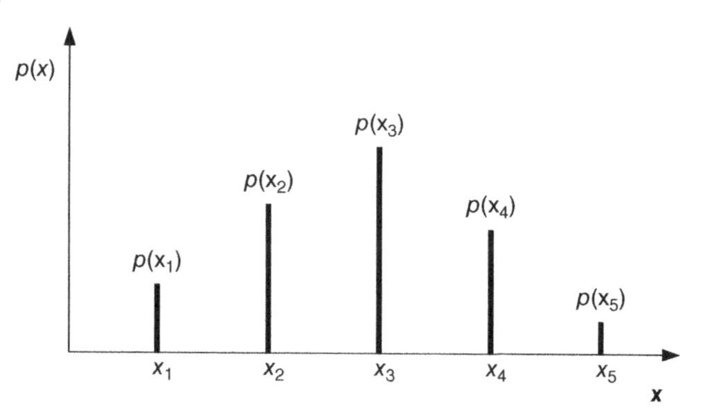

Figure 1.2 Discrete probability distribution.

1.4.1.1 Example: 10 Colored Balls

Let's suppose that a box contains 10 balls:

- Five of the balls are **red**
- Two balls are **green**
- Two balls are **blue**
- One ball is **yellow**

Suppose we take one ball out of the box. Let X be the random variable that represents the ball color. As 5 of the balls are red, and there are 10 balls, the probability that a red ball is drawn from the box is $p(x = \text{red}) = 5/10 = 1/2$.

Similarly, there are two green balls, so the probability that x is green is 2/10. Similar calculations for the other colors yield the probability density function given by Table 1.1 and Figure 1.3.

Table 1.1 Example of a discrete probability distribution.

Ball color	Probability
Red	5/10
Green	2/10
Blue	2/10
Yellow	1/10

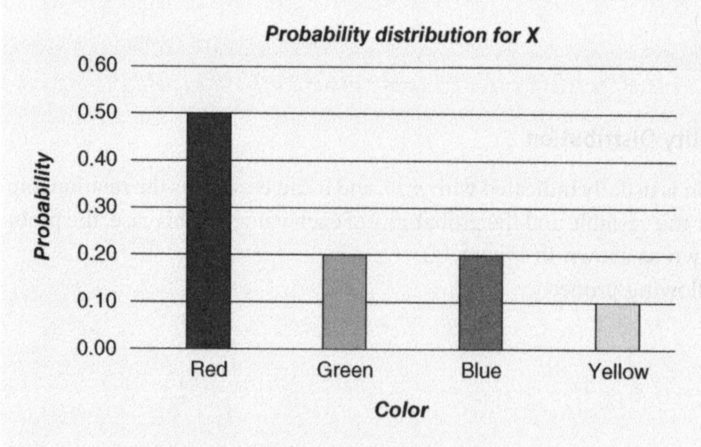

Figure 1.3 Example of a discrete probability distribution.

1.4.1.2 Example: 2 Dice

Let's now suppose that we have two fair six-sided dice. We roll both dice at the same time and add the two numbers that are shown on the upward faces. The discrete probability density function (PDF) is given in Table 1.2 and Figure 1.4.

Table 1.2 Example of a discrete probability distribution for two dice.

Outcome	Sum	Probability
(1.1)	2	1/36
(1.2), (2.1)	3	2/36
(1.3), (2.2), (3.1)	4	3/36
(1.4), (2.3), (3.2), (4.1)	5	4/36
(1.5), (2.4), (3.3), (4.2), (5.1)	6	5/36
...		
(6.6)	12	1/36

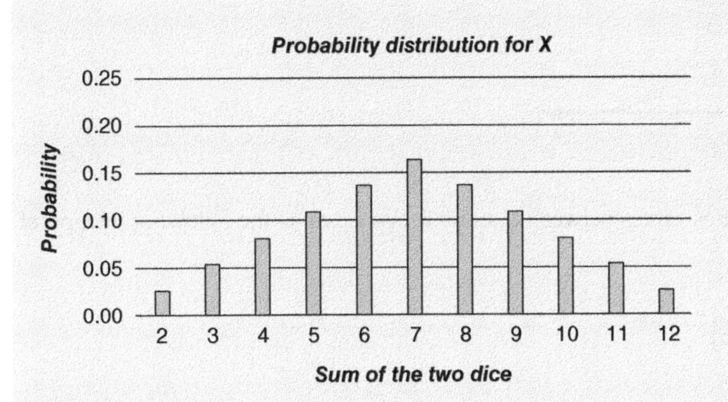

Figure 1.4 Example of a discrete probability distribution for two dice.

1.4.2 The Probability Density Function *f*(*x*)

A continuous probability distribution is indicated with *f*(*x*) and is usually called **PDF**. It is expressed by a function, and it can be represented as in Figure 1.5. The bell curve is just an example of a possible PDF.

The main property of a PDF is that:

$$\int_{-\infty}^{+\infty} f(x)dx = 1 = \mathbb{P}(\Omega)$$

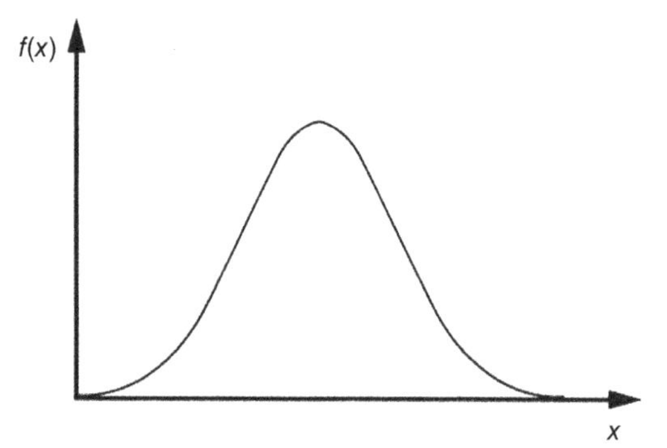

Figure 1.5 Probability density function.

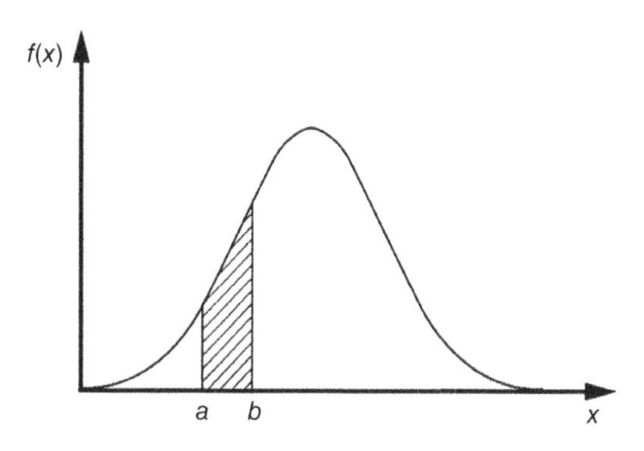

Figure 1.6 The area under the probability density function represents the probability that x assumes values between a and b.

The probability that x assumes values between a and b is evaluated as the following integral of the **PDF:**

$$\int_{a}^{b} f(x)dx = \mathbb{P}\{a \leq X \leq b\}$$

This probability is shown in Figure 1.6.

The PDF is also called **Failure Density** or also **Life Distribution.**

1.4.2.1 Example

A uniform PDF over the interval 0–1 for a random variable x is given by $f(x) = 1$ in the range between 0 and 1, and 0 otherwise.

The function $f(x)$, shown in Figure 1.7, is a valid PDF, since it is non-negative and integrates to one. What is the probability of an outcome in the range 0.2–0.8?

$$PDF(0.2 \text{ to } 0.8) = \int_{0.2}^{0.8} 1 \cdot dx = 0.8 - 0.2 = 0.6$$

Meaning that the probability of failure is 60%.

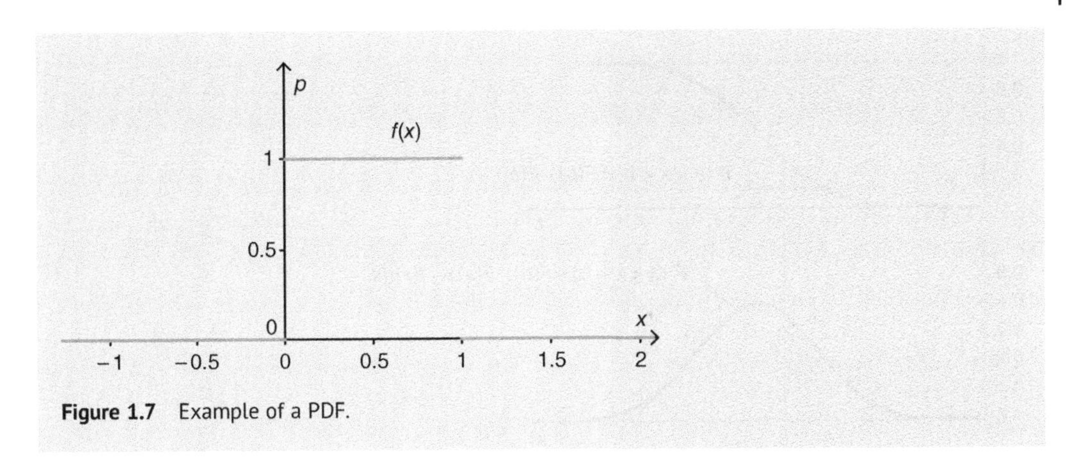

Figure 1.7 Example of a PDF.

1.4.3 The Cumulative Distribution Function *F*(*x*)

The distribution of a continuous variable can be described by the **Cumulative Distribution Function** as well. That gives the probability that the random variable will assume a value smaller or equal to x. Its expression is:

$$F(x) = \mathbb{P}(X \leq x) = \int_{-\infty}^{x} f(\xi)d\xi$$

For $-\infty < x < +\infty$.

$F(x)$ is a non-decreasing function: $F(-\infty) = 0$ and $F(+\infty) = 1$, thus:

$$\int_{-\infty}^{+\infty} f(\xi)d\xi = 1$$

The derivative of the cumulative distribution function is the **PDF** (or failure density) of the random variable x:

$$f(x) = \frac{dF(x)}{dx}$$

The relationship between the **Cumulative distribution function *F*(*x*)** and the **PDF *f*(*x*)** is in Figure 1.8.

These definitions for $F(x)$ allow to express $\mathbb{P}\{a \leq X \leq b\}$ as follows:

$$\mathbb{P}\{a \leq X \leq b\} = \int_{a}^{b} f(x)dx = F(b) - F(a)$$

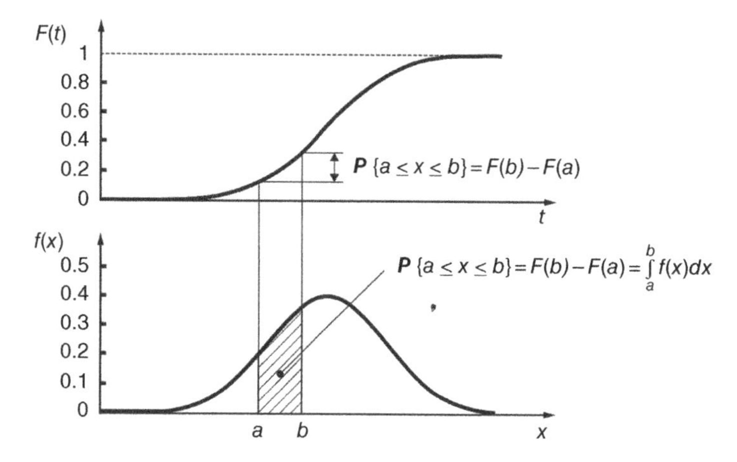

Figure 1.8 $F(t)$ and $f(t)$.

Since **we reason in terms of time** and time is a positive random variable failure time, the **Cumulative Distribution Function** can be written in the following way:

$$F(t) = \int_0^t f(x)dx$$

It is represented graphically as in Figure 1.9.

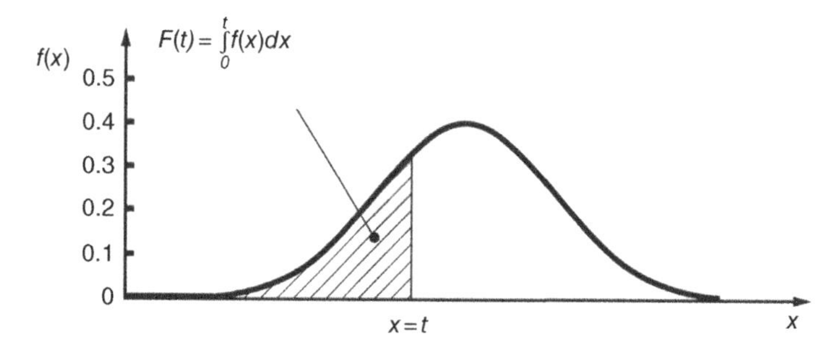

Figure 1.9 $F(t)$ and $f(t)$ for a continuous random variable.

and please consider that

$$\int_0^{+\infty} f(x)dx = 1$$

1.4.4 The Reliability Function *R(t)*

$R(t)$ is the probability that no failure of the item occurs in the interval $(0, t]$. It is represented in Figure 1.10.

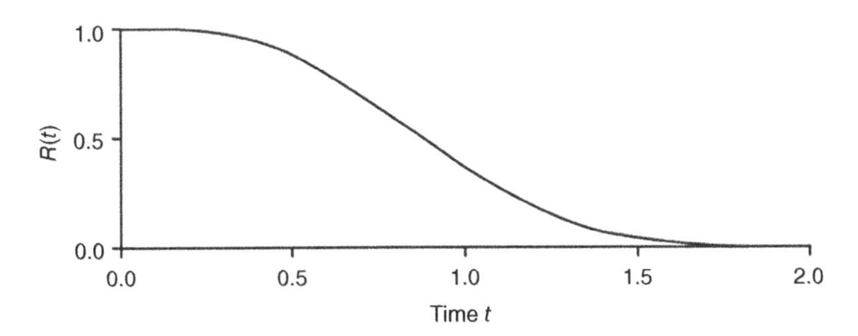

Figure 1.10 The reliability function $R(t)$.

In other terms, $R(t)$ is the probability that an item will operate "failure-free" in time interval $(0, t]$, while the failure will occur in $(t, +\infty)$. Known the PDF $f(x)$ can be represented graphically as in Figure 1.11, we have:

$$R(t) = \int_{t}^{+\infty} f(x)dx$$

$R(t)$ can be represented graphically as the area under $f(x)$ starting from $x = t$, as in Figure 1.11.

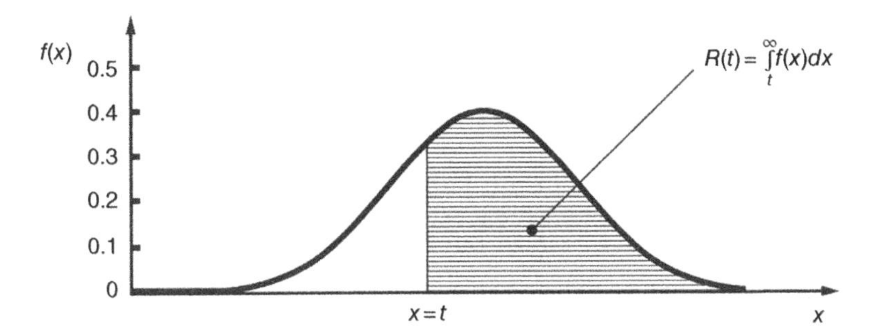

Figure 1.11 $R(t)$ for a continuous random variable.

If the system can be found in two states only, either correct functioning or failure, we can define the function of unreliability $F(t)$ as complementary to $R(t)$, that means:

$$F(t) = 1 - R(t) = \int_{0}^{t} f(x)dx$$

The density function $f(t)$ can now be expressed as:

$$f(t) = \frac{dF(t)}{dt} = -\frac{dR(t)}{dt} \quad \text{[Equation 1.4.4]}$$

1.5 Failure Rate λ

The failure rate is the basis of the Functional Safety theory.

> ***[IEC 61508-4] 3.6 Fault, failure and error***
>
> ***3.6.16 Failure Rate**. Reliability parameter $\lambda(t)$ of an entity (single components or systems) such that $\lambda(t)dt$ is the probability of failure of this entity within $[t, t+dt]$ provided that it has not failed during $[0, t]$*

Mathematically, $\lambda(t)$ is the conditional probability of failure per unit of time over $[t, t + dt]$. It is possible to demonstrate that the instantaneous failure rate is:

$$\lambda(t) = \frac{f(t)}{R(t)}$$

Using the Equation 1.4.4, it is possible to obtain:

$$\lambda(t) = \frac{-dR(t)/dt}{R(t)} = -\frac{d}{dt}\ln R(t)$$

Integrating the upper equation in time:

$$R(t) = exp\left(-\int_0^t \lambda(\tau)d\tau\right)$$

Failure rates and their uncertainties can be estimated from field feedback using conventional statistics.

The most diffuse and widely known model for the failure rate is the "bathtub" curve, represented in Figure 1.12. In the **initial phase** of the component lifetime, $\lambda(t)$ decreases rapidly with time; this fact derives from the existence of a "weak" fraction of the population whose defects cause a failure within a short period of time from the moment they are produced.

In the period called **useful life**, $\lambda(t)$ is approximately constant, in case for example, of electronic components. For electromechanical components, $\lambda(t)$ is a function of time and, in this interval, it constantly increases.

The last period is characterized by **wear out**, with a rapidly increasing failure rate $\lambda(t)$ caused by wearing out, aging, and fatigue.

During the useful life of a component with a **constant failure rate**, considering as an initial condition that Reliability at time 0 is at a maximum and it is equal to 1, we have:

$$R(t) = exp\left(-\int_0^t \lambda d\tau\right) = e^{-\lambda t}$$

The Reliability function $R(t)$ is shown in Figure 1.13a and the PDFs $f(t)$ in Figure 1.13b, in the case $\lambda =$ constant.

Table 1.3 shows a summary of the four functions described so far.

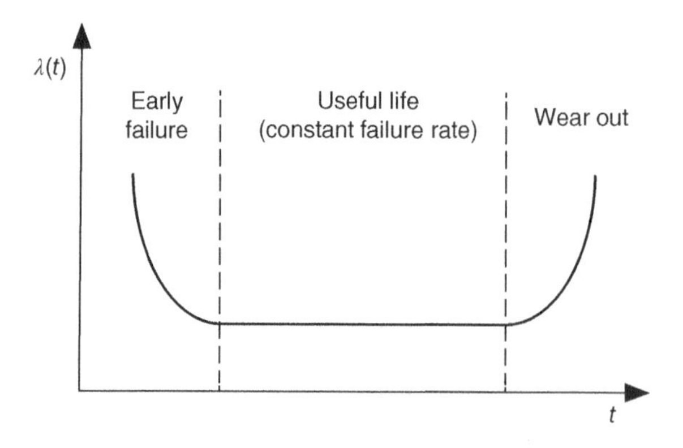

Figure 1.12 The bathtub graph.

Figure 1.13 (a) $R(t)$ and (b) $f(t)$.

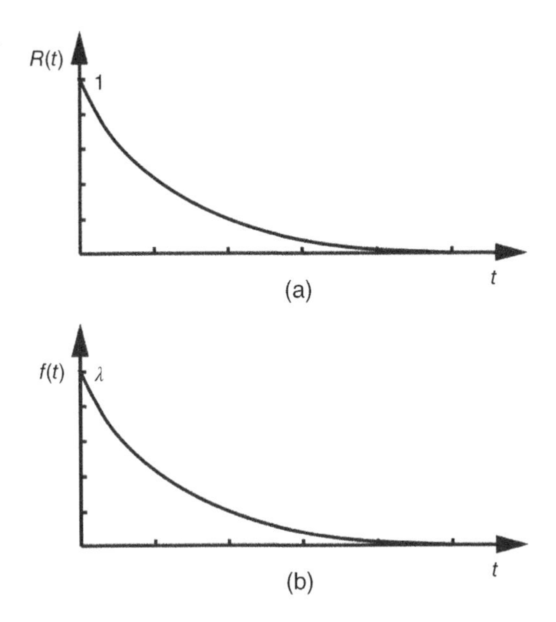

Table 1.3 Summary of the key functions used in Reliability theory.

	$F(t)$	$f(t)$	$R(t)$	$\lambda(t)$	λ = constant
$F(t)$	—	$\int_0^t f(\tau)d\tau$	$1-R(t)$	$1-exp\left(-\int_0^t \lambda(\tau)d\tau\right)$	$1-e^{-\lambda t}$
$f(t)$	$\frac{d}{dt}F(t)$	—	$-\frac{d}{dt}R(t)$	$\lambda(t)\cdot exp\left(-\int_0^t \lambda(\tau)d\tau\right)$	$\lambda \cdot e^{-\lambda t}$
$R(t)$	$1-F(t)$	$\int_t^{+\infty} f(\tau)d\tau$	—	$exp\left(-\int_0^t \lambda(\tau)d\tau\right)$	$e^{-\lambda t}$
$\lambda(t)$	$\frac{dF(t)/dt}{1-F(t)}$	$\frac{f(t)}{\int_t^{+\infty} f(\tau)d\tau}$	$-\frac{d}{dt}ln\,(R(t))$	—	—

Figure 1.14 shows the relationship between $F(t)$ and $R(t)$.

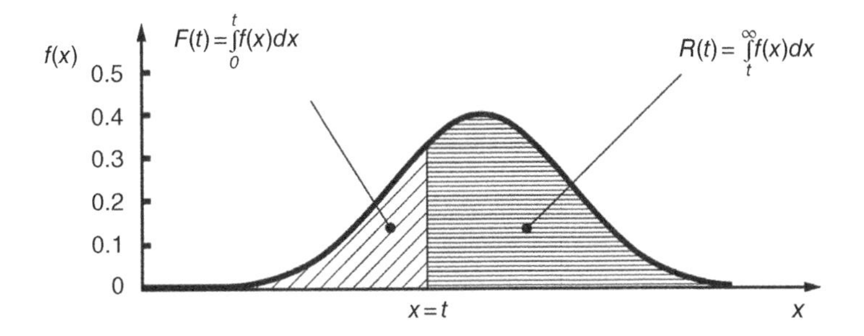

Figure 1.14 $R(t)$ and $F(t)$.

1.5.1 The Maclaurin Series

Mathematically, it can be shown that certain functions can be approximated by a series of other functions. In particular, e^x can be developed as a so-called Maclaurin series:

$$e^x = 1 + x + \frac{x^2}{2!} + \frac{x^3}{3!} + \ldots$$

In case $x \ll 1$,

$$e^x \approx 1 + x$$

That means the Reliability $R(t)$ and Unreliability $F(t)$ functions can be approximated to

$$R(t) = e^{-\lambda t} \approx 1 - \lambda t$$
$$F(t) = 1 - R(t) \approx \lambda t$$

1.5.2 The Failure in Time or FIT

The failure rate λ has a unit of inverse time: it is a common practice to use the unit of "failures per billion (10^9) hours". This unit is known as **FIT: Failure in Time**.

1.5.2.1 Example

A component has a failure rate of 1000 FITs (10^{-6} h^{-1}).

Question: What is its probability of failure after 10^5 hours (about 11 years)?

$$F(t) = 1 - e^{-\lambda t} = 1 - e^{-10^{-6}t} \Rightarrow F\left(10^5\right) = 1 - e^{-10^{-6} \cdot 10^5} = 1 - e^{-0.1} \cong 0.095$$

By using the approximated formula

$$F\left(10^5\right) = 1 - e^{-\lambda t} \approx \lambda t = 10^{-6} \cdot 10^5 = 0.1$$

Answer: the probability of failure after approximately 11 years is 10%.

1.6 Mean Time to Failure

In case of **non-repairable devices**, for example an incandescent lamp, it is common to define the MTTF as an indication of its Reliability.

We need to find the expression of the mean of the continuous random variable t_f with density $f(t)$. Defining E, the mean operator, MTTF is:

$$MTTF = E(t) = \int_{-\infty}^{+\infty} t \cdot f(t)dt$$

Since the time is a positive random variable, the previous equation can be reduced to:

$$MTTF = E(t) = \int_{0}^{+\infty} t \cdot f(t)dt$$

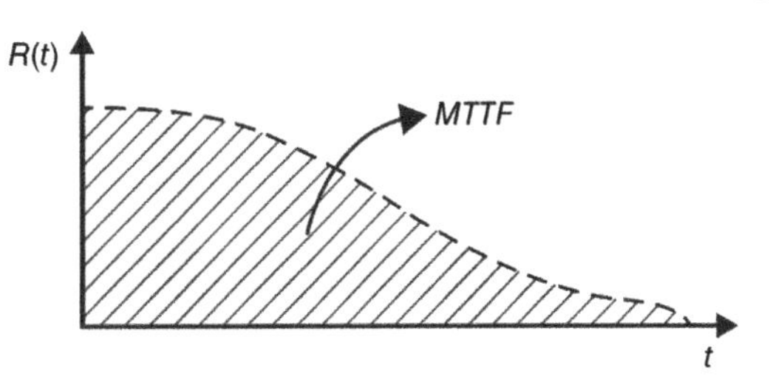

Figure 1.15 MTTF and $R(t)$.

$$MTTF = \int_0^{+\infty} t\frac{dF(t)}{dt}dt = \int_0^{+\infty} -t\frac{dR(t)}{dt}dt = -[t \cdot R(t)]_0^{\infty} + \int_0^{+\infty} R(t)dt = \int_0^{+\infty} R(t)dt$$

MTTF represents the area under the Reliability function $R(t)$, as shown in Figure 1.15.

Based upon the previous considerations and assuming a **constant failure rate,** it follows that:

$$MTTF = \int_0^{+\infty} R(t)dt = \int_0^{+\infty} e^{-\lambda \cdot t} = \frac{1}{\lambda}$$

1.6.1 Example of a Non-Constant Failure Rate

A component has the following Reliability function:

$$R(t) = \frac{1}{(0.2t+1)^2}$$

where t represents the months. The PDF is:

$$f(t) = -\frac{dR(t)}{dt} = \frac{0.4}{(0.2t+1)^3}$$

The failure rate is a function of time:

$$\lambda(t) = \frac{f(t)}{R(t)} = \frac{0.4}{0.2t+1}$$

All three functions are represented in Figure 1.16. MTTF can be calculated as:

$$MTTF = \int_0^{+\infty} R(t)dt = \int_0^{+\infty} \frac{1}{(0.2t+1)^2}dt = 5 \text{ months}$$

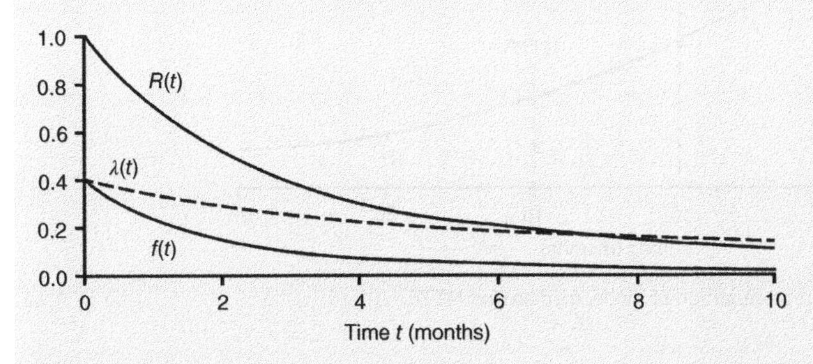

Figure 1.16 $R(t)$, $f(t)$ and $\lambda(t)$.

1.6.2 The Importance of the MTTF

The MTTF is one of the most important parameters used in Functional Safety. Let's look at the value of unreliability $F(t)$ of a component with constant failure rate when it reaches its MTTF value.

$$F(t) = 1 - e^{-\lambda t} \Rightarrow F(t = MTTF) = F\left(t = \frac{1}{\lambda}\right) = 1 - e^{-\lambda \cdot \frac{1}{\lambda}} = 1 - e^{-1} \cong 0.63$$

Therefore, when a component **having a constant failure rate** reaches its MTTF time, its unreliability is about 63% or, in other terms, its Reliability is 37%.

1.6.3 The Median Life

The MTTF is just one way of representing what is also called a "life distribution $f(t)$." Another method is the **Median Life** t_m defined as

$$R(t_m) = 0.5$$

The median divides the distribution in two halves. The component will fail before time t_m with 50% probability.

1.6.4 The Mode

The **Mode** t_{mode} of a life distribution $f(t)$ is the most likely failure time; in other terms, it is the time when the probability density $f(t)$ attains its maximum.

$$f(t_{mode}) = \max_{0 \le t \le \infty} f(t)$$

Figure 1.17 shows the location of the three parameters for a distribution skewed to the right.

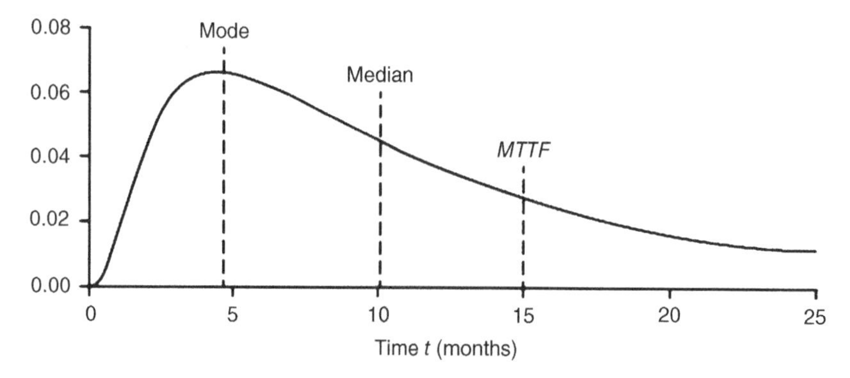

Figure 1.17 Graphical representation of mode, median and MTTF.

1.6.4.1 Example

A component has a constant failure rate $\lambda = 2.5{\cdot}10^{-5}\ h^{-1}$

1. What is the probability that it survives a period of two months, without failures?

$$R(t) = e^{-\lambda t} \Rightarrow R(t = 1460) = e^{-2.5 \cdot 10^{-5} \cdot 1460} = e^{-0.0365} = 0.96$$

   which means **96%**

2. What is its MTTF value?

$$MTTF = \frac{1}{\lambda} \cong 4.6 \text{ years}$$

3. What is the probability the component will survive till its MTTF?

$$R(t) = e^{-\lambda t} \Rightarrow R\left(t = \frac{1}{\lambda}\right) = e^{-1} \cong 0.37$$

   which means **37%**

1.6.4.2 Example

A hydraulic solenoid valve, with constant failure rate λ, will survive a period of two years without failure, with a probability of 90%.

1. Calculate the valve MTTF

$$R(t) = e^{-\lambda t} \Rightarrow 0.9 = e^{-\lambda \cdot 17\,520}$$

   Therefore $\lambda = 6.01{\cdot}10^{-6}$; that means ***MTTF* = 19 years**

2. Find the probability that the valve will have a failure during the time of 10 years.

$$F(t) = 1 - e^{-\lambda t} \Rightarrow F(t = 10\,\text{years}) = 1 - e^{-\lambda t} = 1 - e^{-6.01 \cdot 10^{-6} \cdot 87\,600} = 0.40$$

   which means **40%** failure probability (or 60% survival probability).

3. What is the probability that the valve will fail after 10 years, knowing that it did not fail after 5 years?
   Let's first calculate the survival probability

$$\frac{R(t_1 + t_2)}{R(t_1)} = \frac{e^{-(6.01 \cdot 10^{-6} \cdot 87\,600)}}{e^{-(6.01 \cdot 10^{-6} \cdot 43\,800)}} \cong \frac{0.6}{0.77} \cong 0.78$$

   which means a survival probability of 78% and therefore a failure probability of 22%.
   Please notice that the fact that being 5 years in the way, with the valve that did not fail, gives a higher probability of survival in the 10 years period.

1.7 Mean Time Between Failures

Considering an item or a system that, following a failure, can be **restored**, (these are called **Repairable systems**), the MTBF is the parameter normally used to indicate its level of Reliability.

The MTBF **is defined as the average time period of a Failure + Repair cycle**. It includes the time to failure, any time required to detect the failure, and the actual **repair or restoration time**.

Hereafter the key definitions

[ISO 13849-1] 3 Terms, definitions, symbols and abbreviated terms
3.1.33 Mean Time Between Failures (MTBF)
Expected value of the operating time between consecutive failures.

[IEC 61508-4] 3.6 Fault, failure and error
***3.6.21 Mean Time to Restoration (MTTR)**. Expected time to achieve restoration.*

MTTR encompasses (please see Figure 1.18):

a) the time to detect the failure and,
b) the time spent before starting the repair and,
c) the effective time to repair and,
d) the time before the component is put back into operation.

The sum of the b + c + d time is called **Mean Repair Time** (MRT).

The relationship then becomes:

$$MTBF = MTTF + MTTR$$

If a system is non-repairable, $MTBF = MTTF$. That is the reason why, where only MTBF values are available, **if** the RDF (Ratio of Dangerous Faults) is assumed as 50% of all failures, a conversion to $MTTF_D$ values can be done by

$$MTTF_D \approx 2^* MTBF$$

Otherwise, the worst case is supposing all failures to be dangerous, in that case:

$$\boldsymbol{MTTF_D \approx MTBF}$$

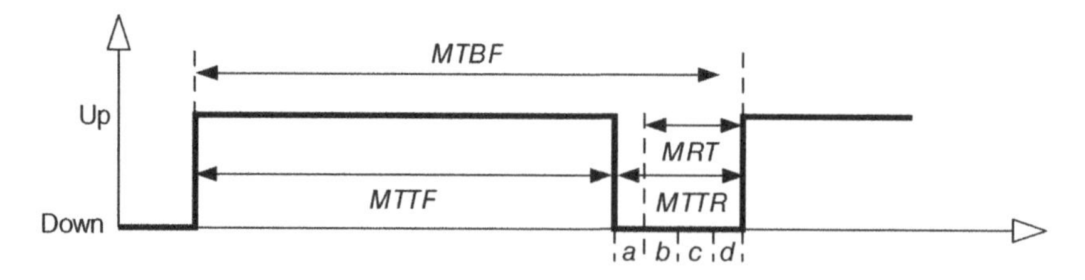

Figure 1.18 Relationship between MTTF and MTBF.

1.8 Frequency Approach Example

In order to make it clear what the Reliability and Unreliability functions stand for, hereafter a simple example. Using the concept of probability based on **relative frequency**, the probability of any event E is defined as:

$$P(E) = \frac{\textit{number of elements of E}}{\textit{number of all possible outcomes}} = \frac{n_E}{n}$$

Combining the $P(E)$ frequency approach with the following formula (§ 1.4.2)

$$\mathbb{P}\{a \leq X \leq b\} = \int_a^b f(x)dx$$

It is possible to define the experimental histogram of relative frequency:

$$f(x) \cdot \Delta x = \frac{n(x)}{n}$$

From a practical point of view, this means that after repeating the experiment n times and after counting the tests $n(x)$, the relationship $x \leq X \leq x + \Delta x$, involving the random variable X is valid. Thinking in histogram terms, Δx represents the width of the classes.

With all these considerations there is an **empirical definition** of Reliability, Unreliability, Density function, and Failure rate, extracted from the **observation and analysis of failure data**.

1.8.1 Initial Data

Let's consider $n = 172$ identical elements, which are statistically independent. We put them into operation under same conditions at time $t = 0$. We obtain the experimental data reported in the table

Time interval (h)	No. of failures at the end of interval
0–1000	59
1000–2000	24
2000–3000	29
3000–4000	30
4000–5000	17
5000–6000	13
Total	**172**

1.8.2 Empirical Definition of Reliability and Unreliability

We will try to arrive at a definition of Reliability and Unreliability in "empirical" terms.

$n_h(t)$ indicates the subset of elements n which have not yet failed at time t, while $n_f(t)$ indicates the number of elements that have failed in time t, considering that $n_h(t) + n_f(t) = n$.

It is now possible to obtain the following definitions:

$$R(t) = \frac{n_h(t)}{n}$$

$$F(t) = \frac{n_f(t)}{n} = \frac{n - n_h(t)}{n} = 1 - R(t)$$

Considering the experimental data, we can evaluate $R(t)$ and $F(t)$ both numerically and graphically (please refer to Figures 1.19 and 1.20).

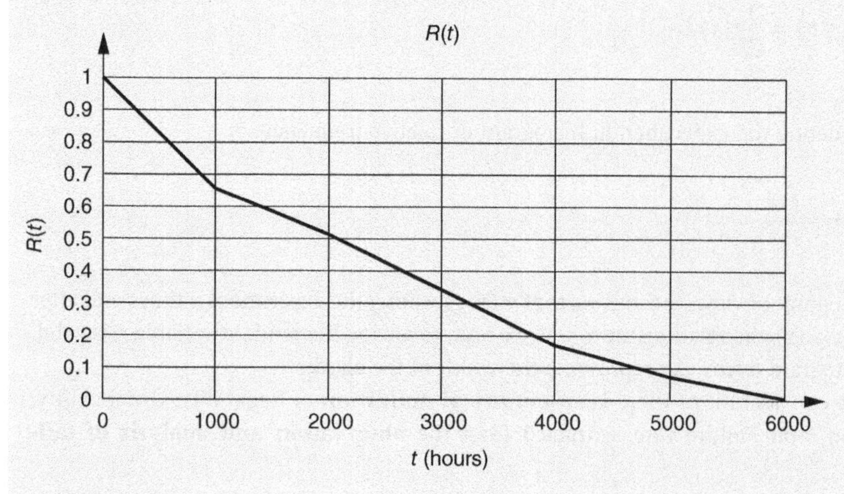

Figure 1.19 *R*(*t*).

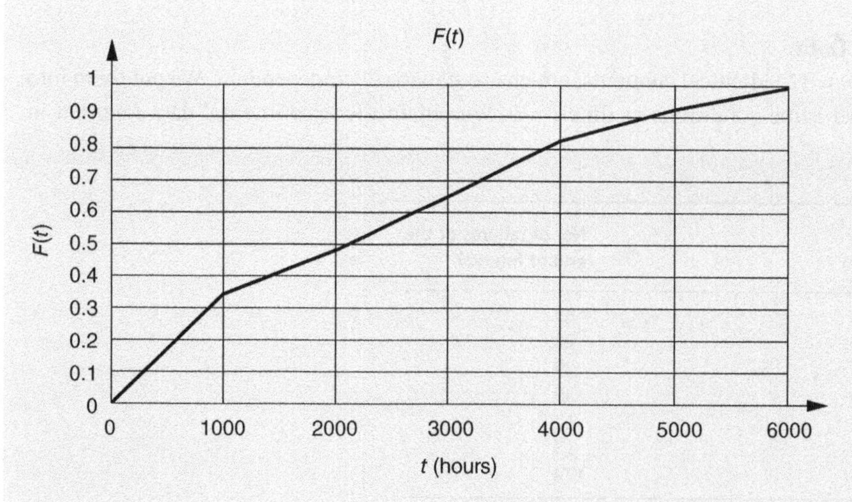

Figure 1.20 *F*(*t*).

t (h)	n_h (t)	R(t)	n_f (t)	F(t)
0	172	1.000	0	0.000
1000	113	0.657	59	0.343
2000	89	0.517	83	0.483
3000	60	0.349	112	0.651
4000	30	0.174	142	0.826
5000	13	0.076	159	0.924
6000	0	0.000	172	1.000

In this case, the experimental histogram of relative frequency can be expressed in the following terms:

$$f_N(t) = \frac{n_f(t+\Delta t) - n_f(t)}{n}\frac{1}{\Delta t} = \frac{F_N(t+\Delta t) - F_N(t)}{\Delta t}$$

We also have the possibility to express the failure rate as the ratio between elements that have broken down in the interval $(t, t + \Delta t]$ and the number of elements functioning at time t, that is:

$$\lambda_N(t) = \frac{n_f(t+\Delta t) - n_f(t)}{n_h(t)}\frac{1}{\Delta t} = f_N(t)\frac{n}{n_h(t)} = \frac{f_N(t)}{R_N(t)}$$

Time interval (h)	f (time interval) 10^{-3}	λ (time interval) 10^{-3}
0–**1000**	0.343	0.343
1001–**2000**	0.140	0.212
2001–**3000**	0.169	0.326
3001–**4000**	0.174	0.500
4001–**5000**	0.099	0.567
5001–**6000**	0.076	1.000

The $f(t)$ and $\lambda(t)$ presented in the above table, can be represented graphically in Figures 1.21 and 1.22.

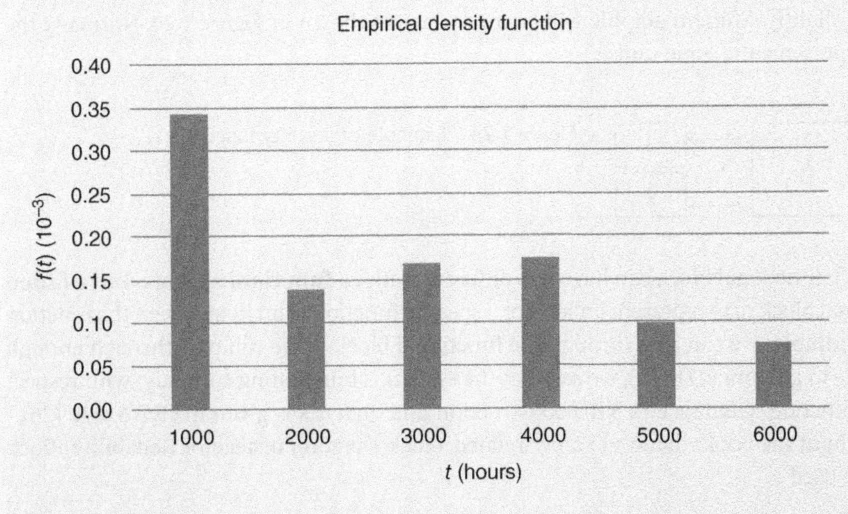

Figure 1.21 $f(t)$.

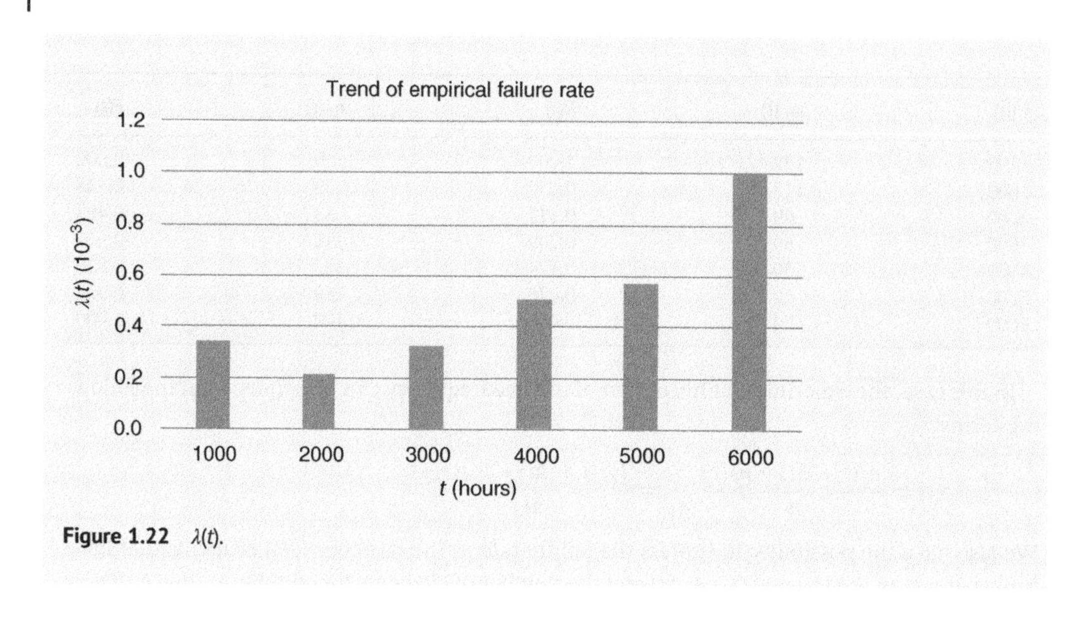

Figure 1.22 $\lambda(t)$.

1.9 Reliability Evaluation of Series and Parallel Structures

1.9.1 The Reliability Block Diagrams

A Reliability block diagram (RBD) illustrates the state of a system with several items. The diagram is made up of **functional blocks**, represented as rectangles, connected by lines. The RBD has a single starting point (a) and a single ending point (b), as shown in Figure 1.23.

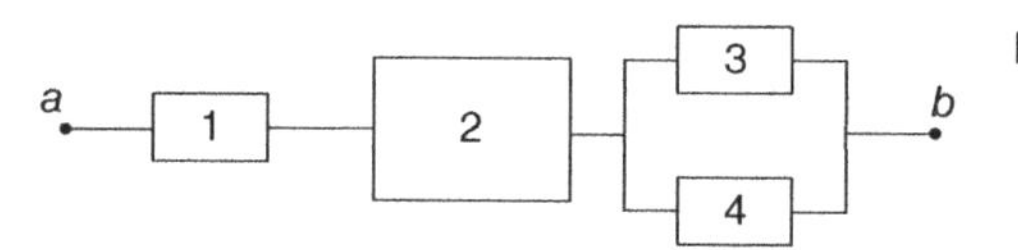

Figure 1.23 Example of a safety function.

In the book, a slightly different graphic will be used: the one shown in Figure 1.24. Normally the biggest block represents the logic unit.

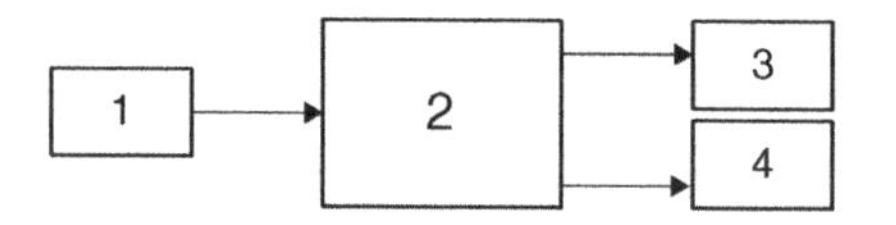

Figure 1.24 Example of a safety function.

In general, each functional block can have two different states, a **functioning state and a failed state**. A functional block may represent an item or a specific function of an item. When the function of the item is **available**, we can pass through the functional block. If we can pass through enough functional blocks to go from (a) to (b), we say that the system is functioning correctly, with respect to its specified function. Guidance to RBD construction and analysis is given in IEC 61078 [26].

Note: Throughout the book, the term Safety-related Block Diagram instead of Reliability Block Diagram will be used.

1.9.2 The Series Configuration

The series functional configuration, whose block diagram is shown in Figure 1.25, represents the simplest and most common Reliability model: a system *S*, composed of n elements, each one with Reliability $R_i(t)$.

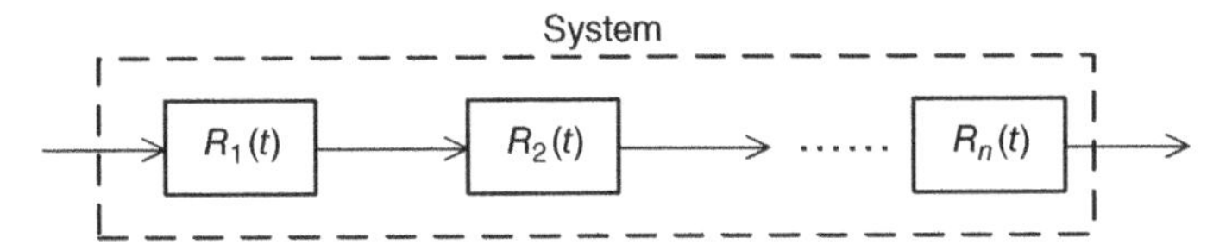

Figure 1.25 RBD for the series configuration.

In the simplified hypothesis of independent events, for which we can assume that the performance of every element, in terms of correct functioning or failure, does not depend on the condition assumed by other elements, the Reliability of the system corresponds to the product of the Reliability of single blocks:

$$\boldsymbol{R_s(t)} = R_1(t) \cdot R_2(t) \cdots R_n(t) = \prod_{i=1}^{n} \boldsymbol{R_i(t)}$$

Assuming:

$$\boldsymbol{R_i(t) = e^{-\lambda_i t}}$$

$$R_s(t) = \prod_{i=1}^{n} R_i(t) = e^{-\left(\sum_{i=1}^{n} \lambda_i\right) t}$$

Therefore, the failure rate of the system λ_S is the sum of the failure rates of the constituent elements λ_i (Figure 1.26). Consequently, the MTTF for the system, in hours, is:

$$MTTF_s = \int_0^{+\infty} R(t)dt = \left[\frac{e^{-\left(\sum_{i=1}^{n} \lambda_i\right) t}}{-\left(\sum_{i=1}^{n} \lambda_i\right)} \right]_0^{\infty} = \frac{1}{\sum_{i=1}^{n} \lambda_i} = \frac{1}{\lambda_S}$$

$$\lambda_S = \sum_{i=1}^{n} \lambda_i$$

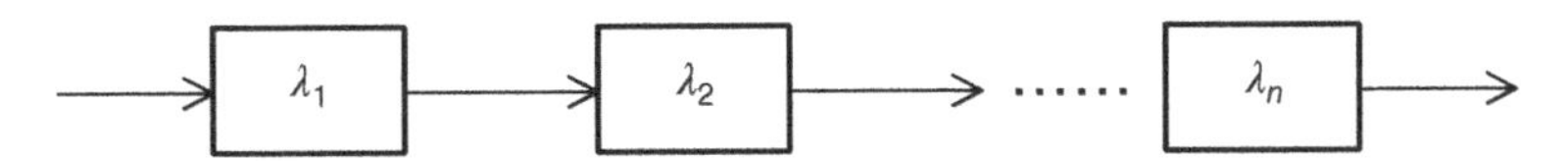

Figure 1.26 The failure rates of reliability functions $R_i(t)$.

In case of **two systems in series** have the same failure rate:

$$\lambda_S = 2\lambda$$

$$MTTF_s = \frac{1}{\lambda_S} = \frac{1}{2\lambda}$$

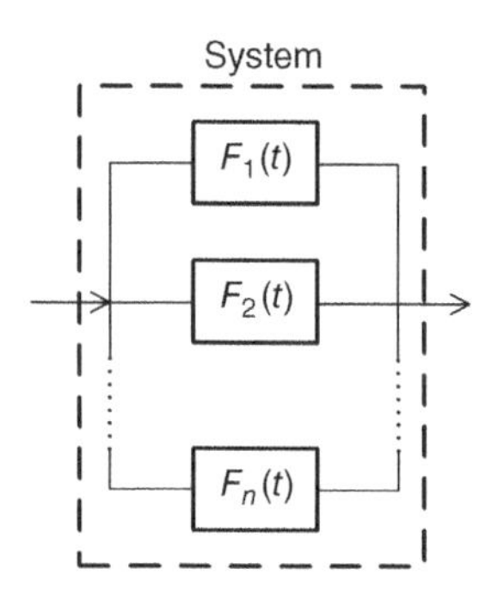

Figure 1.27 RBD for the parallel configuration.

1.9.3 The Parallel Configuration

The parallel function configuration, also called **redundant configuration**, is important when it is necessary to increase the Reliability of a system. The RBD for such a configuration is shown in Figure 1.27.

The system is not functioning when all the elements are faulty, therefore the unreliability of a system corresponds to the product of the unreliability of the constituting elements.

$$\boldsymbol{F_s(t)} = F_1(t) \cdot F_2(t) \cdots F_n(t) = \prod_{i=1}^{n} \boldsymbol{F_i(t)}$$

The Reliability function is, therefore

$$R_s(t) = 1 - F_s(t) = 1 - \prod_{i=1}^{n} F_i(t) = 1 - \prod_{i=1}^{n} \left(1 - e^{-\lambda_i \cdot t}\right)$$

Considering two systems in parallel:

$$R_s(t) = 1 - \left(1 - e^{-\lambda_1 \cdot t}\right) \cdot \left(1 - e^{-\lambda_2 \cdot t}\right) = e^{-\lambda_1 \cdot t} + e^{-\lambda_2 \cdot t} - e^{-(\lambda_1 + \lambda_2) \cdot t}$$

The MTTF for the system, in hours, is:

$$MTTF_s = \int_0^{+\infty} R(t)dt = \left[\frac{e^{-\lambda_1 \cdot t}}{-\lambda_1}\right]_0^{\infty} + \left[\frac{e^{-\lambda_2 \cdot t}}{-\lambda_2}\right]_0^{\infty} - \left[\frac{e^{-(\lambda_1 + \lambda_2) \cdot t}}{-(\lambda_1 + \lambda_2)}\right]_0^{\infty} = \frac{1}{\lambda_1} + \frac{1}{\lambda_2} - \frac{1}{\lambda_1 + \lambda_2}$$

In case the two systems have the same failure rate:

$$MTTF_s = \frac{3}{2 \cdot \lambda}$$

which is a 50% improvement compared with the single element. However, in the parallel case λ_s is a function of time.

1.9.3.1 Two Equal and Independent Elements

Let's consider an element having a constant failure rate λ

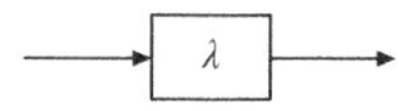

$$R_E(t) = e^{-\lambda t}$$

Let's now compare the configurations with two equal and independent elements functioning in series and in parallel, and plot the Reliability with a single element having the same constant failure rate.

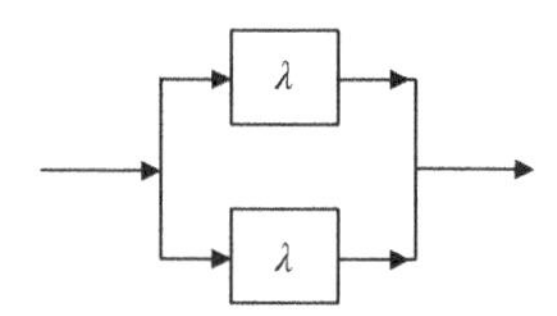

$$R_P(t) = 1 - \left(1 - e^{-\lambda t}\right)^2 = 2e^{-\lambda t} - e^{-2\lambda t}$$

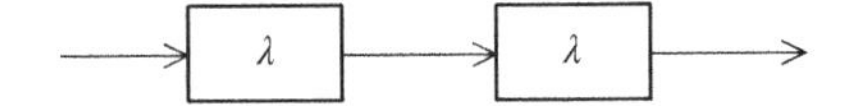

$$R_s(t) = \left(e^{-\lambda t}\right)^2 = e^{-2\lambda t}$$

In Figure 1.28 the three functions are compared.

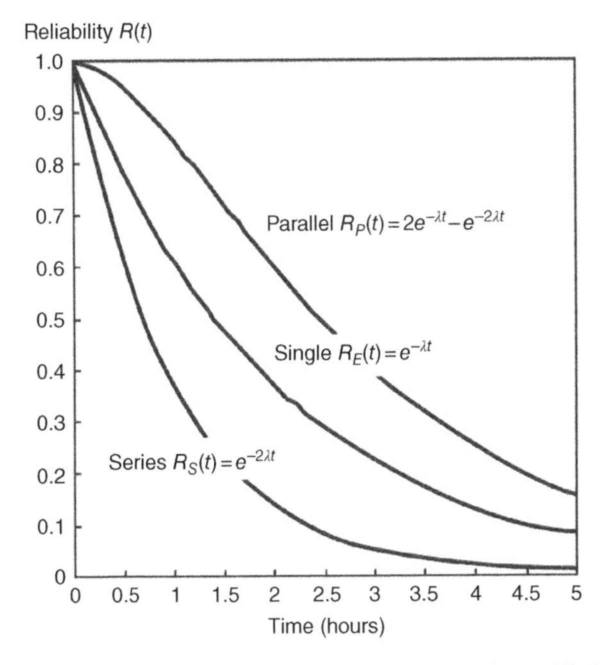

Figure 1.28 Comparison among basic configurations with the same failure rate.

Example 1 Elements in Series.

Please refer to Figure 1.29.

If the values of Reliability of each element, at generic time t, are 0.4, 0.7, and 0.9, respectively, the probability of the system functioning at time t is equal to:

$$R_s(t) = \prod_{i=1}^{n} R_i(t) = 0.4 \cdot 0.7 \cdot 0.9 = 0.252$$

$E_1 = 0.4$ → $E_2 = 0.7$ → $E_3 = 0.9$

Figure 1.29 RBD for a system made of three elements in a series configuration.

Example 2 Elements in Parallel.

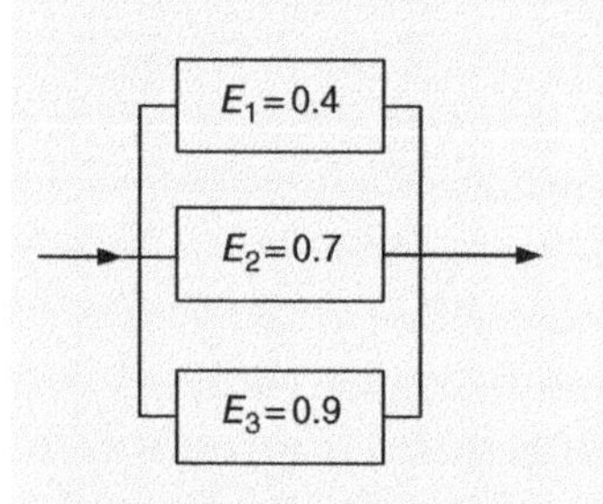

Figure 1.30 RBD for a system made of three elements in a parallel configuration.

Please refer to Figure 1.30.

If the values of Reliability of each element, at time t, are 0.4, 0.7, and 0.9, respectively, the probability of the system functioning at the time t becomes:

$$R_s(t) = 1 - F_s(t) = 1 - \prod_{i=1}^{n} F_i(t) = 1 - (0.6 \cdot 0.3 \cdot 0.1) = 0.982$$

Example 3 Elements Both in Parallel and in Series.

Please refer to Figure 1.31.

Applying the preceding relations in which active redundancy is positioned on the element E_1, with a Reliability value of 0.4, we obtain:

$$R_{S_{PARALLEL}}(t) = 1 - F_s(t) = 1 - \prod_{i=1}^{n} F_i(t) = 1 - (0.6 \cdot 0.6) = 0.64$$

$$R_s(t) = \prod_{i=1}^{n} R_i(t) = 0.64 \cdot 0.7 \cdot 0.9 = 0.4032$$

It is possible to note that redundancy of the less reliable element (E_1) increases the Reliability of the entire system. The Reliability of the entire system is now equal to 0.4, with an increase of 60% with respect to the value of 0.252 in the first example.

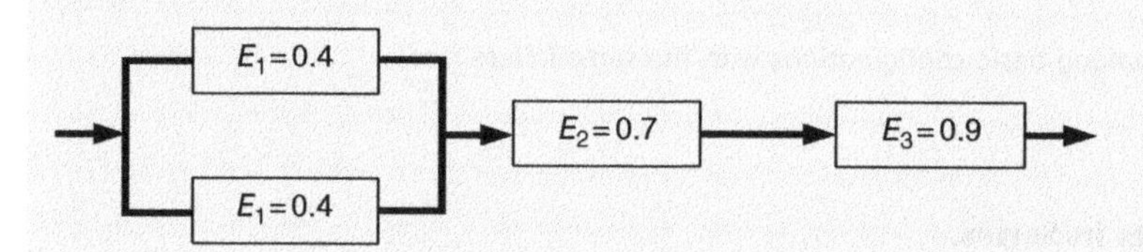

Figure 1.31 RBD for a system made of both parallel and serial configurations.

1.9.4 *M* Out of *N* Functional Configurations

A particular configuration is represented by a system where **at least *M* number of elements, out of a total of *N*, are functioning normally**. The configuration is also called M-out-of-N redundancy with $M \leq$ N or MooN. Here is the definition according to [16].

> ***[IEC 61511-1] 3.2 Terms and definitions***
> ***3.2.41 MooN**. SIS, or part thereof, made up of "N" independent channels, which are so connected, that "M" channels are sufficient to perform the SIF.*

For this configuration, we can assume a structure in which M elements are in active redundancy and the remaining elements (N–M) are in stand-by. A typical example is **a steel cable** formed of N strands that can withstand foreseen stress if at least M numbers of strands are intact.

When the concept is applied to a safety system:

- a **1oo1** is a single channel subsystem, for example a low pressure switch. When the sensor detects low pressure, it shuts down the process.
- a **1oo2** is a subsystem with redundant channels; for example, two pressure switches both detecting a low value. In case one triggers, the safety function will trigger.
- A **1oo3** is a subsystem with three channels; again, in case of a risk given by low pressure, each of the sensors have the same setting; as soon as one of the three detects low pressure, the safety system shuts down the process or the machine.
- In a **2oo3** subsystem, instead, we need two of the three sensors to trigger, in order to trigger the safety function.

The following table summarizes the main configurations used in Functional Safety.

Configuration	RBD	Reliability model	$MTTF_S$
Single element **(1oo1)**	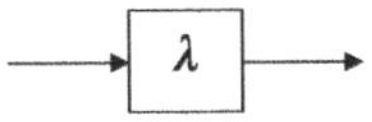 → λ →	$R(t) = e^{-\lambda t}$	$\frac{1}{\lambda}$
1-out-of-2 **(1oo2)**	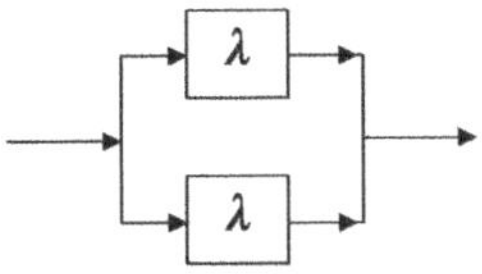 two parallel λ blocks	$R(t) = 2e^{-\lambda t} - e^{-2\lambda t}$	$\frac{9}{6 \cdot \lambda}$
1-out-of-3 **(1oo3)**	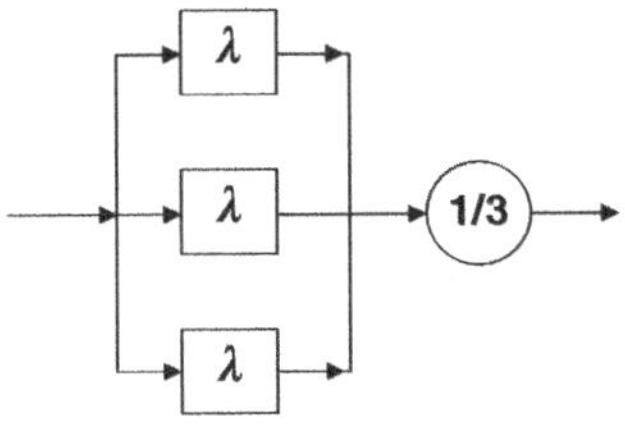 three parallel λ blocks → 1/3 →	$R(t) = 3e^{-\lambda t} - 3e^{-2\lambda t} + e^{-3\lambda t}$	$\frac{11}{6 \cdot \lambda}$
2-out-of-3 **(2oo3)**	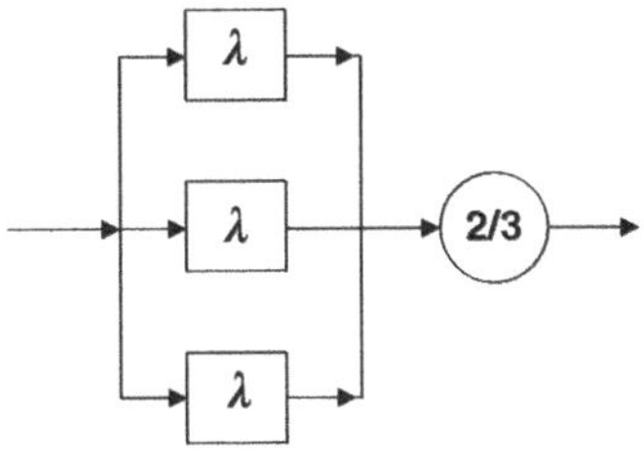 three parallel λ blocks → 2/3 →	$R(t) = 3e^{-2\lambda t} - 2e^{-3\lambda t}$	$\frac{5}{6 \cdot \lambda}$

1.10 Reliability Functions in Low and High Demand Mode

Functional safety was born having in mind the Reliability aspects of Safety-related Control Systems, designed to be activated upon hazardous process deviations; the latter is a process demand generating a Demand Rate of the safety system that protects people, the environment, and material assets.

The parameter used to indicate the Reliability of a Safety-related Control System is the **Unreliability function $F(t)$**. More precisely, there are two functions used, depending if the safety system is working in Low or in High demand mode. Just to give an example, the car airbag safety system is operating in low demand mode since it may remain inactive for years, until a demand occurs (due to a car crash).

In low demand mode, the parameter used to indicate the (un)reliability of a safety function is PFD_{avg}, while in high demand it is PFH(T).

> ***[IEC 61508-4] 3.6 Fault, failure and error***
> ***3.6.18 Average probability of dangerous failure on demand*** *(PFD_{avg}). Mean unavailability (see IEC 60050-191) of an E/E/PE safety-related system to perform the specified safety function when a demand occurs from the EUC or EUC control system.*

$$\boldsymbol{PFD_{avg}} = \frac{1}{T_i}\int_0^{T_i} PFD(t)dt$$

In high demand mode safety systems, the parameter used is PFH(t):

> ***[IEC 61508-4] 3.6 Fault, failure and error***
> ***3.6.19 Average frequency of a dangerous failure per hour*** *(PFH). Average frequency of a dangerous failure of an E/E/PE safety related system to perform the specified safety function over a given period of time.*

$$\boldsymbol{PFH(T)} = \frac{1}{T}\cdot\int_0^{T} w(t)dt$$

1.10.1 The PFD

The PFD(t) is the unreliability function $F(t)$ used in low demand mode. Hereafter its definition, supposing a constant failure rate λ:

> ***[IEC 61508-4] 3.6 Fault, failure and error***
> ***3.6.17 Probability of dangerous failure on demand*** *(PFD). Safety unavailability (see IEC 60050-191) of an E/E/PE safety-related system to perform the specified safety function when a demand occurs from the EUC or EUC control system.*

$$\boldsymbol{PFD} = 1 - e^{-\lambda t} \qquad \text{[Equation 1.10.1]}$$

Therefore, the instantaneous unreliability PFD(t) describes the probability that a safety system is not in a state to perform its required function, under given conditions, at a given instant of time, assuming that the required external resources are provided. Again, it is what we called so far $\boldsymbol{F(t)}$.

$$PFD = 1 - e^{-\lambda t}$$

Considering, for example, a valve with a $\lambda = 50.000$ FIT, its PFD(t) is shown in Figure 1.32. As you can see, the unreliability increases with time; after two years (17 520 hours), $PFD \approx 58\%$. After four years (35 040 hours), $PFD \approx 83\%$.

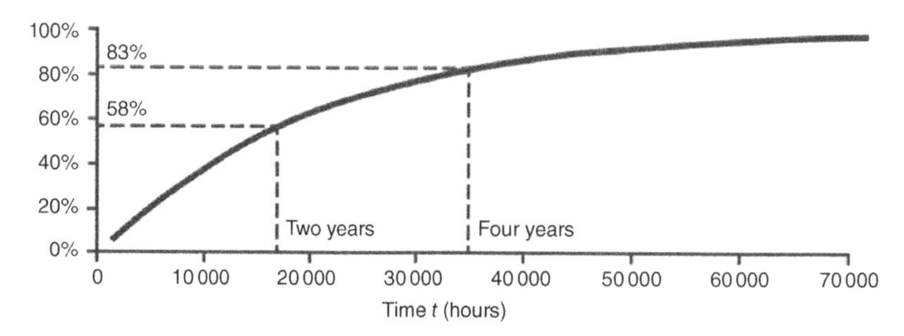

Figure 1.32 PFD over eight years for λ = 50 000 FIT.

Considering a $\lambda = 5.000$ FIT, a more realistic value, its PFD(t) is shown in Figure 1.33. First of all, the PFD has improved: after two years, PFD $\approx 8\%$ and after four years, PFD $\approx 16\%$. **Moreover**, the function can be approximated to a linear one, in case $\lambda{\cdot}t \ll 1$.

$$PFD = PFD(t) = 1 - e^{-\lambda t} \approx \lambda \cdot t$$

As it can be seen from both graphs, the System unreliability increases with time. Going back to the example of the airbag, that means its probability of failure will be very low when the car is new, and it will increase month by month. That is valid for all the elements of a **Safety Instrumented System** (SIS), that is made by one or more sensors, a logic unit, and one or more actuators.

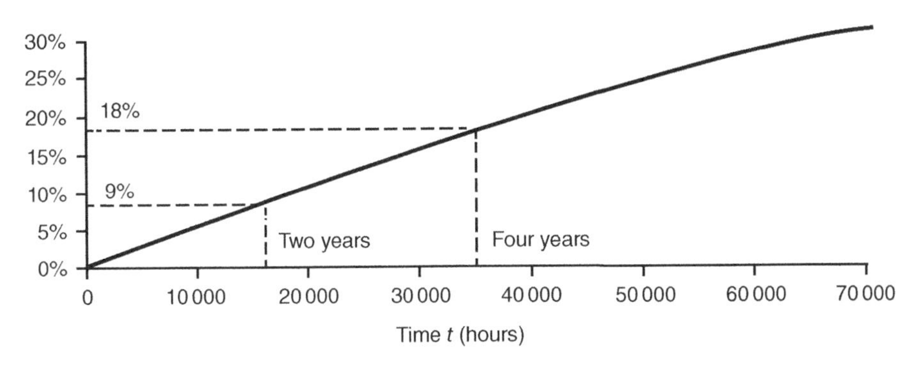

Figure 1.33 PFD over eight years for λ = 5000 FIT.

1.10.1.1 The Protection Layers

A SIS may fail while in passive state and the failure may remain hidden until a demand occurs from the process or until the system is tested.

Let's suppose the pressure in a vessel is controlled by a pressure transmitter and the process control system has to keep the value around a certain set point.

In case the pressure increases above a certain threshold (PSH), an alarm is generated (PAH). In case the value goes "out of control," a safety pressure switch, set at PSHH, shuts down the process (Figure 1.34).

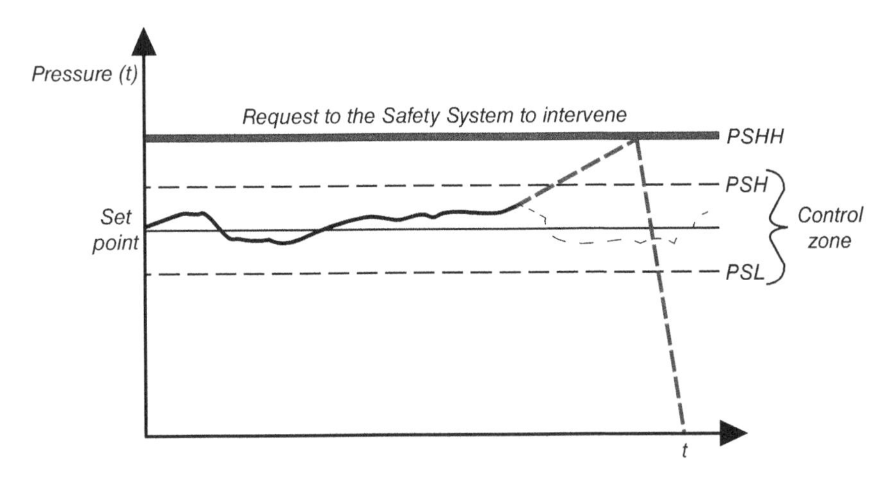

Figure 1.34 Request upon the safety system.

Figure 1.35 EUC, the process control system, and safety system.

We see that there are two protection layers: a Control one and a Safety one. They normally do not share the same field components. In our example, the pressure transmitter belongs to the Control Layer, while the safety pressure switch to the Safety Layer.

Normally, the process control system keeps the pressure around the set point. Very rarely the pressure goes out of control, and at that moment the Safety System intervenes: the issue is that it may have failed in the meantime. A SIS is so-called an **independent protection layer.** It is installed to mitigate the risk associated with the operation of a process that is normally hazardous, and it is called the **Equipment Under Control or EUC** (Figure 1.35).

A Safety Instrumented Function (SIF) is implemented with a SIS that is intended to achieve or maintain a safe state for the EUC with respect to a specific process demand (a high pressure for example). A SIS may consist of one or more SIFs.

1.10.1.2 Testing of the Safety Instrumented System

SISs are normally dormant and their failure may remain undetected, or hidden, until there is a demand upon them (a high temperature or pressure, for example) or until the system is tested if it is still working properly.

There are two types of tests that can be done on such systems.

Diagnostic Testing. They are done automatically by the component itself, or by the logic solver, or by other elements of the safety system. The extent to which this automatic testing reveals a failure is called Diagnostic Coverage (DC). The failures that can be detected in this way are defined as Detectable, the remaining failures are called Undetectable.

Function Testing. The objective of the function testing is especially to reveal the **undetectable failures** and to verify that the system is still able to perform its required function, in case a process demand occurs. Function testing, defined in IEC 61508 as **Proof Test**, is normally done manually, or initiated manually. The time interval between two function tests is indicated as T_i and, in case of a **perfect Proof Test**, the item is considered "as new," after such a test. Please refer to Chapter 3 for the definition of Proof Test and further details.

Do not get confused between Function Test and Functional Test. In literature, you may find that the Proof test is defined as a Function Test, as well as Periodic Test, while the DC is also defined as a Functional Test.

1.10.2 The PFD$_{avg}$

For a single channel subsystem, the Average PFD is defined as

$$\boldsymbol{PFD_{avg}} = \frac{1}{T_i}\int_0^{T_i} PFD(t)dt = \frac{1}{T_i}\int_0^{T_i}\left(1-e^{-\lambda\cdot t}\right)dt \cong \frac{1}{T_i}\int_0^{T_i}\lambda\cdot tdt = \frac{\lambda\cdot T_i^2}{2\cdot T_i} = \frac{\boldsymbol{\lambda\cdot T_i}}{\boldsymbol{2}}$$

[Equation 1.10.2]

T_i is the time when the system is function tested. The PFD(T) of a SIF, that is periodically tested, is represented by a saw tooth curve, with a probability ranging from low, just after a test, to a maximum, just before the next test.

Its average value, or **PFD$_{avg}$**, is represented in Figure 1.36.

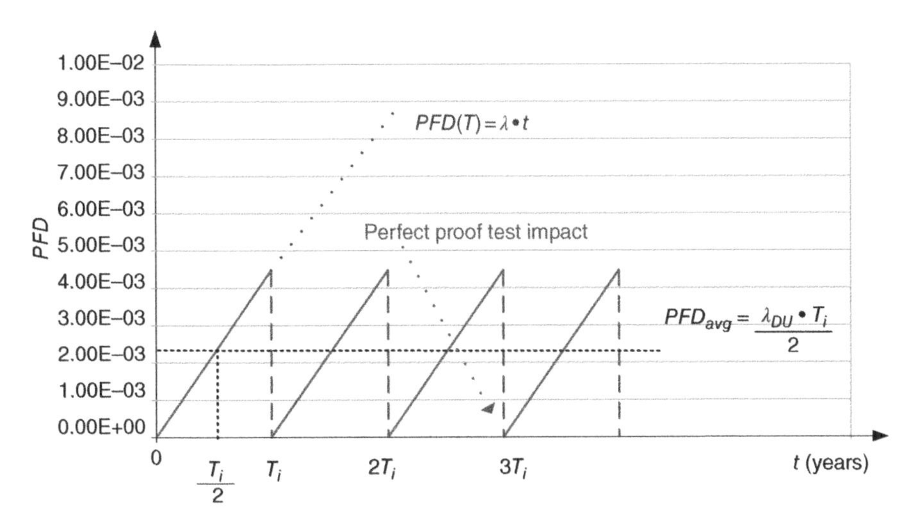

Figure 1.36 PFD(t).

1.10.2.1 Dangerous Failures

When dealing with **Safety Critical Systems**, the important failures are the dangerous ones. Those can be divided into Dangerous Detectable by the Diagnostic tests and Dangerous Undetectable.

Dangerous Undetected Failures (DU) prevent the activation, on demand, of the safety system and are also called **dormant failures**.

Dangerous Detected Failures (DD) may be found immediately when they occur, for example, by an automatic built-in self-test. A short circuit on a normally closed free voltage contact can be revealed with the so-called "trigger" function, now available in almost all Safety-related Control Systems (Chapter 3).

1.10.2.2 How to Calculate the PFD$_{avg}$

In low demand mode, **Dangerous Detected failures** do not play a role in the Unreliability of a Safety System, since, often, they are detected as soon as they appear, and the process is immediately shut down. Therefore, the only significant failures that influence the value of the PFD$_{avg}$ are the DU failures. Therefore, Equation 1.10.2 can be written as:

$$PFD_{avg} = \frac{\lambda_{DU} \cdot T_i}{2} \qquad \text{[Equation 1.10.3]}$$

The test interval T_i is decided based upon the demand rate, so that there is a fair chance that a Dangerous Undetected fault is revealed and corrected before a demand occurs, such that a hazardous event is avoided.

1.10.3 The PFH

The starting point for the calculation of the PFH is the **Failure Frequency**. Hereafter its definitions [29]:

> ***[ISO/TR 12489] 3.1 Basic Reliability concepts***
> ***3.1.22 Failure Frequency*** *(or Unconditional Failure Intensity) w(t). Conditional probability per unit of time that the item fails between t and t + dt, provided that it was working at time 0.*

In high demand mode, the unreliability value used is the **Average Failure Frequency**. Here its definitions [29]:

> ***[ISO/TR 12489] 3.1 Basic Reliability concepts***
> ***3.1.23 Average Failure Frequency*** $\overline{w}(t_1, t_2)$,$\overline{\mathbf{w}}(\boldsymbol{T})$,$\overline{w}$. *Average value of the time-dependent failure frequency over a given time interval.*

$$\overline{w}(T) = \frac{1}{T} \cdot \int_0^T w(t)dt$$

The average failure frequency is also called "**Probability of Failure per Hour**" (PFH) by the standards related to functional safety of safety-related instrumented systems:

$$PFH = PFH_D = \overline{w}(T)$$

However, the correct term for PFH is **Average Failure frequency**. That is the reason why, in the new edition of IEC 62061, PFH is defined as the following [12]:

> ***[IEC 62061] 3.2 Terms and definitions***
> ***3.2.29 Average Frequency of a Dangerous Failure Per Hour PFH or*** $\boldsymbol{PFH_D}$. *Average frequency of dangerous failure of an SCS to perform a specified safety function over a given period of time.*

$$PFH(T) = \frac{1}{T} \cdot \int_0^T w(t)dt$$

Don't be confused by the fact the PFH_D is sometimes written without the subscript *D*. **IEC 62061 uses PFH while ISO 13849 uses** $\mathbf{PFH_D}$**, but they mean exactly the same thing.**

Moreover, ISO 13849-1 kept the old terminology: **Probability of Failure per hour**, even if it is not properly correct.

1.10.3.1 Unconditional Failure Intensity *w(t)* vs Failure Density *f(t)*

The average of the **unconditional failure intensity** $\boldsymbol{w(t)}$ is different from the **failure density** $\boldsymbol{f(t)}$. Let's now understand what the difference is.

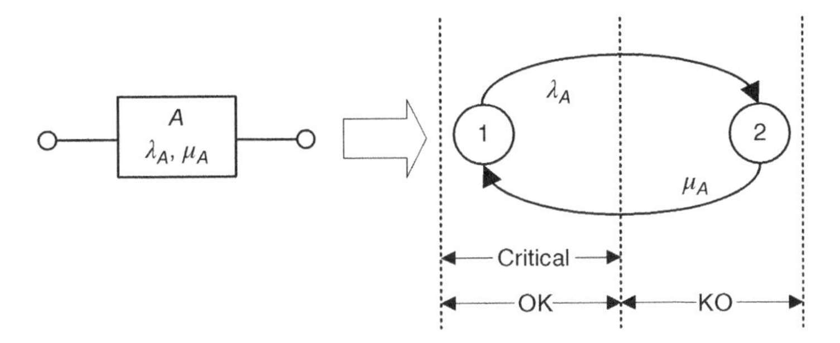

Figure 1.37 Single repairable component.

Please consider a single component that can be in two possible states: OK (1) and KO (2). Figure 1.37 represents the situation and the corresponding Markov Graph failure rate.

where:

- λ_A: is the failure rate of the component.
- μ_A: is the component restoration rate. Please consider that the restoration rate has the same mathematical properties of the failure rate.

Since the model includes the restoration transition, the system is considered repairable; in other terms, it can be brought to an "as new status" after a repair or a Proof Test. In general, the unconditional failure intensity $w(t)$ is a saw-teeth curve while $f(t)$ is decreasing and goes to 0 when t goes to infinity.

Considering the following data (example taken from [29] annex C):

- $\lambda_A = 5{\cdot}10^{-4}$ (h^{-1})
- $\mu_A = 0.01$ (h^{-1})
- $\tau = 2160$ hours

the graphs are shown in Figure 1.38:

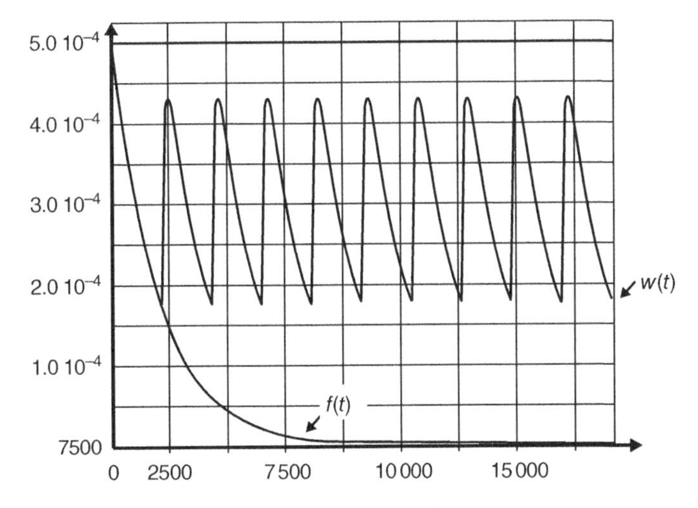

Figure 1.38 Comparison between $w(t)$ and $f(t)$.

1.10.3.2 Reliability Models Used to Estimate the PFH

In high demand mode, the two standards, ISO and IEC, use different models to come to the estimation of the unreliability function.

IEC 62061 uses the RBD method and it assumes the systems (Architectures) as **non-repairable.** **ISO 13849-1** uses Markov Chains and it assumes the systems (Categories) as **repairable**; please refer to § 6.2.5.

That seems a major difference that makes the two approaches irreconcilable. In reality that is not the case and the reason is that in high demand, normally, **the safety control system is the ultimate safety layer:** that is the assumption in both ISO 13849-1 and IEC 62061. Where a safety-related control system is working in high demand or in continuous mode, and it is the ultimate safety layer then, the overall safety-related control system failure **will lead directly to a potentially hazardous situation,** regardless if it is considered repairable or non-repairable.

In case the safety-related control system is the ultimate safety layer $w(t) = f(t)$; that means:

$$PFH(T) = \frac{1}{T} \cdot \int_0^T f(t)dt = \frac{F(T)}{T}$$

Therefore, PFH(T) is the average unavailability of $F(t)$.

Supposing a constant failure rate λ and $\lambda \cdot t \ll 1$:

$$PFH = \frac{F(T)}{T} = \frac{1-R(T)}{T} = \frac{1-e^{\left(-\int_0^T \lambda(t)dt\right)}}{T} = \frac{1-e^{-\lambda t}}{T} \approx \frac{\lambda \cdot T}{T} = \lambda = \frac{1}{MTTF}$$

1.11 Weibull Distribution

It is now clear that, in Functional Safety, **the failure rate of any component has to be constant:** the issue are components subject to wear, like contactors and solenoid valves, since their failure rates are usually not constant. Therefore, **the exponential curve is not helpful to model their life distribution: that is where the Weibull distribution comes in.**

The Weibull distribution [28] is one of the most widely used Life Distributions in Reliability analysis. The distribution is named after the Swedish professor Waloddi Weibull (1887–1979), who developed the distribution for modeling the strength of materials.

The Weibull distribution is very flexible and can, through an appropriate choice of parameters, model many types of failure rate behaviors. **It is therefore used to model the failure behavior of electromechanical components.**

1.11.1 The Probability Density Function

The Weibull **PDF** is the following:

$$f(t) = \beta \cdot \frac{t^{\beta-1}}{\eta^{\beta}} \cdot e^{-\left(\frac{t}{\eta}\right)^{\beta}}$$

- η is called the **Life Characteristics**
- β is referred as the **shape** parameter

Please notice that when $\beta = 1$, the Weibull becomes the Exponential distribution,

In Figure 1.39, the Weibull distribution is plotted for $\eta = 1$ and for some values of β.

Figure 1.39 Weibull $f(t)$.

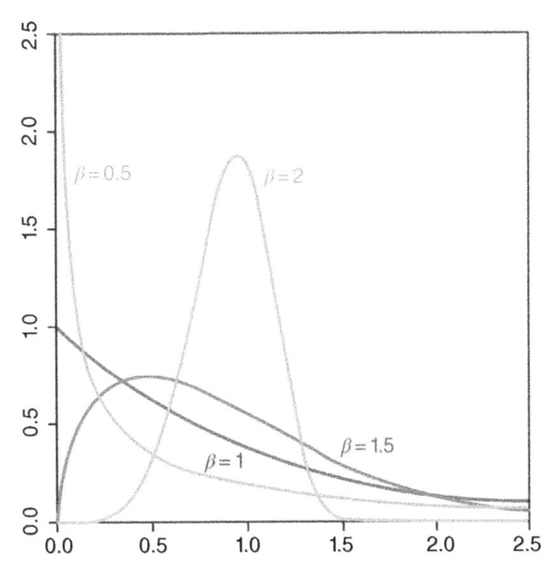

1.11.2 The Cumulative Density Function

The Weibull **Cumulative Density Function** is the following:

$$F(t) = 1 - e^{-\left(\frac{t}{\eta}\right)^{\beta}}$$

Please notice that when $t = \eta$

$$F(\eta) = 1 - e^{-\left(\frac{\eta}{\eta}\right)^{\beta}} = 1 - e^{-(1)^{\beta}} = 1 - e^{-1} = 0.63$$

Therefore, regardless of the distribution shape parameter β, when $t = \eta$, the Probability of unavailability $F(t)$ of the component = 63%.

The parameter $\boldsymbol{\eta}$ is defined as the **characteristic** lifetime of the distribution. Please refer to Figure 1.40.

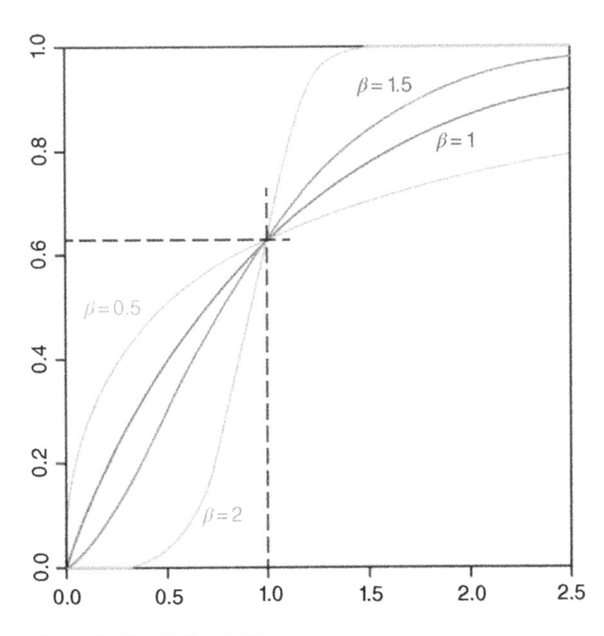

Figure 1.40 Weibull $F(t)$.

1.11.3 The Instantaneous Failure Rate

Finally, the **Instantaneous Failure Rate** is the following:

$$\lambda(t) = \frac{f(t)}{1 - F(t)} = \beta \cdot \frac{t^{\beta-1}}{\eta^{\beta}}$$

When $\beta = 1$, the failure rate is constant and equal to:

$$\lambda = \frac{1}{\eta}$$

In this case, the Weibull distribution is identical to the exponential one. Please refer to Figure 1.41.

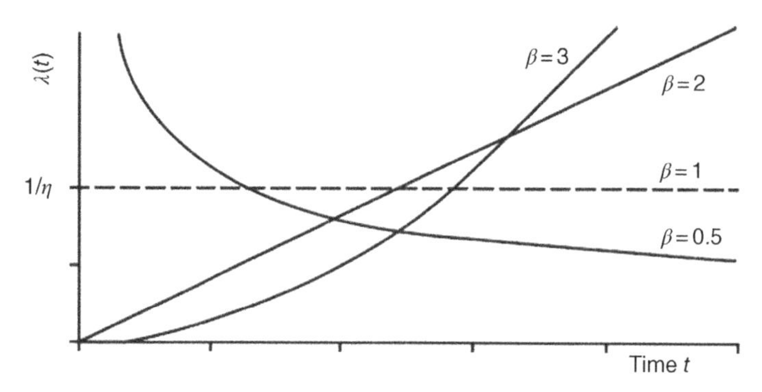

Figure 1.41 Weibull $\lambda(t)$.

When $\beta < 1$, the failure rate decreases with time. Both electronic and mechanical systems may initially have high failure rates. Manufacturers conduct production process control, production acceptance tests, "burn-in," or reliability stress screening (RSS) to prevent early failures before delivery to customers. Therefore, shape parameters of less than one indicate the following:

- lack of adequate process control;
- inadequate burn-in or stress screening;
- production problems, dis-assembly, poor quality control;
- overhaul problems;
- mixture of populations;
- run-in or wear-in.

Many electronic components during their useful life show a decreasing instantaneous failure rate, thus featuring shape parameters less than 1. Preventive maintenance on such a component is not appropriate, as old parts are better than new. Please refer to Figure 1.42.

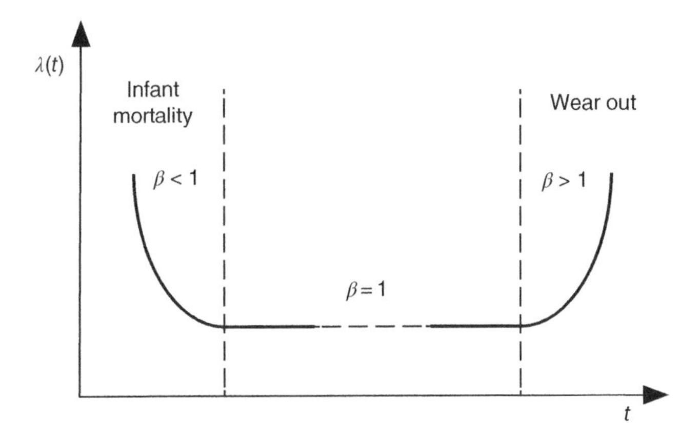

Figure 1.42 Weibull and the bathtub.

When $\beta > 1$, the failure rate increases with time. That behavior is attributed, first of all, to components in the wear-out, or end-of-life, phase. Some typical examples of these cases are:

- wear;
- corrosion;
- crack propagation;
- fatigue;
- moisture absorption;
- diffusion;
- evaporation (weight loss);
- damage accumulation.

Design measures have to ensure that those phenomena do not significantly contribute to the probability of product failure during the expected operational life; however, that is typically the behavior of Contactors and Solenoid valves during their entire life.

1.11.4 The Mean Time to Failure

The MTTF of Weibull distribution is not an easy function. It can be demonstrated that:

$$MTTF = \int_{0}^{+\infty} R(t)dt = \eta \cdot \Gamma\left(\frac{1}{\beta} + 1\right)$$

where Γ is the **Gamma Function,** whose parameters are shown in Table 1.4.

$\Gamma(1 + x) = x \cdot \Gamma(x)$ and $\Gamma(x) \approx \frac{1}{x} + \sum_{k=0}^{7} b_k \cdot x^k$

where $0 \leq x \leq 1$.

Table 1.4 Parameters of gamma function.

Coeff	Value
b_0	−0.577 191 652
b_1	0.988 205 891
b_2	−0.897 056 937
b_3	0.918 206 857
b_4	−0.756 704 078
b_5	0.482 199 394
b_6	−0.193 527 818
b_7	0.035 868 343

1.11.4.1 Example

Let's consider a choke valve, the example is taken from [57], that is assumed to have a Weibull distribution with:

- shape parameter $\beta = 2.25$ and
- scale parameter $1/\eta = 1.15\ 10^{-4}\ \mathrm{h}^{-1}$.

The probability that the valve survives six months (4380 hours) in continuous operation is:

$$R(t) = e^{-\left(\frac{t}{\eta}\right)^{\beta}} = e^{-(1.15\cdot 10^{-4}\cdot 4380)^{2.25}} \cong 0.81$$

Which means 81% probability. The **MTTF** is

$$MTTF = \eta\cdot\Gamma\left(\frac{1}{\beta}+1\right) = \frac{\Gamma\left(\frac{1}{2.25}+1\right)}{1.15\cdot 10^{-4}} = \frac{\Gamma(1.44)}{1.15\cdot 10^{-4}} = \frac{0.886}{1.15\cdot 10^{-4}} \cong 7704\ \text{hours}$$

If the shape parameter $\beta \neq 1$, the MTTF is not equal to η and the MTTF of Weibull distribution is not the time when 63% of the component under test has failed. In case $\beta = 1$, the MTTF is equal to η and the MTTF of Weibull distribution is the time when 63% of the component under test has failed.

$$MTTF = \eta\cdot\Gamma\left(\frac{1}{1}+1\right) = \eta\cdot\Gamma(2) = \eta\cdot 1! = \eta$$

The Median Life (50% failure probability) is

$$t_m = \eta\cdot(\ln 2)^{\frac{1}{\beta}} \cong 7389\ \text{hours}$$

A Valve that has survived the first six months, will again survive six months with a probability of:

$$\frac{R(t_1+t_2)}{R(t_1)} = \frac{e^{-(1.15\cdot 10^{-4}\cdot 8760)^{2.25}}}{e^{-(1.15\cdot 10^{-4}\cdot 4380)^{2.25}}} \cong \frac{0.36}{0.81} \cong 0.44$$

which means 44% probability, which is significantly less than the probability that a new valve would survive six months, of course. That fact the failure rate is increasing plays an important role!

1.12 B_{10D} and the Importance of T_{10D}

Once a Safety-related block diagram has been defined, for each safety-related system, the technique used to calculate the probability of hardware failure is based upon specific Markovian formulae obtained from Taylor's series and slightly conservative underlying hypotheses, among which **a constant failure rate**.

The starting point of the analysis is a probabilistic model of the Safety-related control system. That consists in:

- Identifying all the states of the system.
- Analyzing the transitions of the system from state to state, according to events that may arise during its life, like failures, repairs, and tests.

One of the advantages of the Markov models is that they can be modeled with equations.

1.12.1 The $B_{X\%}$ Life Parameter and the B_{10D}

For components having mechanical wear, a straight MTTF cannot be defined. That is the reason why the concept of **$B_{X\%}$ Life** was introduced.

The $B_{X\%}$ life metric originated in the ball and roller bearing industry but has become a product lifetime metric used across a variety of industries. The $B_{X\%}$ life is the lifetime metric that takes to fail $X\%$ of the units in a population. For example, if an item has a B_{10} life of 1000 km, this means that 10% of the population will have failed by the time it reached 1000 km in operation. Other percentages can be defined; however, in high demand mode applications, B_{10} is the one used.

B_{10D} is the mean number of cycles until 10% of the components failed dangerously and is used for components having mechanical wear.

If only the B_{10} of a component were available, B_{10D} can be estimated as twice the B_{10} (50% dangerous failure).

Sometimes, the component manufacturer provides the B_{10} value and the **Ratio of dangerous failures** (RDF). The relation among those parameters is the following.

$$B_{10D} = \frac{B_{10}}{RDF}$$

Linked to B_{10} is the **number of operations** a component does **in a year**: n_{op}

The lower the RDF, the higher is the B_{10D}, and therefore the longer is the T_{10D}. However, both ISO 13849-1 and IEC 62061 limit such a value; in particular, if the ratio of dangerous failure is estimated to be less than 0.5 (50% dangerous failure), the T_{10D} of the component **is limited to $2 \times T_{10}$**.

T_{10D} is linked to the number of operations n_{op} by the following formula:

$$T_{10D} = \frac{T_{10}}{RDF} = \frac{B_{10}}{RDF \cdot n_{op}} = \frac{B_{10D}}{n_{op}}$$

The ratio of dangerous failure is estimated as 50% of dangerous failures if no information is available.

1.12.1.1 Example

Let's assume a component has a $B_{10} = 10^6$, an RDF = 30%, and is used six times per hour (n_{op} = 6·8·240 = 11 520).

$$T_{10} = \frac{B_{10}}{n_{op}} = 87 \text{ years}$$

$$B_{10D} = \frac{B_{10}}{RDF} = 3.3 \cdot 10^6$$

Therefore T_{10D}:

$$T_{10D} = \frac{B_{10D}}{n_{op}} = 286 \text{ years}$$

However, given the standard limitation, T_{10D} is limited to 87·2 = 174 years.

1.12.2 How λ_D and $MTTF_D$ are Derived from B_{10D}

In Functional Safety and, in particular, in high demand Mode of Safety-related Control Systems, **B_{10D} is used to indicate the Reliability of components that do not have a constant failure rate.**

As it will be described later in the chapter, since an "approximated" constant failure rate will be associated to those components, in order to limit the error on the calculation of the PFH_D of the safety Function, **the usage of the component will be limited to when it reaches the B_{10D} number of operation**. That means the component must be replaced when B_{10D} is reached, or earlier, if its Mission Time is shorter.

Since the cycle duration corresponds to the reciprocal of the operating frequency $\boldsymbol{n_{op}}$, the point in time $\boldsymbol{T_{10D}}$, at which the element has completed B_{10D} cycles is:

$$T_{10D} = \frac{B_{10D}}{n_{op}}$$

Given a component with the unreliability function $F(t)$,

$$F(t) = 1 - e^{-\lambda \cdot t}$$

the probability of dangerous failures that a component has, when T_{10D} is reached is

$$F(t = T_{10D}) = 1 - e^{-\lambda_D \cdot T_{10D}}$$

But we know that the probability $F(t = T_{10D}) = 10\%$, therefore:

$$\frac{1}{10} = 1 - e^{-\lambda_D \cdot T_{10D}}$$

$$\lambda_D = \frac{1}{T_{10D}} \cdot ln\,\frac{10}{9} \cong \frac{1}{10 \cdot T_{10D}} = \frac{n_{op}}{10 \cdot B_{10D}}$$

In case of a constant failure rate,

$$\lambda = \frac{1}{MTTF}$$

Therefore, the following is the formula to be used to calculate the $MTTF_D$ from B_{10D}:

$$MTTF_D = \frac{B_{10D}}{0.1 \cdot n_{op}}$$ [Equation 1.12.2]

1.12.3 The Importance of the Parameter T_{10D}

If a component has a B_{10D}, that means its failure rate is not constant over time. The B_{10D} was chosen not by chance! It can be demonstrated that by limiting the product life to T_{10D}, there is no significant error in the estimation of the PFH_D by assuming a constant failure rate, compared to the use of a Weibull distribution.

Let's calculate the error, comparing three Weibull distributions: with $\beta = 1$ (constant failure rate), $\beta = 2$, and with $\beta = 3$. Please refer to Table 1.5.

Table 1.5 Parameters comparison for three Weibull distributions.

	$\beta = 1$	$\beta = 3$	$\beta = 4$
$F(t)$	$1 - e^{-\frac{t}{\eta}}$	$1 - e^{-\left(\frac{t}{\eta}\right)^3}$	$1 - e^{-\left(\frac{t}{\eta}\right)^4}$
$F(T_{10D}) = 0.1$	$\eta = \frac{T_{10D}}{0.1}$ (Note 1)	$\eta = \frac{T_{10D}}{\sqrt[3]{0.1}}$ (Note 2)	$\eta = \frac{T_{10D}}{\sqrt[4]{0.1}}$ (Note 3)
$\lambda_D(t)$	$\frac{0.1 \cdot n_{op}}{B_{10D}} = constant$	$\beta \cdot \frac{t^{\beta-1}}{\eta^\beta}$	$\beta \cdot \frac{t^{\beta-1}}{\eta^\beta}$
$\boldsymbol{\lambda_D(t)}$	$\boldsymbol{\frac{0.1 \cdot n_{op}}{B_{10D}} = \frac{1}{\eta}}$	$\boldsymbol{3 \cdot \frac{t^2}{\eta^3}}$	$\boldsymbol{4 \cdot \frac{t^3}{\eta^4}}$
$\lambda_D(t)$ with η as per Note 1, 2 or 3	$\frac{0.1}{T_{10D}}$	$3 \cdot \frac{t^2}{\left(\frac{T_{10D}}{\sqrt[3]{0.1}}\right)^3}$	$4 \cdot \frac{t^3}{\left(\frac{T_{10D}}{\sqrt[4]{0.1}}\right)^4}$
$T_{10D} \cdot \lambda_D(t)$	0.1	$3 \cdot \frac{t^2}{\left(\frac{T_{10D}}{\sqrt[3]{0.1}}\right)^3} \cdot T_{10D}$	$4 \cdot \frac{t^3}{\left(\frac{T_{10D}}{\sqrt[4]{0.1}}\right)^4} \cdot T_{10D}$
$\boldsymbol{T_{10D} \cdot \lambda_D(t)}$	**0.1**	$\boldsymbol{0.3 \cdot \left(\frac{t}{T_{10D}}\right)^2}$	$\boldsymbol{0.4 \cdot \left(\frac{t}{T_{10D}}\right)^3}$

Hereafter, in the three Notes, the detailed calculations of "η":

Note 1: $\boldsymbol{\beta = 1} \rightarrow F(t = T_{10D}) = 0.1 \Rightarrow 1 - exp\left[-\left(\frac{t}{\eta}\right)^1\right] = 0.1 \Rightarrow exp\left[-\left(\frac{t}{\eta}\right)^1\right] = 0.9 \Rightarrow \left[-\left(\frac{t}{\eta}\right)^1\right] = -0.1 \Rightarrow$
$t = \boldsymbol{0.1\,\eta = T_{10D}}$

Note 2: $\boldsymbol{\beta = 3} \rightarrow F(t = T_{10D}) = 0.1 \Rightarrow 1 - exp\left[-\left(\frac{t}{\eta}\right)^3\right] = 0.1 \Rightarrow exp\left[-\left(\frac{t}{\eta}\right)^3\right] = 0.9 \Rightarrow \left[-\left(\frac{t}{\eta}\right)^3\right] = -0.1 \Rightarrow$
$t = \boldsymbol{\sqrt[3]{0.1}\,\eta = T_{10D}}$

Note 3: $\boldsymbol{\beta = 4} \rightarrow F(t = T_{10D}) = 0{,}1 \Rightarrow 1 - exp\left[-\left(\frac{t}{\eta}\right)^4\right] = 0{,}1 \Rightarrow exp\left[-\left(\frac{t}{\eta}\right)^4\right] = 0{,}9 \Rightarrow \left[-\left(\frac{t}{\eta}\right)^4\right] = -0{,}1 \Rightarrow$
$t = \boldsymbol{\sqrt[4]{0{,}1}\,\eta = T_{10D}}$

Figures 1.43 and 1.44 show the three dangerous failure rates, standardized at T_{10D}.

$$\lambda_D(t) = \frac{0.1 \cdot n_{op}}{B_{10D}} = constant$$

$$\lambda_D(t) = 3 \cdot \frac{t^2}{\eta^3}$$

$$\lambda_D(t) = 4 \cdot \frac{t^3}{\eta^4}$$

At the beginning, the "surrogate" failure rate (the one with $\beta = 1$) is greater but then the "real" failure rate (the one with $\beta = 3$ or with $\beta = 4$) becomes greater. At the point $t = T_{10D}$ the area under both curves is equal and its value is **0.1**: it means that the two failure rates in the interval from $t = 0$ to $t = T_{10D}$ provide the same estimation.

That also shows that for $t > T_{10D,}$ the estimation no longer works. If the component is used longer than T_{10D}, then the surrogate $MTTF_D$ does not reflect the real behavior, and the failure rate would be under-estimate in a dangerous way.

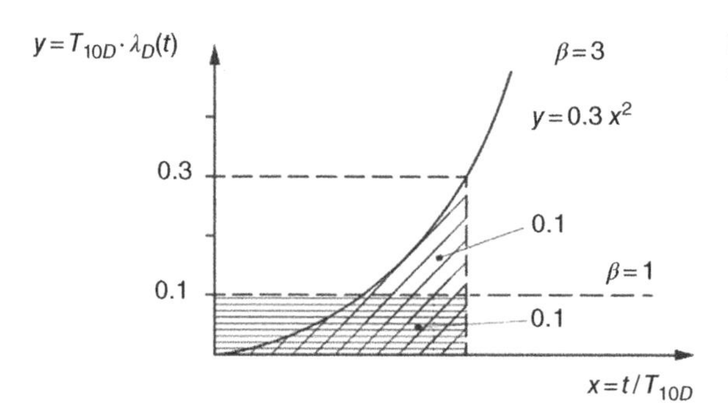

Figure 1.43 Different failure rate functions for β = 3.

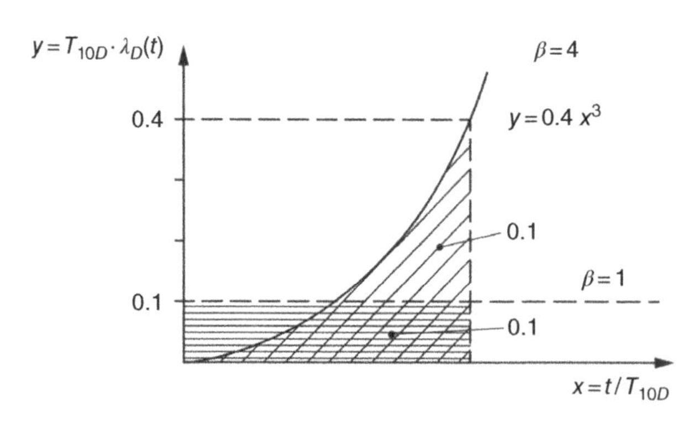

Figure 1.44 Different failure rate functions for β = 4.

That is the meaning of the following note in IEC 62061:

> ***[IEC 62061] 7.3.4 Failure rate of subsystem element***
> ***7.3.4.2 Relationship of relevant parameters** [...]*
>
> ***Note 4:** For electronic systems, the exponential distribution is applicable. For non-electronic systems, the exponential distribution is not applicable. The Weibull distribution (see also IEC 61649) is more appropriate, but parameters and calculations are difficult to apply. However, when using exponential distribution for non-electronic components within the limits of T_{10D}, then the results of the calculations are pessimistic and the formula with $1 - e^{-\lambda t}$ could be applied as a simplified method.*

1.12.4 The Surrogate Failure Rate

Since Markov only works with a constant failure rate, a **Surrogate failure rate** (also called **Substitute failure rate**) is defined for components subject to wear that do not show such characteristics.

Actually, the first step is to use the Weibull distribution to estimate the best fit value of B_{10}. Examples of those methods are given in standards like ISO 19973-1 [21] and ISO 19973-2 [50], valid for pneumatic components.

The useful life of a component is represented in Figure 1.45 whereby, even if the component has a failure rate that increases with the time, since its lifetime is limited to T_{10D}, its failure rate can be assumed as constant during this period.

Please refer to IEC 60947-1 annex K [78] for further insights.

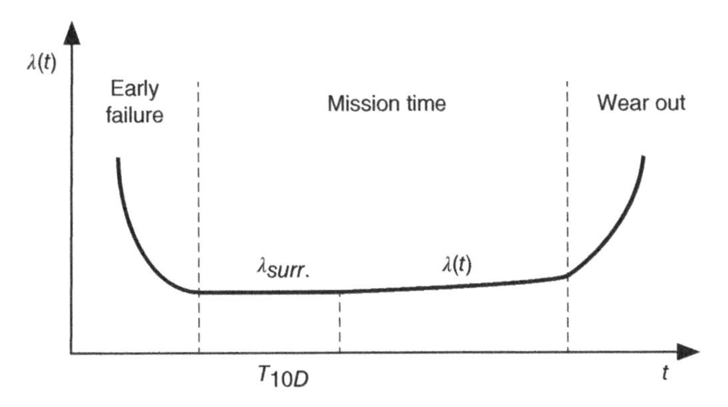

Figure 1.45 Surrogate failure rate.

1.12.5 Markov

The Markov approach is one of the recommended approaches in IEC 61508-6 for Reliability assessment of a Safety Instrumented System, and it is used to analyze **how the state of a system changes with time**.

It is applicable to **Stationary systems** only. That means their behavior must be the same at any moment under consideration, and consequently, the probability of a transition between two states must remain the same during the specific time interval: $\boldsymbol{\lambda} =$ **constant**.

Finally, the systems must be **without memory**: the future random behavior of a system depends only on its actual state and not on preceding states or the way in which the actual state has occurred.

With those assumptions, the transition probability due to random failures is given by

$$P_f = \lambda \cdot \Delta t$$

where λ is the failure rate, and Δt is the time interval the transition probability is related to. The Markov model for this simple failure process is shown in Figure 1.46:

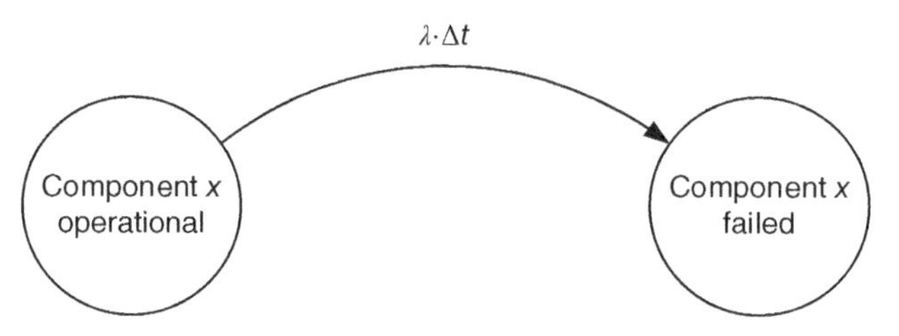

Figure 1.46 Failure of a component with constant failure rate.

In practice, the time is omitted from the graph. Markov processes used in Functional Safety have a finite number of states, for example:

- **State 0:** Functioning State
- **State 1:** Fault state
- λ is the failure rate of the component
- μ is the repair rate of the component

Please refer to Figure 1.47.

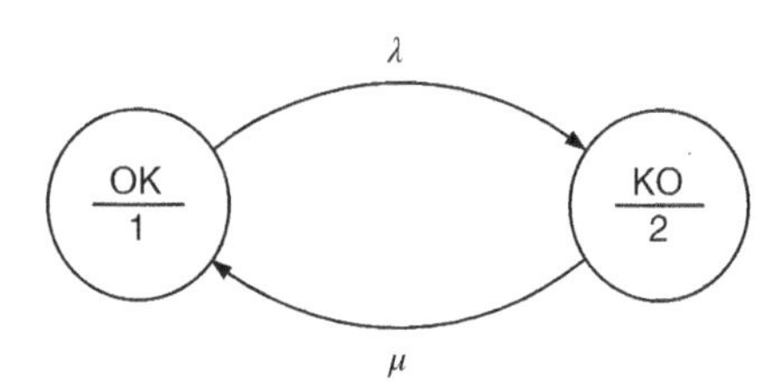

Figure 1.47 A two states transition model.

This Markov model is described by the following differential equation system:

$$\dot{P_1}(t) = -\lambda \cdot P_1(t) + \mu \cdot P_2(t)$$

$$\dot{P_2}(t) = \lambda \cdot P_1(t) - \mu \cdot P_2(t)$$

with the initial conditions:

- Probability that the system is in state 1 at time 0: $P_1(0) = 1$
- Probability that the system is in state 2 at time 0: $P_2(0) = 0$

The solution is as follows:

$$P_1(t) = \frac{\mu}{\lambda + \mu} + \frac{\lambda}{\lambda + \mu} \cdot e^{-(\lambda + \mu)t}$$

$$P_2(t) = \frac{\lambda}{\lambda + \mu} - \frac{\lambda}{\lambda + \mu} \cdot e^{-(\lambda + \mu)t}$$

That was a simple example. A transition model may have three states like in Figure 1.48:

- **State 0:** the channel is functioning correctly
- **State 1:** the channel has a safe fault
- **State 2:** the channel has a dangerous fault

Supposing the components are non-repairable, the transition model is described in Figure 1.48.

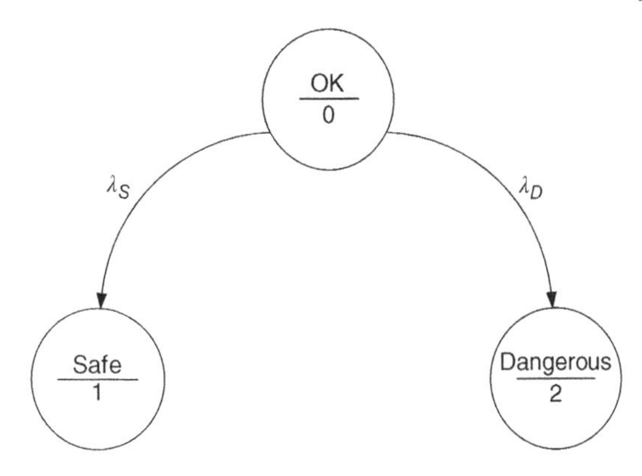

Figure 1.48 Three state transition diagram.

1.13 Logical and Physical Representation of a Safety Function

The Blocks used to represent a Safety Function are a **logical view** of the subsystem architectures.

Blocks can be in a **Series configuration** (i.e. any failure of a block causes the failure of the relevant safety function) or in a **Parallel configuration** (i.e. coincident block failures are necessary for the relevant safety function to fail). However, they do not necessarily represent a specific physical connection scheme.

A Hardware Fault Tolerance of 1 is represented by parallel subsystem elements or blocks, but the corresponding physical connections will depend upon the application of the subsystem.

1.13.1 De-energization of Solenoid Valves

As a first example, we consider an output subsystem whose **mission is to remove power** from a pneumatic cylinder.

Let's suppose the removal of air from the pneumatic cylinder is the safe state condition. In order to implement a redundant pneumatic subsystem ($HFT = 1$), two pneumatic valves (A and B) must be connected in series (Figure 1.49).

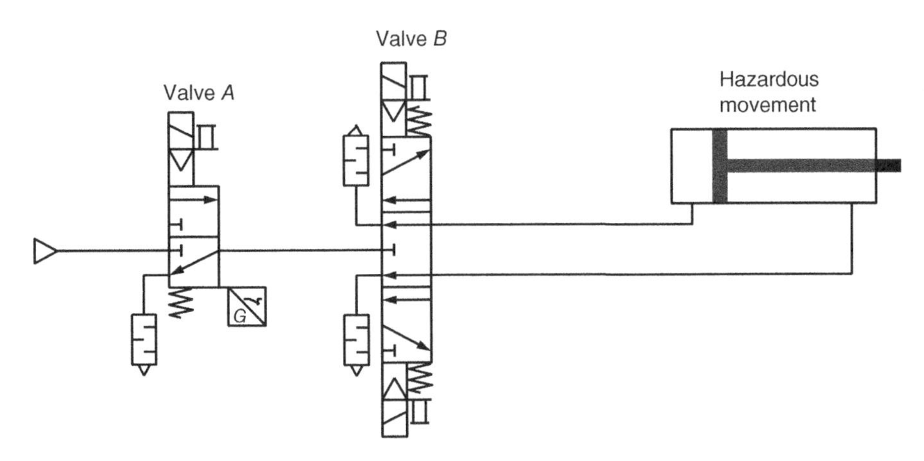

Figure 1.49 Physical representation of a safety output subsystem.

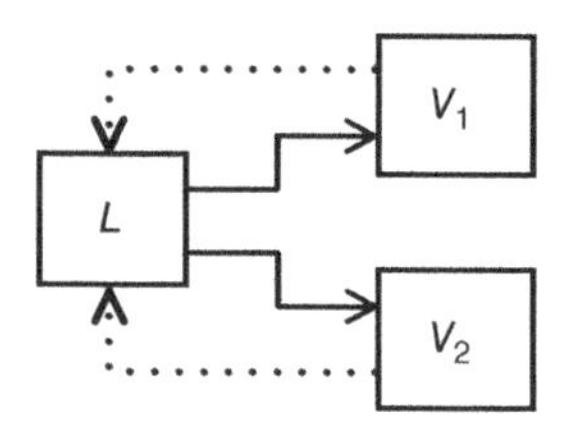

Figure 1.50 Logical representation of a safety 1oo2D output subsystem.

While the physical architecture has two valves in series, the logical one shows the two elements in parallel (Figure 1.50) since we are dealing with a so-called **redundant system** ($HFT = 1$ or 1oo2D).

1.13.2 Energization of Solenoid Valves

Let's consider a continuous casting line. The ladle is positioned above the tundish. At the bottom of the ladle, there is a drawer. When the drawer is opened, the steel flows into the tundish, and the continuous casting can produce steel.

A safe state of the system is when the ladle drawer is closed. The drawer movement is guaranteed thanks to a hydraulic cylinder that is attached to the cylinder stem.

The cylinder is moved thanks to a Solenoid valve that is normally de-energized. Its energization closes the drawer and places the continuous casting machine into a safe state.

In order to implement a redundant subsystem ($HFT = 1$ or 1oo2D), two hydraulic valves in parallel must be installed. The valves must be energized in order to close the drawer. The way they are installed, if one fails, the second one can guarantee the cylinder stem movement and, therefore, the drawer closure.

While the logical system is the same as in Figure 1.50, the physical one is shown in Figure 1.51 two valves in parallel (*A* and *B*) are installed, so that, in case one fails, the other guarantees oil to the cylinder and therefore the drawer closure. Please consider that the circuit was simplified compared with the real one implemented in a continuous casting plant.

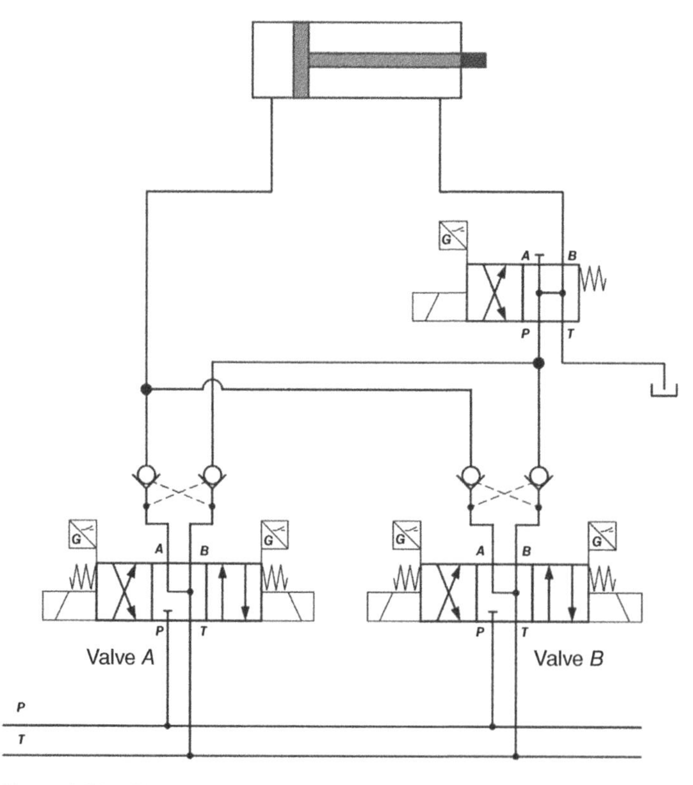

Figure 1.51 Physical representation of a safety output subsystem (with the cylinder as shown in the figure, the drawer is open).

2
What is Functional Safety

2.1 A Brief History of Functional Safety Standards

In machinery, one of the first safety standards was BS 5304 "Code of Practice for Safety of Machinery," first published in 1975. Design features in this code of practice were essentially qualitative. The guide existed, with various revisions, until well into the 1990s, where such guidance was then provided by the European Standard **EN 954-1**:1996 "Safety of machinery – Safety related parts of control systems. General principles for design" [2]: **the safety of Machinery Control Systems was formally born**.

At that time, the world of programmable electronics, despite already heavily used in process safety, was kept out of Machinery standards. Here is what the 1998 version of IEC 60204-1 was stating about emergency stops:

> ***[IEC 60204-1: 1997] 9.2.5.4 Emergency operations (emergency stop, emergency switching off)***
> ***9.2.5.4.2 Emergency Stop.*** *[...] Where a Category 0 stop is used for the emergency stop function,* ***it shall have only hardwired electromechanical components****. In addition, its operation shall not depend on electronic logic (hardware or software) or on the transmission of commands over a communications network or link. Where a Category 1 stop is used for the emergency stop function, final removal of power to the machine actuators shall be ensured and carried out by means of electromechanical components.*

The reason for the skepticism towards Electronics was that an Electromechanical component has clearly defined failure modes. For example, a power contactor (§ 4.12.2) can fail open or fail closed.

EN 954-1 had, what was later called, a **Deterministic Approach**. Safety components were accepted only if electromechanical or made with simple electronics. Safety was relying mainly on the so-called "Architectures": Single or Double channels. In case of need for a low-risk reduction, a single interlocking device and a single contactor that stops the motor would be enough. In case the motor were moving a high-risk element, like a saw, two contactors with monitoring function (Fault Detection) would be needed to stop the same motor.

For Software and Programmable Electronics used in safety applications, some countries had local technical standards. In Germany, in the nineties, the standard DIN VDE 0801 for processor and software-based Safety-related Control Systems (SCS) was sometimes used in addition to EN 954-1. Basic principles of the DIN VDE standard went into the IEC 61508 series.

Functional Safety of Machinery: How to Apply ISO 13849-1 and IEC 62061, First Edition. Marco Tacchini.

In the 1990s, IEC started writing what later became the IEC 61508 [4] series of standards that officially defined the term **Functional Safety**. Since IEC 61508 series only consider electrical/electronic/programmable electronic (E/E/PE) safety-related systems, this is its definition of Functional Safety.

> ***[IEC 61508-4] 3 Definitions and abbreviations***
> ***3.1.12 Functional Safety.*** *Part of the overall safety relating to the EUC and the EUC control system that depends on the correct functioning of the E/E/PE safety-related systems and other risk reduction measures.*

With the acronym EUC, we mean the machinery or process whose risk we want to reduce.

> ***[IEC 61508-4] 3 Definitions and abbreviations***
> ***3.2.1 Equipment Under Control*** *(EUC). Equipment, machinery, apparatus or plant used for manufacturing, process, transportation, medical or other activities.*

However, Functional Safety can be achieved with other technologies, like Pneumatic or Hydraulic; therefore, this is another definition [1]:

> ***[Electropedia] Functional Safety****: part of the overall safety that depends on functional and physical units operating correctly in response to their inputs.*

2.1.1 IEC 61508 (All Parts)

This is the "mother" of all Functional safety standards used in several industries worldwide. It was written in order to allow the use of Electronic components in safety critical systems.

In 1985, the International Electrotechnical Commission (IEC) set up a Task Group to assess the viability of developing a **generic standard for programmable electronic systems to be used for safety applications.** A working group had previously been set up to deal with safety-related software. These two working groups collaborated on the development of an international standard that became known as the IEC 61508 series, published at the end of the 90s.

The original scope of the Task Group, programmable electronic systems used for safety applications, was extended to include all types of electro-technical based technologies, electrical, electronic, and programmable electronic systems: the so-called **E/E/PE systems**.

Parts 1–7 of IEC 61508 were published during the period 1998–2000. In 2005 IEC/TR 61508-0 was published. A review process to update and improve the standard was initiated in 2002, and it was completed with the publication of IEC 61508 Edition 2 in April 2010.

The overall title of IEC 61508 is "Functional Safety of electrical, electronic and programmable electronic (E/E/PE) safety-related systems." It has eight parts.

- Part 0: Functional Safety and IEC 61508.
- Part 1: General requirements.
- Part 2: Requirements for electrical/electronic/programmable electronic safety-related systems.
- Part 3: Software requirements.
- Part 4: Definitions and abbreviations.
- Part 5: Examples of methods for the determination of safety integrity levels.
- Part 6: Guidelines on the application of parts 2 and 3.
- Part 7: Overview of techniques and measures.

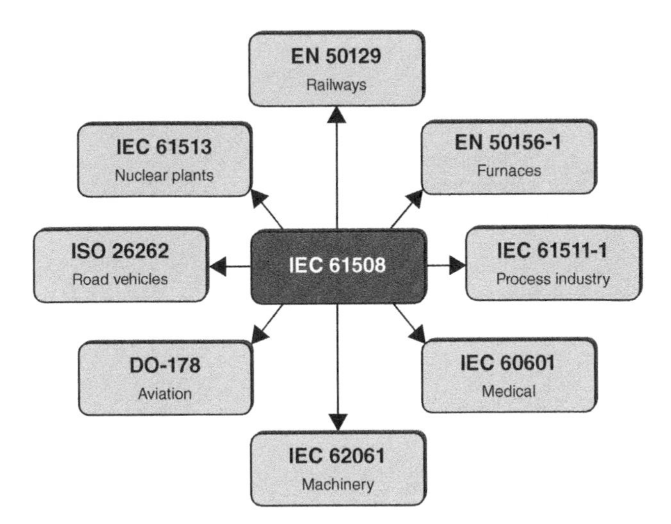

Figure 2.1 Safety standards based upon IEC 61508 series.

Parts 1, 2, and 3 contain the normative requirements and some informative parts. Parts 1, 2, 3, and 4 of IEC 61508 are **IEC Basic Safety Publications (BSP)**. Those are, at the moment, the only BSP on Functional Safety.

IEC 61508 is used as the basis for sector and product standards. It has been used to develop standards for the process, nuclear, and railway industries and for machinery and power drive systems. It has influenced and will continue to influence the development of E/E/PE safety-related systems and products across several sectors. This concept is illustrated in Figure 2.1.

Despite its title, **EN 50156-1** is applicable to water heating systems, steam boiler installations, and heat recovery steam boilers.

Industrial furnaces and associated processing equipment (TPE) follow ISO 13849-1 and IEC 62061 for high demand mode applications and IEC 61511-1 for low demand mode safety systems.

The strategy for achieving Functional Safety is made up of the following key elements:

- Management of functional safety.
- Technical requirements for relevant phases of the applicable safety lifecycle.
- Functional safety assessment (FSA).
- Competence of persons.

The standard covers the whole safety lifecycle: from the initial concept until the system decommissioning or disposal. It proposes three complementary lifecycles:

- The **overall Safety Lifecycle** can be considered as the leading one. One of its phases, Realization, is decomposed in two lifecycles which are executed in parallel:
  - The **E/E/PE system safety lifecycle,** related to hardware; and
  - The **Software safety lifecycle**.

2.1.1.1 HSE Study

Evidence of the need to adopt an approach that covers all phases of a system Safety Lifecycle was illustrated in a study undertaken by the UK Health and Safety Executive [81]. The study analyzed a

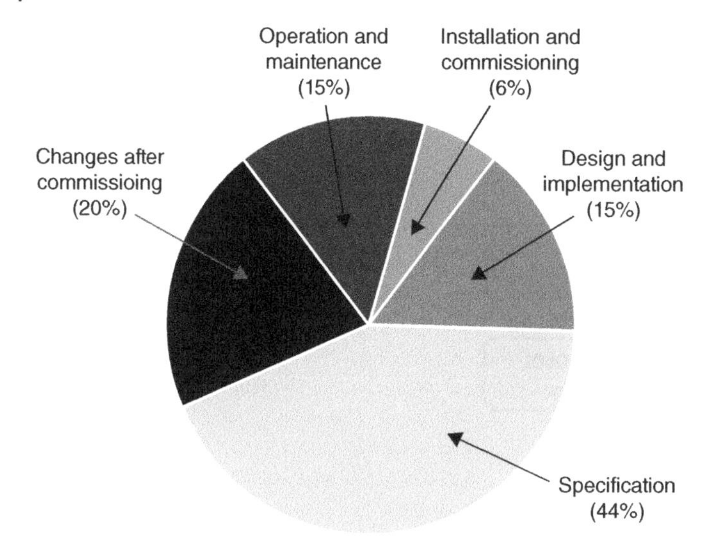

Figure 2.2 Primary cause of system failure for each lifecycle phase.

number of accidents and incidents involving SCS. Figure 2.2 shows the primary causes of failure for each lifecycle phase.

Based on the HSE study, more than 60% of failures were "built in" the safety-related system before being taken into service. Whilst the primary causes by phase will vary, depending upon the sector and complexity of the application, what is self-evident is that it is important that all phases of the lifecycle are addressed if Functional Safety is to be achieved.

That, again, is the reason why IEC 61508 puts so much emphasis on the Safety Lifecycle of the Safety Control System.

2.1.1.2 Safety Integrity Levels

According to IEC 61508, failures can be classified as **either random hardware failures or systematic failures**. The challenge to anyone designing a complex system, such as a programmable electronic system, is to determine how much confidence is necessary for the specified safety level. IEC 61508 tackles this on the following basis:

- that it is possible to quantify the **random hardware** failures;
- that is not usually possible to quantify **systematic** failures.

IEC 61508 series specifies four levels of safety performance for a safety function. These are called **safety integrity levels**. Safety integrity level 1 (SIL 1) is the lowest level, and safety integrity level 4 (SIL 4) is the highest level. The standard details the requirements necessary to achieve each safety integrity level. These requirements are more rigorous at higher levels of safety integrity in order to achieve the required lower likelihood of dangerous failures.

An E/E/PE safety-related system will usually implement more than one safety function. If the safety integrity requirements for these safety functions differ, unless there is sufficient independence of implementation between them, the requirements applicable to the highest relevant safety integrity level shall apply to the entire E/E/PE safety-related system.

If a single E/E/PE system is capable of providing all the required safety functions and the required safety integrity is less than that specified for SIL 1, then IEC 61508 does not apply.

As previously stated, in order to design a reliable safety control system, two aspects have to be considered:

- **Hardware Safety Integrity.** This is achieved through meeting the quantified target failure measures for random failures, together with meeting the Architectural Constraints for the specified SIL.
- **Systematic Safety Integrity.** It is a group of measures used to avoid systematic failure mechanisms; they are in general qualitative measures with increasing rigor, assurance, and confidence, the higher the SIL.

Therefore, safety integrity is made up of **Hardware Safety Integrity,** in relation to random failures, and **Systematic Safety Integrity,** in relation to systematic failures. The above concept is shown in Figure 2.3.

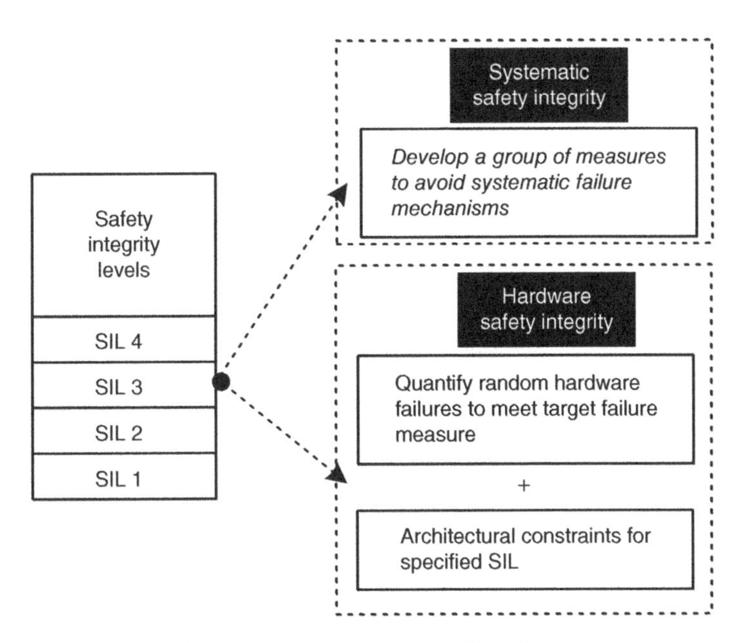

Figure 2.3 Design strategy to achieve a specified SIL.

2.1.1.3 FMEDA

The Failure Modes, Effects, and Diagnostic Analysis (**FMEDA**) is used to calculate the product random failures: it is an extension of the classic FMEA procedure [67]. The technique was first developed for electronic devices, but it is now used for mechanical and electromechanical devices as well. The FMEDA results are the different failure rates described in § 3.1.

A FMEDA is done by examining each component in a product and, for each one, the effect of a random failure on the product is analyzed. Questions asked are: will a failure in a specific resistor cause the product to fail safe, fail dangerous, or lose calibration? If the serial communication line from the A/D to the microprocessor gets shorted, how does the product respond? If this spring fractures, does that cause a dangerous or a safe failure? In this way, the failure rate of each component is analyzed and the various groups are added. The end result is a **product specific set of failure data** that includes failure rates for each failure mode: failure rates that are detected and undetected by diagnostics, Safe Failure Fraction calculations and, often, an explanation on how to use the numbers for safety verification calculations.

A FMEDA is sometimes done by the product manufacturer but, typically, it is done by third parties.

It should be emphasized that a FMEDA provides failure rates, failure modes, and diagnostic coverage effectiveness **for random hardware failures**. It does not include failure rates due to "systematic" causes, including incorrect installation, inadvertent damage, incorrect calibration, or any other human error.

Please also consider that, in order to estimate the component failure rates, it is also possible to use field data [73]. That is what IEC 61508-2 refers to as Route 2_H. Please look at § 7.4.4.3.3 of the mentioned standard.

2.1.1.4 High and Low Demand Mode of Operation

IEC 61508-1 clarifies which target failure measure, or unreliability function $F(t)$, should be used, **depending upon the mode of operation**:

- **PFD_{avg}** should be used for safety systems in **low demand mode**.
- **PFH_D** should be used for safety systems in **high demand or in continuous mode**.

> ***[IEC 61508-1] 7.6 Overall safety requirements allocation***
> *[...] 7.6.2.9 When the allocation has sufficiently progressed, the safety integrity requirements, for each safety function allocated to the E/E/PE safety-related system(s), shall be specified in terms of the safety integrity level in accordance with tables 2 or 3 and shall indicate whether the target failure measure is, either:*
>
> - *the average probability of dangerous failure on demand of the safety function, (**PFD_{avg}**), for a **low demand** mode of operation (table 2), or*
> - *the average frequency of a dangerous failure of the safety function (h^{-1}), (**PFH**), for a **high demand** mode of operation (table 3), or*
> - *the average frequency of a dangerous failure of the safety function (h^{-1}), (PFH), for a continuous mode of operation (table 3).*

The content of table 2 of IEC 61508-1 is the same as shown in Table 2.1; The one of table 3 is in Table 2.2.

Table 2.1 Safety integrity levels – target failure measures for a safety function operating in low demand mode.

Safety integrity level (SIL)	Average probability of dangerous failure on demand (PFD_{avg})
4	$10^{-5} \leq PFD_{avg} < 10^{-4}$
3	$10^{-4} \leq PFD_{avg} < 10^{-3}$
2	$10^{-3} \leq PFD_{avg} < 10^{-2}$
1	$10^{-2} \leq PFD_{avg} < 10^{-1}$

Table 2.2 Safety integrity levels – target failure measures for a safety function operating in high demand mode.

Safety integrity level (SIL)	Average probability of dangerous failure per hour (PFH_D) [h^{-1}]
4	$10^{-9} \leq PFH_D < 10^{-8}$
3	$10^{-8} \leq PFH_D < 10^{-7}$
2	$10^{-7} \leq PFH_D < 10^{-6}$
1	$10^{-6} \leq PFH_D < 10^{-5}$

2.1.1.5 Safety Functions and Safety-Related Systems

IEC 61508 series sees safety as the freedom from unacceptable risk of physical injury or of damage to the health of people, **either directly or indirectly, as a result of damage to property or to the environment.**

Therefore, it considers damage to property but as a risk of indirectly affecting people health.

Functional safety is part of the overall safety that depends on a system or equipment operating correctly in response to its inputs. For example, an overtemperature protection device, using a thermal sensor in the windings of an electric motor to de-energize the motor before it can overheat, is an instance of functional safety. But providing specialized insulation to withstand high temperatures is not an instance of functional safety, although it is still an instance of safety and it could protect against the same hazard.

Neither safety nor functional safety can be determined **without considering** the systems as a whole and **the environment with which they interact**.

Generally, the significant hazards for equipment and any associated control system in its intended environment have to be identified by the developer via a **risk assessment and a risk reduction process. The analysis determines whether functional safety is necessary to ensure adequate protection against each significant hazard**. Therefore, functional safety is just one method of dealing with hazards; other means for their elimination or reduction, such as **inherent safety through design**, remain of primary importance.

The term **safety-related, used in all functional safety standards,** describes systems that are required to perform a specific function or functions to ensure risks are kept at an acceptable level. Such functions are, by definition, safety functions.

Two types of requirements are necessary to achieve functional safety:

- **Safety function requirements:** what the function does; and
- **Safety integrity requirements:** the likelihood of a safety function being performed satisfactorily.

The **safety function requirements** are derived from the hazard analysis and the **safety integrity requirements** are derived from a risk assessment. The higher the level of safety integrity, the lower the likelihood of dangerous failure.

Any system, implemented in any technology, which carries out safety functions is a safety-related system. A safety-related system may be separate from any equipment control system, or the equipment control system may itself carry out safety functions. In the latter case, the equipment control system will be safety-related.

2.1.1.6 An Example of Risk Reduction Through Functional Safety

Consider a machine with a rotating blade that is protected by a hinged solid cover. The blade is accessed for routine cleaning by lifting the cover. The cover is interlocked so that whenever it is lifted, an electromechanical or electronic circuit de-energizes the motor and applies a brake. In this way, the blade is stopped before it could injure the operator. In order to ensure that safety is achieved, a **risk assessment and risk reduction** are necessary.

a) The first step is to identify the hazards associated with cleaning the blade. For this machine, it might show that it should not be possible to lift the hinged cover more than 5 mm without the brake activating and stopping the blade. Therefore, the risk assessment established that we need to reduce the risk. Further analysis could reveal that the time for the blade to stop shall be 1 s or less. Therefore we decided that the risk has to be reduced, and we will use a safety-related control system.
b) At this point, we need to determine the performance requirements of the safety function. The aim is to ensure that the safety integrity of the safety function is sufficient to ensure that no one is exposed to an unacceptable risk associated with this hazard.

The harm resulting from a failure of the safety function could be the amputation of the operator's hand or could be just a bruise. The risk also depends on how frequently the cover has to be lifted, which might be many times during daily operation, or it might be less than once a month.

The level of safety integrity required increases with the severity of the injury and the frequency of exposure to the hazard.

The safety integrity of the safety function will depend on all the equipment that is necessary for the safety function to be carried out correctly: that means, **the interlock, the associated electromechanical or electronic circuit, and the braking system**. Both the safety function and its safety integrity specify the required behavior for the systems as a whole within a particular environment.

To summarize, these two elements, "What the safety function shall do," the safety function requirements, and "What degree of certainty is needed for the safety function," the safety integrity requirements, **are the foundations of functional safety.**

2.1.1.7 Why IEC 61508 was Written

Back in the 1990s, Safety functions were more and more carried out by electronic or programmable electronic systems. These systems are usually complex, making it impossible, in practice, to fully determine every failure mode or to test all possible behaviors.

The challenge was to design the system in such a way as to prevent dangerous failures or to control them when they arise. **Dangerous failures may arise from:**

- Incorrect specifications of the safety-related control system.
- Omissions in the safety requirements specification (e.g. failure to develop all relevant safety functions during different modes of operation).
- Random hardware failure mechanisms.
- Systematic hardware failure mechanisms.
- Software errors.
- Common cause failures.
- Human errors.
- Environmental influences (e.g. electromagnetic, temperature, mechanical phenomena).

IEC 61508 contains requirements to minimize these failures and build a reliable safety-related control system. Its aim was:

- Release the potential of E/E/PE technology to improve machinery and process safety.
- Enable technological developments to take place within an overall safety framework.
- Provide a technically sound, system-based approach, with sufficient flexibility for the future.
- Provide a risk-based approach for determining the required performance of SCS.
- Provide a generically-based standard that can be used directly by industry but can also help with developing sector standards (e.g. machinery, process chemical plants, medical or rail) or product standards (e.g. power drive systems).
- Provide a means for users and regulators to gain confidence when using computer-based technology.

2.1.2 ISO 13849-1

It is one of the two standards used in machinery and, at least in Europe, it is the most used of the two, the other being IEC 62061. It is divided into two parts:

- **ISO 13849-1:** Safety of machinery – Safety-related parts of control systems – Part 1: General principles for design.
- **ISO 13849-2:** Safety of machinery – Safety-related parts of control systems – Part 2: Validation.

The first edition of ISO 13849-1 was published in 1999, and it was identical to EN 954-1:1996. ISO 13849-1 was just the corresponding ISO number for EN 954-1; therefore, the real first edition (officially, it was the second one) was in 2006.

The stakeholders who edited the EN 954-1 saw the need to include Programmable Electronics in machinery safety systems. Actually, Electronics were already included in EN 954-1, but without any detailed software requirements. The standard needed to go through a so-called **probabilistic approach**, the same used by IEC 61508 series. Therefore, the revision leading to the second edition of ISO 13849-1 combined the deterministic aspects of EN 954-1 with the probabilistic approach of IEC 61508 and included software requirements for the first time. A few Mathematicians from IFA [59] designed the Markov models for the 2006 edition of the standard.

The third edition was issued in 2015, while the latest one, the fourth, was published in 2023. This new edition is based upon the same principles as the previous one. Hereafter we mention two key changes:

1. **The Validation process**, detailed in ISO 13849-2 is now included in the first part. The main reason is that people were not focused enough on the validation process. Manufacturers normally run the number crunching and fail to check if, once the machine is installed and commissioned, the safety system works as expected: validation is key to confirm and guarantee the level of safety required.
2. It is now clear that **the Category is a characteristics of the Safety Subsystem and not of the whole safety function**. The input subsystem can be Category 1 (single channel), while the output subsystem of the same safety function can be Category 3 (double channel). The confusion was due to the EN 954-1 heritage. The fact that a Safety Function can be made of different subsystem categories means **its reliability level is represented by its Performance level only**. In EN 954-1, the reliability was represented by a category level only. When we moved to ISO 13849-1, Type C standards kept giving both the PL and the Category requirements for Safety Functions. From this fourth edition, it is clear that **only the PL represents the reliability**

level of a Safety Function: the category is only a way to reach it. The same is valid for IEC 62061, whereby a Safety Function is characterized by a SIL level only and not by which architectures are used for the various subsystems.

2.1.3 IEC 62061

This is the second standard applicable to machinery and, so far, the least used for the reasons explained in the book. However, there is a confidence that, with the publication of the 2021 edition, there will be more and more machinery manufacturers who will choose it.

The first edition of IEC 62061 [12] was published in 2005. It is part of the approach detailed in IEC 61508. It is addressed to the machinery sector, and it allows the verification of the Reliability level reached by a Safety-related Control System (SCS).

> ***[IEC 62061 DIS: 2020] Introduction***
> *[...] This International Standard is intended for use by machinery designers, control system manufacturers and integrators, and others involved in the specification, design and validation of an SCS. It sets out an approach and provides requirements to achieve the necessary performance.*

Around 2010, a working group was established with the assignment of writing one common standard for Functional Safety of Machinery, called ISO/IEC 17305, combining ISO 13849-1 and ISO 62061. Unfortunately, the new standard did not see the light. When the MT 62061 (maintenance team, as they are called in IEC) met for the first time, they decided that the results of that work should be the starting point for the new edition. That is one of the reasons why the new edition of IEC 62061 is closer in the approach to ISO 13849-1: the team mediated the IEC 61508 approach **with the pragmatism of ISO 13849-1**. These are the main changes compared with the previous edition [15]:

- The new standard is now applicable to non-electrical technologies. That is the reason why it now refers to SCS instead of Safety Related Electrical Control Systems (SRECS).
- The Architectures are now better defined, especially Architecture C, as well as the formulas to be used.
- The Architectural Constraint, previously called SIL Claim, is now defined as the maximum SIL that a Subsystem can reach.
- Requirements on independence for software verification and validation activities were added.
- New important Informative annexes were added; the information contained and the approach described come from ISO 13849-1:
  - **Annex C:** on examples of B_{10D} and $MTTF_D$ values for components.
  - **Annex D:** examples of Diagnostic Coverage values.

2.1.4 IEC 61511

2.1.4.1 Introduction

IEC 61511 (all parts) is intended as the process industry sector implementation of IEC 61508; it addresses the application of SISs (Safety Instrumented Systems) in the process industry.

Safety Instrumented Functions (SIFs) are protective functions implemented in a **Safety Instrumented System** (SIS). A typical SIS is comprised of multiple SIFs; typically, each SIS has process sensors that measure a process deviation, a logic solver that executes the functional logic, and final control elements like on/off valves that bring the process to a safe state. The IEC 61511 series of standards addresses SIS based on the use of electrical, electronic, or programmable electronic technology in the process industry sector. IEC 61511 series also addresses a process Hazard and Risk Assessment (H&RA) from which the specification for SISs are derived.

The standard recognizes that systematic failures come from human errors, and they **can be reduced with the implementation of solid organizational processes**.

The first edition of the standard was issued in 2003. The second edition was published in 2016. The standard has four parts:

- **Part 1** – It is the only **normative part** of the series. It includes terminology, requirements for specification, hardware design and application programming, commissioning, validation, operation, maintenance, and testing of SIS components.
- **Part 2** – It is an **Informative guidance** on Part 1. It contains annex A that provides guidance and implementation examples of requirements outlined in Part 1.
- **Part 3 –** It is an informative part of the series that provides information on available methods to conduct H&RA to determine integrity requirements, i.e. Safety Integrity Level (SIL).
- **Part 4** – It is actually a **Technical Report**. It contains an explanation and rationale for changes from Editions 1 to 2.

Compliance with the IEC 61511 standard series will help assure reliable and effective implementation of SIS to achieve risk reduction objectives and thereby improving process safety.

IEC 61511 is recognized as a good engineering practice in most countries and a regulatory requirement in an increasing number of countries. End users in the process industry should use this standard series to develop their internal procedures, work processes, and management systems. Implementing a SIS lifecycle management system provides a framework for managing people, processes, and systems to improve overall safety and operational performance.

The standard applies when devices that meet the requirements of IEC 61508 series are integrated into an overall safety-related control system, **to be used in a process sector application**. It does not apply to manufacturers wishing to claim that their devices are suitable for use in SISs for the process sector; for this purpose, IEC 61508-2 [6] and IEC 61508-3 [7] have to be used.

The normative part does not contain any formula. The reason is that the spirit of IEC 61511-1 **is to define what has to be achieved and not how to achieve it.**

2.1.4.2 The Second Edition

The Second Edition reinforces the need **to design for Functional Safety management rather than a narrow focus on calculations,** and it can be used to manage the actual performance of the SIS over time. IEC/TR 61511-4 was written to provide a brief introduction to the above issues, with more detailed content remaining in the main parts of the standard. **Management of Functional Safety addresses systematic failures, mostly caused by humans**, that are not quantifiable with mathematical models. These activities, covering the whole safety lifecycle, are applied through processes and procedures.

In this second edition, there is the idea that Safety is not only based upon reliable components, but it comes from a holistic approach given by the concept of the **Safety Life Cycle**. To ensure that Functional Safety can be achieved, several activities (done by different stakeholders, like end users, engineering companies, vendors, etc.) need to be done. They are all connected to each other like a chain, and the strength of this chain will be only as strong as the weakest link. **It is crucial to consider Functional Safety as a lifecycle, which starts with hazard identification and ends**

with the decommissioning of SIS: all activities in the safety lifecycle are impacted by upstream and downstream activities.

Every SIS project has clear roles and responsibilities. All involved parties are aware of their responsibilities and are competent to fulfill the related activities necessary for Functional Safety. Competencies are kept up to date. All necessary activities in a project are described in a **safety plan** which can be project specific or a general company-specific document. For all relevant activities, an FSA is carried out to demonstrate that SIF fulfills all requirements and it is compliant to the agreed standards. Performance management during operation is done by collecting field data for SIS Reliability and SIS process demand information. **Functional Safety audits** are done at regular intervals to demonstrate that the organization remains capable of fulfilling the defined Functional Safety requirements. Assessment and auditing activities are done by individuals independent of the project team. Meaningful documentation of the assessment and audit results are generated, and recommendations are tracked for effective closure.

2.1.4.3 Designing a SIS

Designing a SIS means:

1. Controlling the effects of random hardware failures and
2. Avoiding or controlling systematic failures.

The activity can be summarized in the following four parts:

a) Select devices appropriately, based on prior use or in accordance with IEC 61508.
b) Ensure minimum redundancy determined by HFT, either in accordance with the process sector approach, defined in IEC 61511-1 or in IEC 61508
c) Design the architecture and application program to meet the requirements of the Safety Requirements Specification and verify that the specified performance objectives for Integrity, Reliability, and Systematic error control have been met; including aspects such as human capabilities, bypass management, diagnostic coverage, common cause failures, Proof Test interval, MTTR, etc.
d) Ensure adequate demarcation between the SIS and the BPCS for both hardware and application program, so that the overall risk reduction performance is achieved.

2.1.4.4 Three Methods

The Second Edition allows **three different methods** to determine the required HFT of an SIS:

1. **Route 1_H** of IEC 61508-2 [6], based on FMEDA analysis and conformance with the related clauses in IEC 61508-2.
2. Use of the concept of **Prior Use** (§ 5.3). That means the use of table 6 of IEC 61511-1 [16], same as Table 2.3, in conjunction with the requirements in IEC 61511-1 clauses 11.5–11.9.
3. **Route 2_H** of IEC 61508-2, based on product returns to the manufacturer and conformance with related clauses in IEC 61508-2.

Table 2.3 clarifies that, in case SIL 1 has to be reached, no redundancy is necessary; the same is valid in the case of SIL 2 in low demand mode. However, if the system is working in high demand mode, in order to reach an SIL 2 level, an $HFT = 1$ is required.

Table 2.3 Minimum HFT requirements according to SIL.

SIL	Mode of operations	Minimum required *HFT*
1	Any mode	0
2	Low demand mode	0
2	High demand mode or continuous mode	1
3	High demand mode or continuous mode	1
4	Any mode	2

2.1.4.5 The Concept of Protection Layers

An intuitive way of representing the risk reduction capability of a SIS is to use the concept of the Protection layer as described in IEC 61511-1 [16].

> ***[IEC 61511-1] 3.2 Terms and definitions***
> ***3.2.57 Protection Layer****: any independent mechanism that reduces risk by control, prevention or mitigation.*
>
> ***Note 1 to entry****: It can be a process engineering mechanism such as the size of vessels containing hazardous chemicals, a mechanical mechanism such as a relief valve, a SIS or an administrative procedure such as an emergency plan against an imminent hazard. These responses may be automated or initiated by human actions (see figure 9).*

In a typical chemical process, **various protection layers are in place** to lower the frequency of undesired consequences: the process design (including inherently safer concepts), the basic process control system (BPCS), SIS, active devices (such as relief valves), passive devices (such as dikes and blast walls), human intervention, etc.

Protection layers that perform their function with a high degree of reliability may qualify as independent protection layers (IPL). Figure 9 of IEC 61511-1 is similar to Figure 2.4 hereafter.

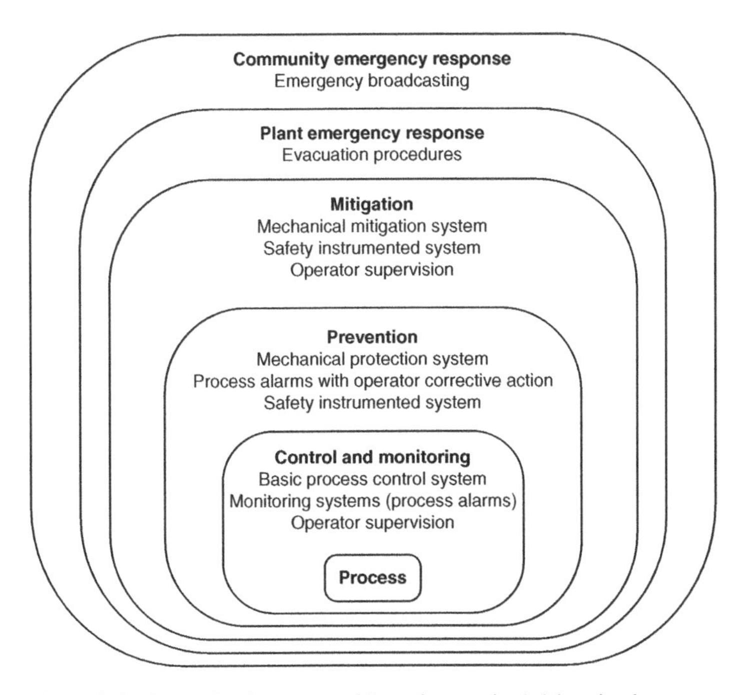

Figure 2.4 Protection layers to achieve the required risk reduction.

2.1.4.6 The Different Types of Risk

The starting point in the risk assessment is the process itself, also called **Equipment Under Control** (EUC). The process has a certain "**Process Risk,**" that is, the risk without the implementation of any Risk Reduction Measure (machinery safety terminology) or Protection layer (process terminology).

Figure 2.5 illustrates the general concepts of risk reduction. The general model assumes that:

- there is a process and an associated BPCS;
- there are associated human factor issues;
- the safety protection layer includes:
  - mechanical protection system;
  - safety instrumented system;
  - non-SIS instrumented system;
  - mechanical mitigation system.

The risk is then reduced thanks to the implementation of protection layers, part of which is also the SIS. What is left is called "**Residual Risk.**"

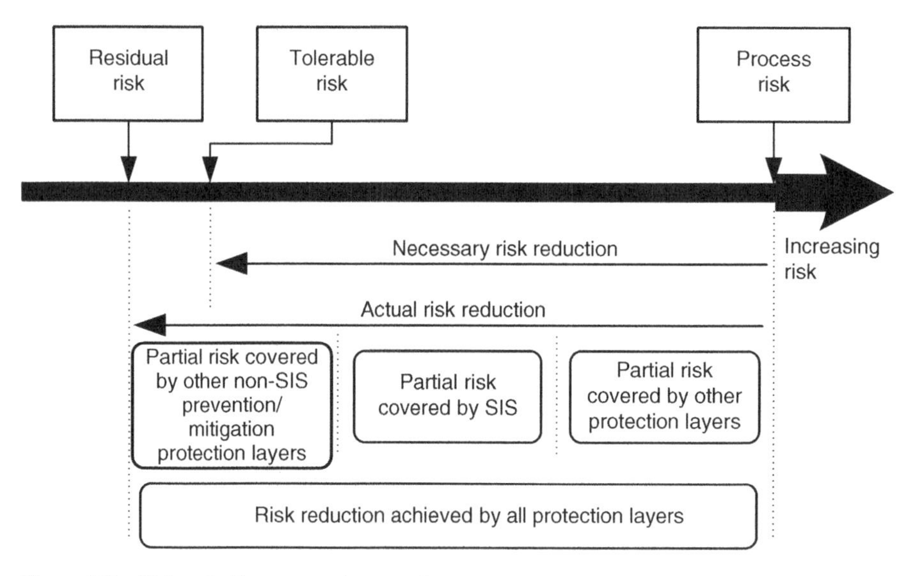

Figure 2.5 Risk reduction: general concepts.

2.1.4.7 The Tolerable Risk

There is no such thing as zero risk. That is because no physical item has zero failure rate, no human being makes zero errors, and no piece of software design can foresee every operational possibility [65].

Nevertheless, public perception of risk, particularly in the aftermath of a major incident, often calls for zero risk. In general, most people understand that is not practicable, as it can be seen from the following examples of the everyday risk of death from various causes.

Table 2.4 Risk reduction: general concepts.

Causes of death	Probability per annum
All causes (mid-life including medical)	1×10^{-3} pa
All accidents (per individual)	5×10^{-4} pa
Accident in the home	4×10^{-4} pa
Road traffic accident	6×10^{-5} pa
Natural disasters (per individual)	2×10^{-6} pa

From Table 2.4, a company may decide that the lower limit of the tolerable risk (called **Residual Risk**) is 10^{-6}. That means, once the risk has been reduced below that level, it is acceptable. Before that assessment, it is important to decide the upper limit of the tolerable risk. That is a more difficult task!

In Figure 2.6 there are three regions:

- The **Broadly Acceptable Region** is where the risk is "Socially acceptable." For example, below the 10^{-6} threshold.
- The **Unacceptable Region** is where the risk cannot be accepted, and it must be reduced, regardless of the cost involved.
- The in-between region, called **Tolerable Region**, is where the concept of ALARP, as low as Reasonably Practicable, comes in.

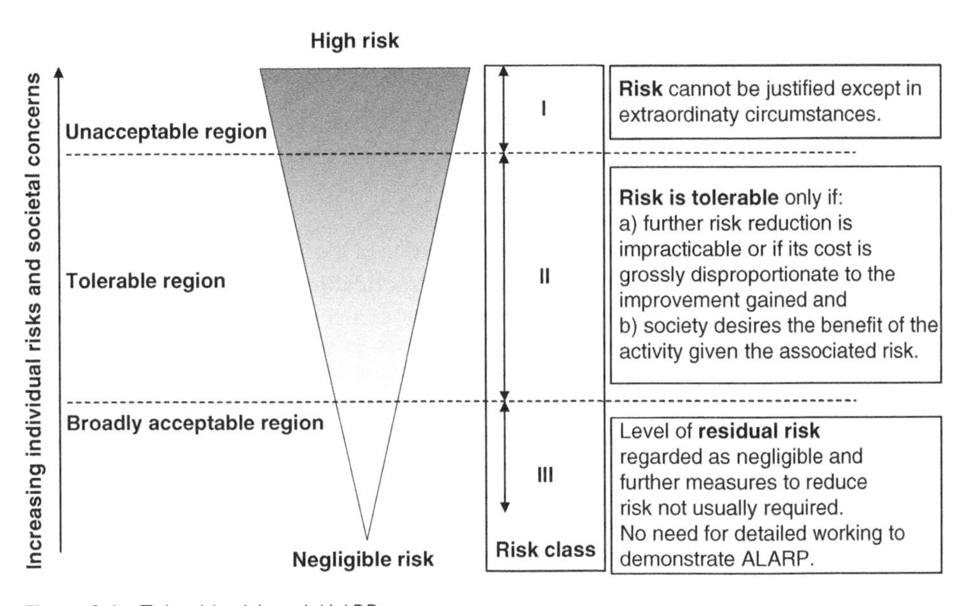

Figure 2.6 Tolerable risk and ALARP.

HSE in [66] indicates the limit of Residual risk, or the **Broadly Acceptable Region**, at 10^{-6} death of one person in a year:

> ***Boundary between the "broadly acceptable" and "tolerable" regions for risk entailing fatalities***
> *HSE believes that an individual risk of death of one in a million per annum for both workers and the public corresponds to a very low level of risk and should be used as a guideline for the boundary between the broadly acceptable and tolerable regions. As is very apparent from tables 1–4 at appendix 4, we live in an environment of appreciable risks of various kinds which contribute to a background level of risk – typically a risk of death of one in a hundred per year averaged over a lifetime.* ***A residual risk*** *of one in a million per year is extremely small when compared to this background level of risk. Indeed many activities which people are prepared to accept in their daily lives for the benefits they bring, for example, using gas and electricity, or engaging in air travel, entail or exceed such levels of residual risk. [...]*

The limit between the Tolerable and Unacceptable region is more difficult to set. Here is what HSE states in the same document:

> ***Boundary between the "tolerable" and "unacceptable" regions for risk entailing fatalities***
> *We do not have, for this boundary, a criterion for individual risk as widely applicable as the one mentioned above for the boundary between the broadly acceptable and tolerable regions. [...] Nevertheless, in our document on the tolerability of risks in nuclear power stations, we suggested that an individual risk of death of one in a thousand per annum should on its own represent the dividing line between what could be just tolerable for any substantial category of workers for any large part of a working life, and what is unacceptable for any but fairly exceptional groups. For members of the public who have a risk imposed on them "in the wider interest of society" this limit is judged to be an order of magnitude lower – at 1 in 10 000 per annum.*

Once the Acceptable and Tolerable risk level have been set, the risk assessment can be done and, case by case, using the so-called ALARP principle, the team can decide if the risk reduction measures are enough.

2.1.4.8 The ALARP Principle

The ALARP principle can be applied during the determination of the tolerable risk and the safety integrity levels. ALARP stands for **As Low As Reasonably Practicable,** and it is a concept that can be applied during the determination of the safety integrity level of a SIS. It is not, in itself, a method for determining safety integrity levels.

A risk analysis can result in three situations:

1. The risk is so great that it must be refused altogether: we are in the **Unacceptable Region** of figure 2.6;
2. The risk is, or has been made, so small as to be negligible: we are in the **Broadly Acceptable Region**;
3. The risk falls between the two states specified in 1 and 2, and it has been reduced to a level that is As Low As Reasonably Practicable: we are in the **Tolerable Region**.

Reasonably Practicable is not easy to quantify. It implies a calculation in **which additional risk reduction** is balanced against the **sacrifice involved in achieving it**. If there is a gross disproportion between them and the benefit is insignificant in relation to the cost, then the risk is considered ALARP.

A safety improvement measure could not eliminate risk but reduce it by an amount: it's important to **assess the risk reduction benefit against the cost of its implementation.**

For this evaluation, some organizations operate a cost-per-life saved target or **CPL**. The cost of preventing fatalities over the life of the plant is compared with the target CPL but considering a correcting factor. The English HSE recommends the use of a gross disproportionality factor **GDF**. The CPL is multiplied by a GDF that depends upon how close the predicted risk is to the target. For predicted risks approaching the Maximum Tolerable Risk, a factor of 10 is applied. This falls, with a logarithmic scale, as the risk moves towards the Broadly Acceptable region. This will be illustrated in the examples which follow. The improvements will be implemented unless costs are grossly disproportionate.

Example

We have a situation whereby a CPL saved of 2 M€ is used in a particular industry. **GDF = 10,** therefore the **Target cost per life saved** becomes 20 M€.

A **maximum tolerable risk target of 10^{-5} pa** has been established for a particular hazard, which is likely to cause 2 fatalities.

The proposed safety system has been assessed and a **risk of 8×10^{-6} pa predicted.** Given that the **Broadly Acceptable negligible Risk is 10^{-6} pa,** then the application of ALARP is required since we are in the Tolerable Region of Figure 2.6.

Let's suppose the cost of the additional instrumentation system is 10 000 € and that would **reduce the risk to 2×10^{-6} pa** over the life of the plant, which is 30 years.

The specific "cost per life saved" is given by the formula;

$$CPL = \frac{Cost\ of\ the\ Risk\ reduction\ measure}{No\ of\ lives\ saved}$$

That means:

$$CPL = \frac{10\,000\ €}{(8 \cdot 10^{-6} - 2 \cdot 10^{-6}) \cdot 2 \cdot 30} = \frac{10\,000\ €}{3.6 \cdot 10^{-4}} = 27.8\ M€$$

The specific CPL is greater than the target CPL of 20 M€ and therefore **the proposal should be rejected**.

Example

Another example can be the situation where a CPL saved of 1 M€ is used in a particular industry. **GDF = 10,** therefore the **Target CPL** becomes 10 M€.

The Acceptable risk target is 10^{-6} and the Tolerable one is 10^{-4}.

With the proposed safety system, a **risk of 8×10^{-5} pa was predicted.** We are in the ALARP Region.

The cost of the additional instrumentation system is 3000 € and that would **reduce the risk to 2×10^{-6} pa** over the life of the plant, which is 30 years. Should the company implement the risk reduction measure?

The specific "cost per life saved" is:

$$Cost\ per\ live\ saved = \frac{3000\ €}{(8 \cdot 10^{-5} - 2 \cdot 10^{-6}) \cdot 1 \cdot 30} = \frac{3000\ €}{2.34 \cdot 10^{-3}} = 1.28\ M€$$

The specific CPL is lower than the target CPL of 10 M€, and therefore **the proposal should be adopted.**

2.1.4.9 Hazard and Operability Studies (HAZOP)

The HAZOP methodology [43] is a structured and systematic technique for examining a process, with the objectives of:

- Identifying risks associated with the operation and maintenance of the system. The hazards or other risk sources involved can include both those essentially relevant only to the immediate area of the process and those with a much wider sphere of influence, for example some environmental hazards;
- Identifying potential operability problems with the system and in particular identifying causes of operational disturbances and production deviations likely to lead to hazards.

An important benefit of HAZOP studies is that the resulting knowledge, obtained by identifying risks and operability problems in a structured and systematic manner, is of great assistance in determining appropriate remedial measures

The basis of a HAZOP study is a "guide word examination," which is a deliberate search for deviations from the design intent. **To facilitate the examination, a system is divided into parts** in such a way that the design intent or function for each part can be adequately defined. The size of the part chosen is likely to depend on the complexity of the system and the potential magnitude and significance of the consequences. In complex systems or those where the level of risk might be expected to be high, the parts are likely to be small in comparison to the system. In simple systems or those where the level of risk might be expected to be low, the use of larger parts will expedite the study.

The design intent for a given part of a system is expressed in terms of properties (Pressure, Temperature, etc.), which convey the essential characteristics of the part and which represent natural divisions of the process.

For each part, the HAZOP study examines each property for deviation from the design intent, which can lead to undesirable consequences. The identification of deviations from the design intent is achieved by using guide words like "more" or "less." The role of the guide word is to stimulate imaginative thinking, maximizing the chances of study completeness.

2.1.4.10 Layer of Protection Analysis (LOPA)

LOPA is a simplified and not fully quantitative form of risk assessment whose primary purpose is to determine if there are sufficient layers of protection against an accident scenario. The final result is a probability or frequency of accidents; this method is based on probability calculations.

LOPA uses the data developed by HAZOP [43] and documents the initiating cause and the protection layers that modify the risk. This can then be used to determine the amount of risk reduction achieved by existing controls and to decide whether further risk reduction is needed.

The method starts with data developed during hazard identification and accounts for each identified hazard by documenting the initiating cause and the protection layers that prevent or mitigate the hazard. The total amount of risk reduction can then be determined, and the need for more risk reduction analyzed. If additional risk reduction is required and if it has to be provided in the form of a SIF, **the LOPA methodology allows the determination of the appropriate SIL for each specific SIF**.

Depending on the process complexity and potential severity of consequences, a scenario may require one or several protection layers. If the estimated risk of a scenario is not acceptable, additional IPLs may be added.

LOPA does not suggest which IPL to add or which design to choose, but it assists in judging between alternatives for risk mitigation. It is limited to evaluating a single cause-consequence pair as scenario.

A protection layer can be a BPCS, as well as a Safety System. Safety Valves can also be a protection layer, and even alarm systems that trigger an operator intervention can be considered protection layers. The analysis starts from the likelihood of an event (for example, the explosion of a tank due to high pressure), and layers of protection are added in order to reduce the probability of that event.

Let's suppose the bare probability is 10%; the presence of a BPCS may reduce it by a factor of 10. We may then add a Safety Control System with a $PFD_{avg} = 3 \cdot 10^{-3}$. Finally, we install a Pressure Relief Valve to which we attribute a Risk Reduction Factor (RRF) of 100. Implementing the three protection layers, the likelihood of an accident goes from 10^{-1} to $10^{-1} \cdot 10^{-1} \cdot 3 \cdot 10^{-3} \cdot 10^{-2} = 3 \cdot 10^{-7}$.

The original probability was multiplied by the reduction factors of each layer. That can be done only if **each layer is independent from all the others**. If that is not the case, Common Cause Failures have to be considered between the two non-independent layers. A common cause failure occurs when there is the potential for the failure of more than one component or system as a result of a single fault.

Typical rules for independence of a Protection Layer from the Initiating Event (IE) and from other IPLs in the same LOPA scenario include [71]:

- **The device** or system implementing each IPL **is independent** of the failure of the equipment that causes the IE and of the failure of other IPLs credited in the scenario. **That is the main reason why**, when dealing with low demand mode Safety Systems, a component used for process control, for example a control valve, cannot be used for the process shut down system: typically, an on/off safety valve is dedicated to the safety protection layer. Please consider that **when dealing with high demand mode Safety Systems**, since a different Risk Analysis and Risk Reduction techniques are used, the same component, like a contactor, can be shared by both the General Purpose PLC and the Safety Control System.
- The utility or support system implementing each IPL is independent of the failure of the equipment that causes the IE as well as of other IPLs. Safety systems are normally under UPS.

Other events that may jeopardize the independence of a Protection Layer may be considered.

2.1.5 PFD_{avg} for Different Architectures

In low demand mode, the average PFD can be calculated by integrating the PFD(t) function in the Proof Test interval.

$$PFD_{AVG} = \frac{1}{T_i}\int_0^{T_i} PFD(t)dt = \frac{1}{T_i}\int_0^{T_i} F(t)dt$$

2.1.5.1 1oo1 Architecture in Low Demand Mode

Let's consider a 1 out of 1 (1oo1) subsystem with Failure Rate λ (Figure 2.7). This architecture consists of a single channel, where any dangerous failure leads to the failure of the safety function when demand arises.

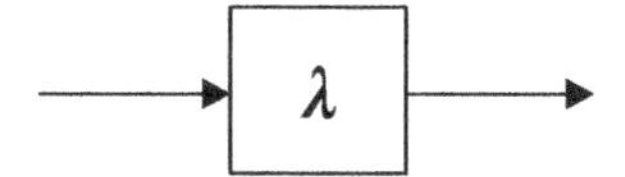

Figure 2.7 1oo1 subsystem.

starting from its Reliability function

$$R(t) = e^{-\lambda t}$$

the unreliability function can be written and approximated as

$$F(t) = 1 - R(t) = 1 - e^{-\lambda t} \cong 1 - (1 - \lambda t) = \lambda t$$

therefore:

$$PFD_{AVG} = \frac{1}{T_i}\int_0^{T_i} F(t)dt = \frac{1}{T_i}\int_0^{T_i}\left(1 - e^{-\lambda t}\right)dt \cong \frac{1}{T_i}\int_0^{T_i}\lambda t dt = \frac{\lambda}{2T_i}T_i^2 = \frac{\lambda T_i}{2}$$

2.1.5.2 Series of 1oo1 Architecture in Low Demand Mode

In case of two 1oo1 Subsystems in series, Figure 2.8, the total average PFD is the sum of the average PFD of the single subsystems.

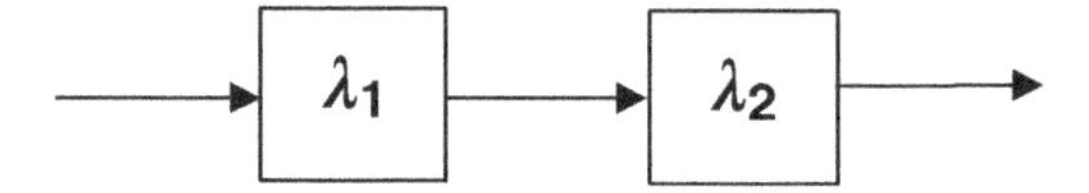

Figure 2.8 1oo1 subsystems.

$$PFD_{AVG-TOT} = PFD_{AVG-1} + PFD_{AVG-2} = \frac{\lambda_1 T_1}{2} + \frac{\lambda_2 T_2}{2}$$

2.1.5.3 1oo2 Architecture in Low Demand Mode

Considering a 1oo2 subsystem, where each element has the same Failure Rate λ, **safety is guaranteed if at least one of them is functioning correctly** (Figure 2.9). In other words, in case of one failure, the safety function is still guaranteed.

This architecture consists of two channels connected in parallel, such that either channel can process the safety function. Thus there should be a dangerous failure in both channels before a safety function fails on demand.

Supposing each element of the subsystem is, for example, a pressure transmitter; if one of them detects a dangerous situation, the subsystem triggers a shuts down.

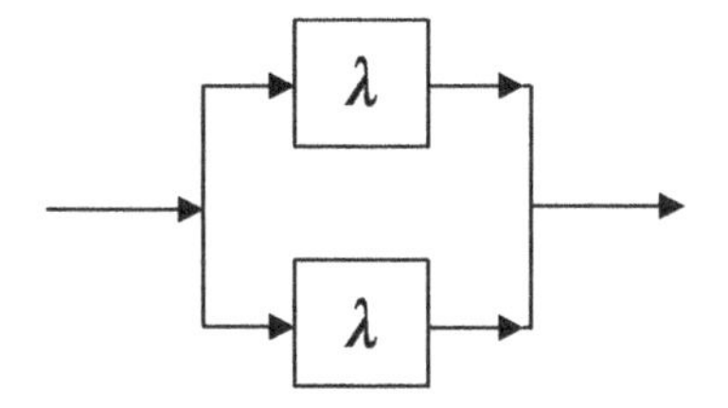

Figure 2.9 1oo2 subsystem.

For devices connected in parallel, knowing their $F(t)$, the total failure probability function $F_{tot}(t)$ is obtained through this formula:

$$F_{tot}(t) = \prod_{i=1}^{2} F_i(t) = F(t)_1 * F(t)_2$$

$$F(t) = 1 - R(t) = 1 - e^{-\lambda t} \cong 1 - (1 - \lambda t) = \lambda t$$

$$F_{tot}(t) = \lambda t * \lambda t = \lambda^2 t^2$$

therefore:

$$PFD_{AVG} = \frac{1}{T_i}\int_0^{T_i} F_{tot}(t)dt \cong \frac{1}{T_i}\int_0^{T_i} \lambda^2 t^2 tdt = \frac{\lambda^2}{3T_i}\int_0^{T_i} 3t^2 tdt = \frac{\lambda^2}{3T_i} T_i^3 = \frac{\lambda^2 T_i^2}{3}$$

2.1.5.4 1oo3 Architecture in Low Demand Mode

Considering a 1oo3 subsystem, where each element has the same Failure Rate λ, **safety is guaranteed if at least one of them is functioning correctly** (Figure 2.10). In other words, in case of two faults, the safety function is still guaranteed.

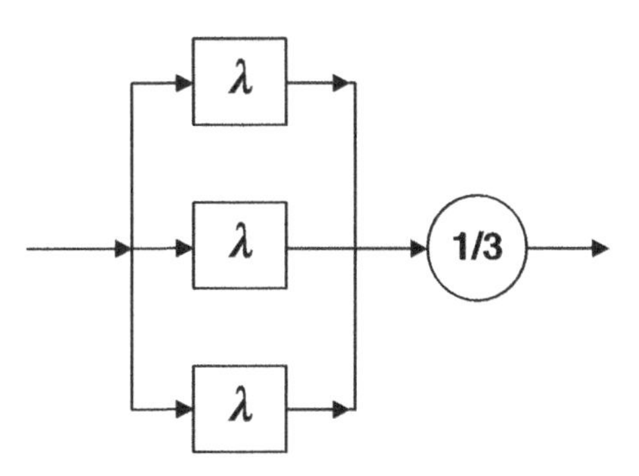

Figure 2.10 1oo3 subsystem.

Supposing each element of the subsystem is, for example, a temperature sensor: even if one of them detects a dangerous situation (for example, a high temperature), the subsystem will activate the Emergency Shutdown System they belong to.

For devices connected in parallel, knowing their $F(t)$, the total failure probability function $F_{tot}(t)$ is obtained through this formula:

$$F_{tot}(t) = \prod_{i=1}^{3} F_i(t) = F(t)_1 * F(t)_2 * F(t)_3$$

Therefore:

$$PFD_{AVG} = \frac{1}{T_i}\int_0^{T_i} F_{tot}(t)dt \cong \frac{1}{T_i}\int_0^{T_i} \lambda^3 t^3 tdt = \frac{\lambda^3}{4T_i}\int_0^{T_i} 4t^3 tdt = \frac{\lambda^3}{4T_i} T_i^4 = \frac{\boldsymbol{\lambda^3 T_i^3}}{\mathbf{4}}$$

2.1.5.5 2oo3 Architecture in Low Demand Mode

This architecture consists of three channels or elements (Figure 2.11) connected in parallel with a majority voting arrangement for the output signals, such that the output state is not changed if only one element gives a different result, which disagrees with the other two elements.

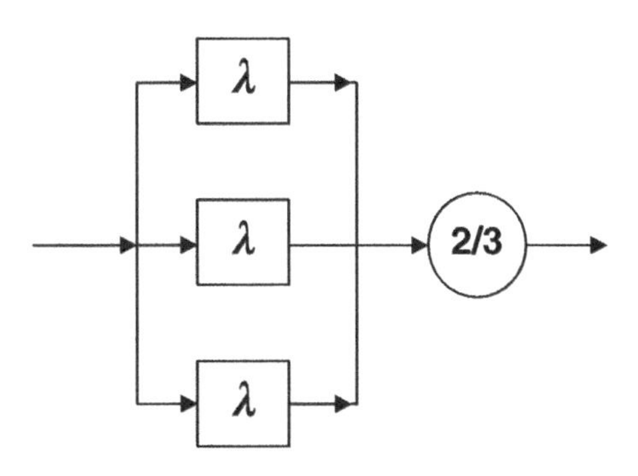

Figure 2.11 1oo3 subsystem.

Supposing each element of the subsystem is, for example, a temperature sensor, if one of them detects a dangerous situation (for example, a high temperature), the subsystem does not trigger the shutdown of the process. In this case, two temperature sensors must detect a high temperature to trigger the system to shut down. This architecture has a high level of both reliability and availability.

In this case, we have to use a more general formula, able to express the R(t) function for architectures with redundancy, knowing "n", total number of devices and "m", number of devices that have to be functioning to ensure that the entire system is functioning correctly:

$$\begin{cases} R_{tot}(t) = \sum_{i=m}^{n} \binom{n}{i} (R(t))^i (1 - R(t))^{n-i} \\ \binom{n}{i} = \frac{n!}{i!(n-i)!} \end{cases}$$

This set of formulas can be used for different combination of devices connected in parallel, including the previous scenario. For the 2oo3 architecture:

$$\begin{aligned} R_{tot}(t) &= \binom{3}{2} \left(e^{-\lambda t}\right)^2 \left(1 - e^{-\lambda t}\right) + \binom{3}{3} \left(e^{-\lambda t}\right)^3 \\ &\cong 3(1-\lambda t)^2(\lambda t) + (1-\lambda t)^3 = [...] = 1 + 2\lambda^3 t^3 - 3\lambda^2 t^2 \\ F(t)_{tot} &= 1 - R(t)_{tot} = 1 - \left(1 + 2\lambda^3 t^3 - 3\lambda^2 t^2\right) = 3\lambda^2 t^2 - 2\lambda^3 t^3 \cong 3\lambda^2 t^2 \end{aligned}$$

Therefore, PFD_{avg} can be calculated using the well-known formula:

$$PFD_{AVG} = \frac{1}{T_i} \int_0^{T_i} F_{tot}(t)dt \cong \frac{1}{T_i} \int_0^{T_i} 3\lambda^2 t^2 tdt = \frac{\lambda^2}{T_i} \int_0^{T_i} 3t^2 tdt = \frac{\lambda^2}{T_i} T_i^3 = \boldsymbol{\lambda^2 T_i^2}$$

2.1.5.6 Summary Table

In case we consider Safety Critical Systems, only the dangerous part of the failure rate is significant. Moreover, **considering non-repairable systems and no common cause failures**, the formula for the Average PFD are summarized in Table 2.5.

Table 2.5 Summary of *M* out of *N* Architectures.

Configuration	RBD	PFD_{avg}
Single element **(1oo1)**	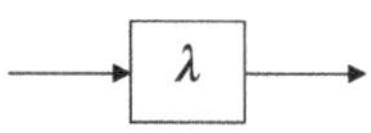 → λ →	$\frac{\lambda_{DU} \cdot T_i}{2}$
1-out-of-2 **(1oo2)**	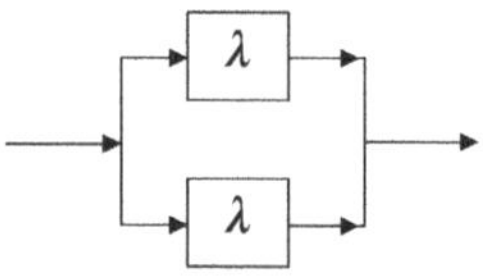 λ, λ (in parallel)	$\frac{\lambda_{DU}^2 \cdot T_i^2}{3}$
1-out-of-3 **(1oo3)**	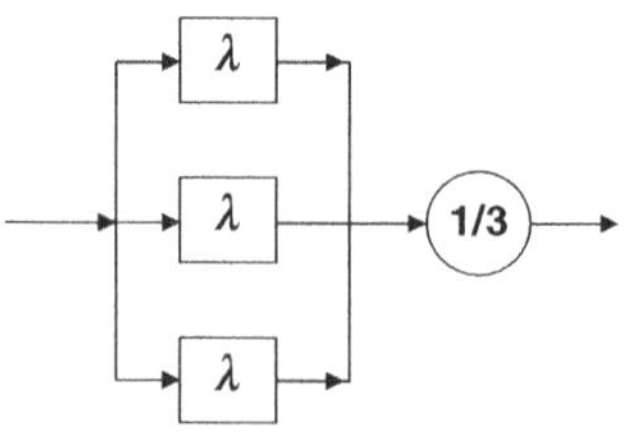 λ, λ, λ (in parallel) → 1/3 →	$\frac{\lambda_{DU}^3 \cdot T_i^3}{4}$

Table 2.5 (Continued)

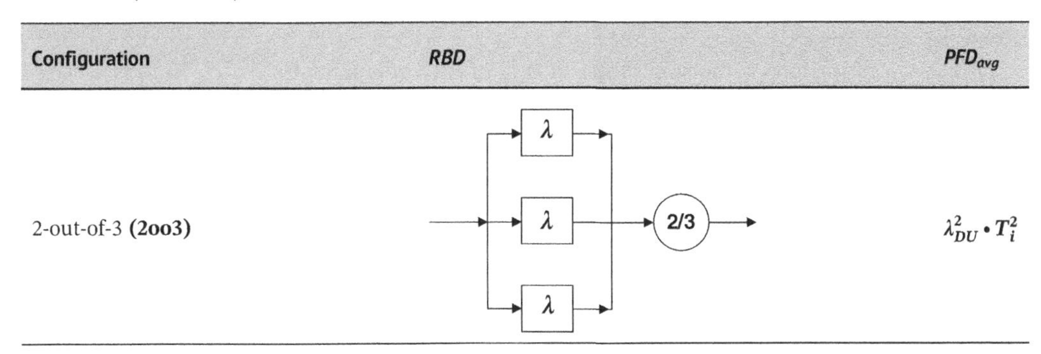

The value of PFD_{avg}, in case of repairable systems and common cause failures for redundant subsystems ($\beta \neq 0$), can be found in IEC 61508-6 annex B.

Just for a quick reference, **for a 1oo2 architecture**, in case only Common Cause Failures are taken into consideration, the formula for the PFD calculation becomes the following:

$$PFD_{avg} = (1-\beta)^2 \cdot \frac{\lambda_{DU}^2 \cdot T_i^2}{3} + \beta \cdot \frac{\lambda_{DU} \cdot T_i}{2}$$

2.1.5.7 Example of PFD_{Avg} Calculation

We want to calculate the PFD_{avg} for a low demand mode Safety Loop involving a pressure transmitter, an analog barrier, two electronic safety modules, and the STO (Safe Torque off) of a Variable Speed Drive. Please refer to Figure 2.12.

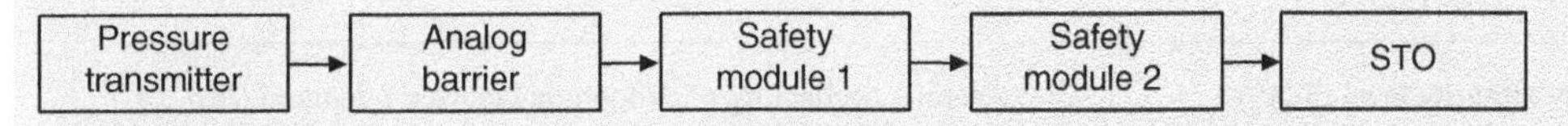

Figure 2.12 Example of a safety instrumented function.

We need to know the safety data from the datasheet of each component; for example, we need to know λ_{DU}, SFF, and the component type. It is then necessary to establish the structure of each subsystem and the Proof test period Ti.

Hereafter, examples of data:

Component	λ_{DU}	SFF (%)	Type	Arch. constr.	Structure	T_i
Pressure transmitter	$9.9 \cdot 10^{-8}$	87	B	2	1oo1	10 years
Analog barrier + converter	$9.13 \cdot 10^{-8}$	93	B	2	1oo1	10 years
Safety module	$1.14 \cdot 10^{-11}$	99	A	3	1oo1	10 years
STO/VSD	$3.65 \cdot 10^{-9}$	99.66	A	3	1oo1	10 years

The PFD_{avg} of the 1oo1 structure can be calculated with the following formula, resulted from the integral in *par. 2.2 Terms and definitions from 61508-4*

$$PFD_{avg} = \frac{\lambda_{DU} \cdot T_i}{2}$$

The SIL architecture of each subsystem has been fixed using tables 2 and 3 of the so-called Route 1_H.

PFD_{avg_TOT} of the loop is the sum of the PFD_{avg} of each component.

The system reaches SIL 2.

Component	PFD_{Avg}
Pressure transmitter	$4.34 \cdot 10^{-3}$
Analogue barrier + converter	$4 \cdot 10^{-3}$
Safety module 1	$5 \cdot 10^{-7}$
Safety module 2	$5 \cdot 10^{-7}$
STO/VSD	$1.6 \cdot 10^{-4}$
Total (SIL 2)	$\mathbf{8.5 \cdot 10^{-3}}$

2.1.6 Reliability of a Safety Function in Low Demand Mode

In low demand mode, the Reliability of a safety function is defined with the parameter PFD_{avg}.

IEC 61508 divides the requirements into four safety integrity levels, SIL 1, SIL 2, SIL 3, and SIL 4, with SIL 4 being the most reliable and SIL 1 being the least reliable. Table 2.6 shows the relationship between the SIL and the PFD_{avg} range.

Table 2.6 SIL and limits of PFD_{avg} values.

Safety integrity level (SIL)	Average probability of dangerous failure on demand (PFD_{avg})
4	$10^{-5} \leq PFD_{avg} < 10^{-4}$
3	$10^{-4} \leq PFD_{avg} < 10^{-3}$
2	$10^{-3} \leq PFD_{avg} < 10^{-2}$
1	$10^{-2} \leq PFD_{avg} < 10^{-1}$

However, it is not enough to have a safety instrumented function with a certain PFD_{avg} value (for example, SIL 2) to decide that the SIF is a SIL 2 Safety instrumented Function.

Consider a SIF that operates in low-demand mode and assume that we have determined the PFD_{avg} to be $5 \cdot 10^{-3}$. Because this value is in the interval from 10^{-3} to 10^{-2}, the system may fulfill the requirements for SIL 2 if it also fulfills the requirements for

- Systematic Safety Integrity.
- Hardware Safety integrity.

Please refer to Figure 2.13.

A SIF will, therefore, not automatically fulfill the SIL 3 requirements, for example, when the PFD_{avg} is within the interval for SIL 3. Let's see an example for that.

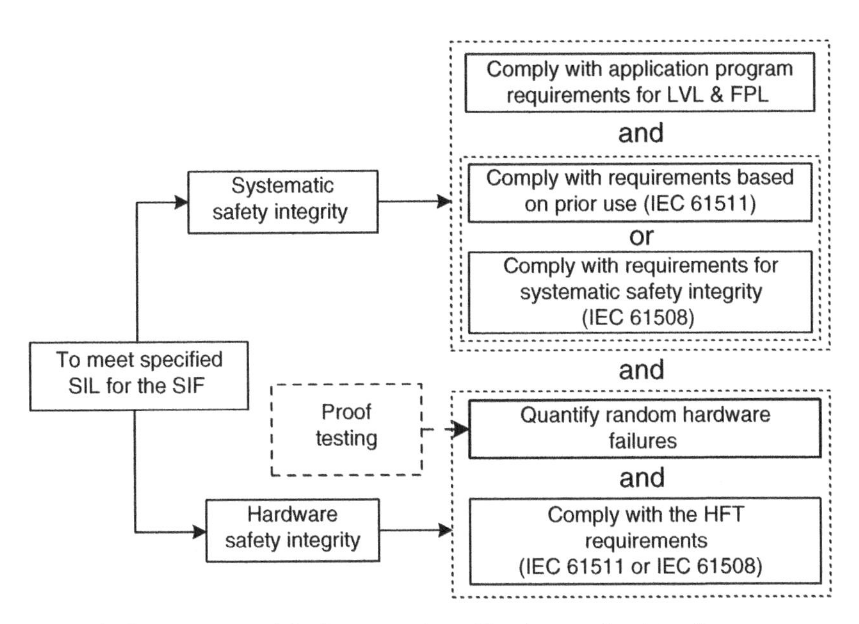

Figure 2.13 How to reach both systematic and hardware safety integrity.

Example

Let's consider an Input subsystem made of a Pressure Transmitter as detailed in Table 2.7. It is an ABB model 2600T, 268 Safety.

It is a Type B Component with $HFT = 0$.

SIL Capability is 2.

Table 2.7 Pressure transmitter reliability data.

Pressure transmitter failure rate data (Source: Exida SERH 2015 – 01 sensors – item 1.6.5)		p-Cap Per 10^9 hours (FITs)
Fail dangerous detected	λ_{DD}	669
Fail dangerous undetected	λ_{DU}	37
Fail safe detected	λ_{SD}	134
Fail safe undetected	λ_{SU}	160
No effect failure	λ_{NE}	181

$$SFF = \frac{\lambda_S + \lambda_{DD}}{\lambda_T} = \frac{294 + 669}{1000} = 96.3\%$$

Let's suppose the dangerous situation happens once every 10 years; we decide that a Proof Test has to be performed at least once every year.

The Reliability level of the subsystem is therefore:

$$PFD_{avg} = \frac{\lambda_{DU} \cdot T_i}{2} = \frac{37 \cdot 10^{-9} \cdot 1 \cdot 8760}{2} = 1.6 \cdot 10^{-4}$$

The reliability level is SIL 3; however, the safety subsystem can be used in a System that reaches SIL 2 as a maximum. There are two reasons for that:

- The first one is that the architectural constrains from Table 2.8 limit the maximum SIL to SIL 2, if the transmitter is used as a single channel ($HFT = 0$)
- The other reason is that the SIL capability limits the component to SIL 2, even if used with HFT higher than zero.

Table 2.8 Architectural constraints for Type B components.

Safe failure fraction of an element	Hardware fault tolerance		
	0	1	2
$SFF < 60\%$	Not allowed	SIL 1	SIL 2
$60\% \leq SFF < 90\%$	SIL 1	SIL 2	SIL 3
$90\% \leq SFF < 99\%$	SIL 2	SIL 3	SIL 4
$SFF \geq 99\%$	SIL 3	SIL 4	SIL 4

2.1.7 A Timeline

In Figure 2.14, a timeline of the Functional Safety Standards discussed in the book is shown.

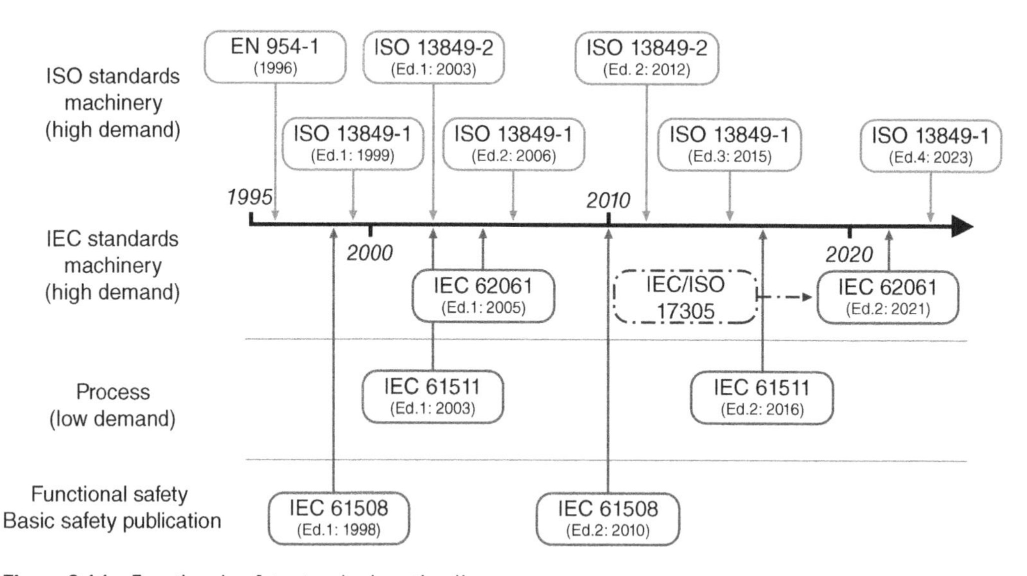

Figure 2.14 Functional safety standards: a timeline.

2.2 Safety Systems in High and Low Demand Mode

From the beginning of Functional Safety, two types of safety systems were identified: they were based upon the frequency of demand of the safety function. For that reason, the characteristics of a safety system and the way its Reliability level is calculated, differ based upon the frequency of demand.

Using an analogy with urban mobility, the characteristics of a car that is used every day should be different from those of a car used once every four years. In both cases we are looking for a reliable car; if we do not specify that we want to use the car once every four years, it will be likely that when we enter the car, after it was parked in our garage for four years and try to start it, it fails, for example because the battery is flat.

That is the reason safety systems are divided into two major kinds of demand mode of operation:

- Safety Systems in **high demand** mode of operation
- Safety Systems in **low demand** mode of operation

Here are the definitions from IEC 61508-4 [8]:

> ***[IEC 61508-4] 3.5 Safety functions and safety integrity***
> ***3.5.16 Mode of Operation****: way in which a safety function operates, which may be either*
>
> - ***low demand mode:*** *where the safety function is only performed on demand, in order to transfer the EUC into a specified safe state, and where the frequency of demands is no greater than one per year; or*
> - ***high demand mode:*** *where the safety function is only performed on demand, in order to transfer the EUC into a specified safe state, and where the frequency of demands is greater than one per year; or*
> - ***continuous mode:*** *where the safety function retains the EUC in a safe state as part of normal operation*

Therefore, if a Safety Control System is activated less than one time per year (or exactly once in a year), it works in low demand mode.

2.2.1 Structure of the Control System in High and Low Demand Mode

2.2.1.1 Structure in Low Demand Mode, Process Industry

In low demand mode of operation, the control and the safety system can be represented as an "onion model." Normally, the process is kept under control thanks to the Process Control System: PID controllers (part of the BPCS) keep the main variables like pressure, temperature, level, or flow, under control. In case a failure occurs in the control loop and a variable goes out of control, the safety system brings the whole system into a safe state.

That is the reason why the BPCS is being part of the overall risk reduction. When a LOPA (Layer of Protection Analysis) is performed, normally a maximum reduction level of 10 is assigned to the Control System.

Looking at Figure 2.15, the **Demand Rate** (DR) is the frequency of request of the safety function; not to be confused with the Initiating **Event Rate** (ER). In case of an overspeed safety function in a gas turbine, there is always a rotation speed control system. In case there is an issue in the process (damaged bearing = Initiating Event), the control system may succeed in bringing the turbine into a safe state: in that case, no demand onto the safety system is generated. Therefore, in general, the DR is lower than the Initiating ER.

For a system operating in low demand mode, the request upon the safety function is less than or equal to once per year; moreover, the **Proof Test interval should be an order of magnitude greater than the expected interval between demands (DR).**

Figure 2.15 The onion model.

2.2.1.2 Structure in High Demand Mode, Machinery

In high demand mode, normally, the Non-Safety-related control system (in this book, the terms "Automation PLC" or "General Purpose PLC" are also used as synonyms) makes the machine performing its tasks. In case there is a request for access into a safeguarded area, the safety-related control system (in the book, the term "Safety System" is also used as a synonym) stops the machine (the function is a "Safety-related stop" § 4.3.1) and the operator can work inside the area, safely.

In this case, the Non-Safety-related control system is not part of the safety function, and therefore it is not taken into account for the evaluation of the Reliability of the safety function itself. The Safety-related I/O cards may also be used for non-safety-related aspects.

In Figure 2.16, the upper "I/O components" are only safety I/O cards; the I/O components shown below the robot are Automation I/O cards on the left and Safety I/O cards on the right.

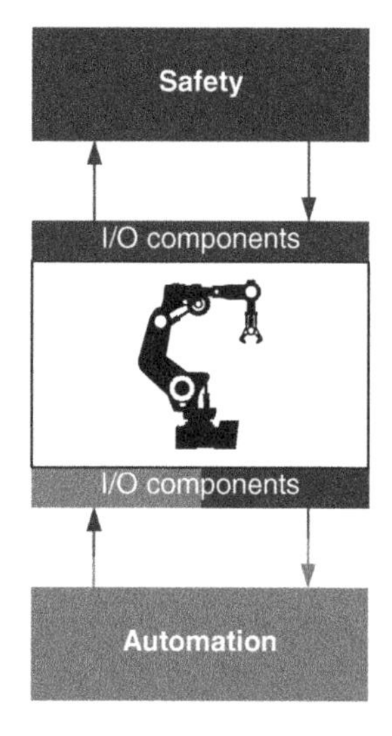

Figure 2.16 The shared model.

2.2.1.3 Continuous Mode of Operation

In continuous mode of operation, the safety function is performed "permanently" and can be used as an operational control function. That means the Machine is constantly under the control of a Safety System.

As an example, consider the control system for a centrifugal compressor. It typically prevents surge by continuously modulating a recycle valve (or surge control valve), which routes part of the compressor discharge to the compressor suction part. A failure of this control system can cause the compressor to go into surge, with significant hazardous effects and hence **it performs both a control and a safety function,** and it can be considered a continuous mode SIF.

2.2.2 The Border Line Between High and Low Demand Mode

2.2.2.1 Considerations in High Demand Mode

The high demand mode approach of ISO 13849-1 and IEC 62061 works well when the demand (number of operations per year) is high, in other words, when the safety control system is running constantly and **the demand upon the safety function is daily or weekly**. An example could be a contactor that is de-energized when an operator enters a safeguarded area. When calculating the reliability of the contactor output subsystem, starting from a B_{10D} value, the more the contactor is used (in other words, the higher is the number of operations in a year or $\boldsymbol{n_{op}}$, the lower is its $MTTF_D$ value and therefore its reliability).

In high demand mode, when the frequency decreases, we are in a situation whereby **the least a component is used, the higher is its reliability**. Considering again the example of a contactor, **the lower is the n_{op}, the higher is its $MTTF_D$.**

Finally, every time there is a request upon the safety function, the correct functioning of its components can be verified: that is also the meaning of the **Diagnostic Coverage** (DC). Let's consider, for example, two pressure switches that monitor a high pressure in a pipe. In case they work in high demand mode (demand of once every week, for example), every time the safety system triggers, in case one of the two pressure switches has failed in the meantime, the logic solver detects the inconsistency and, besides shutting down the machine or the process, it does not allow to restart it until the inconsistency is cleared.

The issue is when the safety system, considered working in high demand mode, initiates a safety function, for example, **twice per year**. Again, looking at the contactor reliability, when used twice in a year ($n_{op} = 2$), its reliability results higher than when it was used once in a week ($n_{op} = 52$). That issue is known and, for example, that is the reason why IEC 62061 requires a test every month for Safety Systems in SIL 3.

Reliability in high demand mode relies upon automatic testing of components. In case electronic components are used (for example, a Type 4 RFID interlocking device with OSSD output), the test can be done more often than the frequency of demand upon the safety function. In the case of electromechanical components, like a pressure switch, the diagnostic can be done only when there is a demand upon the safety function.

2.2.2.2 Considerations in Low Demand Mode

In **low demand mode**, the least the safety function is tested (Proof Test), the lower is its Reliability; let's refer to Equation 1.10.3:

$$PFD_{avg} = \frac{\lambda_{DU} \cdot T_i}{2}$$

where T_i indicates the test interval. The longer the interval, the higher the PFD_{avg} and therefore the lower is the system Reliability. The test interval T_i is called **Proof Test Interval**, and it is normally done "offline."

Reliability in low demand mode relies upon automatic testing of components as well: many sensors contain electronics (for example, a 4–20 mA transmitter), and that allows the diagnostic of all detected failures (λ_{DD}). However, being in low demand mode, it is important to detect the so-called Dangerous Undetected failures (λ_{DU}). That cannot be done in an "automatic way" using the component or the system diagnostic capabilities, and it has to be done "offline."

2.2.2.3 The Intermediate Region

The region where the system is in high demand mode but used few times per year is called the "**intermediate demand mode**" [62]. In this region, the high demand mode formula gives a "too rosy" picture; while using the low demand mode approach, the issue may be the practicality of execution of the Proof Test.

The dividing line of once per year was not based upon specific research: it was arbitrary. Therefore, the strict distinction of the low-demand and high-demand mode, using the not justified criterion of once per year as the borderline, is sometimes creating unclear situations around the design and architecture of safety systems.

Regardless of which approach is the one used, in the case of high demand mode SCS used less than once per month, a **periodic verification** is recommended, especially for electromechanical components. Such verification can be done by forcing the safety function to act or by creating a demand upon it. If that is not possible, inspections can be done to find visible damages that may impair safety, like evidence of overheating or of physical damage to components. Other aspects to consider are for example the presence of pollution agents (dust, water, etc.) or of modifications to the circuit: that becomes a sort of Proof Test or at least **Partial Proof Test**. The issue with Proof Test in high demand mode of operation is that the Proof Test procedure needs to be specified by the component manufacturer, and the ones operating in high demand mode only rarely provide such procedures.

2.3 What is a Safety Control System

A control system is, in general, composed of (Figure 2.17):

- Sensors
- A Logic solver
- Final elements.

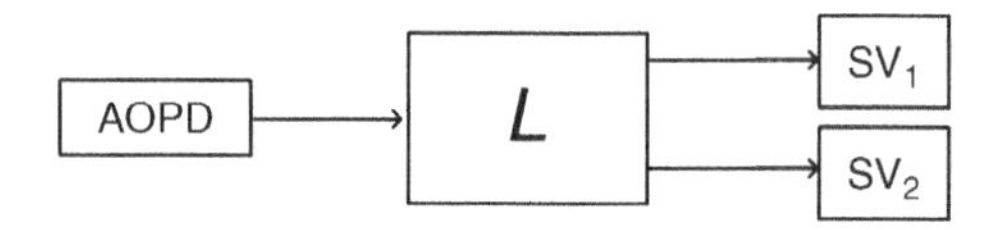

Figure 2.17 RBD of a safety control system.

2.3.1 Control System and Safety System

Let's consider a conveyor transporting metal pieces; we need a control system made of a sensor that detects that a piece has reached the end position and for example activates a cylinder that moves the piece to a certain location. The signal from the sensor goes to a PLC (Programmable Logic Controller), whose logic opens the motor contactor and stops the belt. **This is an example of an open loop control system**.

In process control, closed-loop systems are often used. The sensor can be a pressure or flow transmitter and the final element can be a valve.

In the first case, the control system makes "the iron move," and in the second, it makes "the process flow correctly within certain limits." We could call the "iron" or the piping, the Equipment under Control (EUC). Here its definition [8]:

> ***[IEC 61508-4] 3.2 Equipment and devices***
> ***3.2.1 Equipment Under Control – EUC***. *Equipment, machinery, apparatus or plant used for manufacturing, process, transportation, medical or other activities.*
>
> ***Note:*** *The EUC control system is separate and distinct from the EUC.*

There are situations where the sensor detects a danger. In that case, **the control system has a safety duty** in the sense that it intervenes in order to avoid a dangerous situation.

ISO 13849-1 calls it a Safety-related part of a control system or **SRP/CS:**

[ISO 13849-1] 3.1 Terms and definitions
3.1.1 Safety-related Part of a Control System. SRP/CS. *Part of a control system that performs a safety function, starting from safety-related input(s) to generating safety-related output(s).*

Note 1 to entry: *The safety-related parts of a control system starts at the point where the safety-related inputs are initiated (including, for example, the actuating cam and the roller of the position switch) and end at the output of the power control elements (including, for example, the main contacts of a contactor).*

IEC 62061 used to call it SRECS: safety-related electrical control system. Since the new edition includes non-electrical systems as well, the new definition is:

[IEC 62061] 3.2 Terms and definitions
3.2.3 Safety-related Control System. SCS. *Part of the control system of a machine which implements a safety function.*

Note 1 to entry: *SCS is similar to SRECS of the previous edition of this document.*

IEC 61511-1 [16] calls it a **Safety Instrumented System or SIS**. Actually, this standard starts from the concept of **Safety Function** that can be implemented by a Safety System or by any other way: for example, a containment system for a big tank is a safety system but not done with a safety control system.

[IEC 61511-1] 3.2 Terms and definitions
3.2.65 Safety Function. *Function to be implemented by one or more protection layers, which is intended to achieve or maintain a safe state for the process, with respect to a specific hazardous event.*

Then, it defines the abstract concept of a safety system and calls it Safety Instrumented Function:

[IEC 61511-1] 3.2 Terms and definitions
3.2.66 Safety Instrumented Function SIF. *Safety function to be implemented by a safety instrumented system (SIS).*

Note 1 to entry: *A SIF is designed to achieve a required SIL which is determined in relationship with the other protection layers participating to the reduction of the same risk.*

It finally defines the "physical" level of a safety control system and calls it an SIS:

[IEC 61511-1] 3.2 Terms and definitions
3.2.67 Safety Instrumented System. SIS. *Instrumented system used to implement one or more SIFs.*

***Note 1 to entry:** A SIS is composed of any combination of sensor (s), logic solver (s), and final elements(s) (e.g., see figure 6). It also includes communication and ancillary equipment (e.g., cables, tubing, power supply, impulse lines, heat tracing).*

In essence, the three standards call the same concept in three different ways (please refer to Table 2.9): a sensor placed to detect a dangerous situation, a control system that translates the input signal into an output one. The signal then goes to a final element that, normally by being de-energized, brings the machine or the process to a safe state. We also call it a **Safety Loop**.

Table 2.9 Summary of the key terminology used in the different functional safety standards.

	IEC 61508	IEC 61511-1	IEC 62061	ISO 13849-1
Logical layer	Safety function	Safety Instrumented Function (**SIF**)	Safety-related Control Function (**SCF**)	Safety Function
Physical layer	Safety-related system	Safety Instrumented System (**SIS**)	Safety-related Control System (**SCS**)	Safety-related part of the Control System (**SRP/CS**)

The logical layer is the result of the risk assessment; that means, from the risk assessment comes the need of certain Safety Functions. The above standards help designing the "physical layers," in other words **designing a proper SRP/CS or SCS or SIF**.

2.3.2 What is Part of a Safety Control System

In order to be part of a Safety Control System, components need Reliability data. Not all components have them; therefore the question is: what are the components that belong to an SCS? Figure 2.18, taken from ISO 12100 annex A, shows that Sensors (Ex. interlocking devices), Logic systems (Safety PLC), and Power control elements (Contactors, Valves, etc.) are part of a Safety Control System.

However, **Machine actuators** like Engines and Cylinders are not.

This definition is valid in the case of Emergency Stop functions. Different is the situation in case an **Emergency start function** is implemented. Here are the definitions from IEC 60204-1 [3] annex E:

[IEC 60204-1] Annex E: Explanation of emergency operation functions

***Emergency stop:** An emergency operation intended to stop a process or a movement that has become hazardous.*

***Emergency start:** An emergency operation intended to start a process or a movement to remove or to avoid a hazardous situation.*

There are situations where, in order to bring the machine into a safe state, it is necessary to activate a safety element. An example can be **a motor break.**

Another example is, in a Continuous Casting line, the interruption of the steel flow from the tundish; that can be achieved by energizing a Solenoid valve that activates the movement of a cylinder.

In both cases, the machine actuators (the brake or the cylinder) are part of the safety system; however, that case is more the exception than the rule.

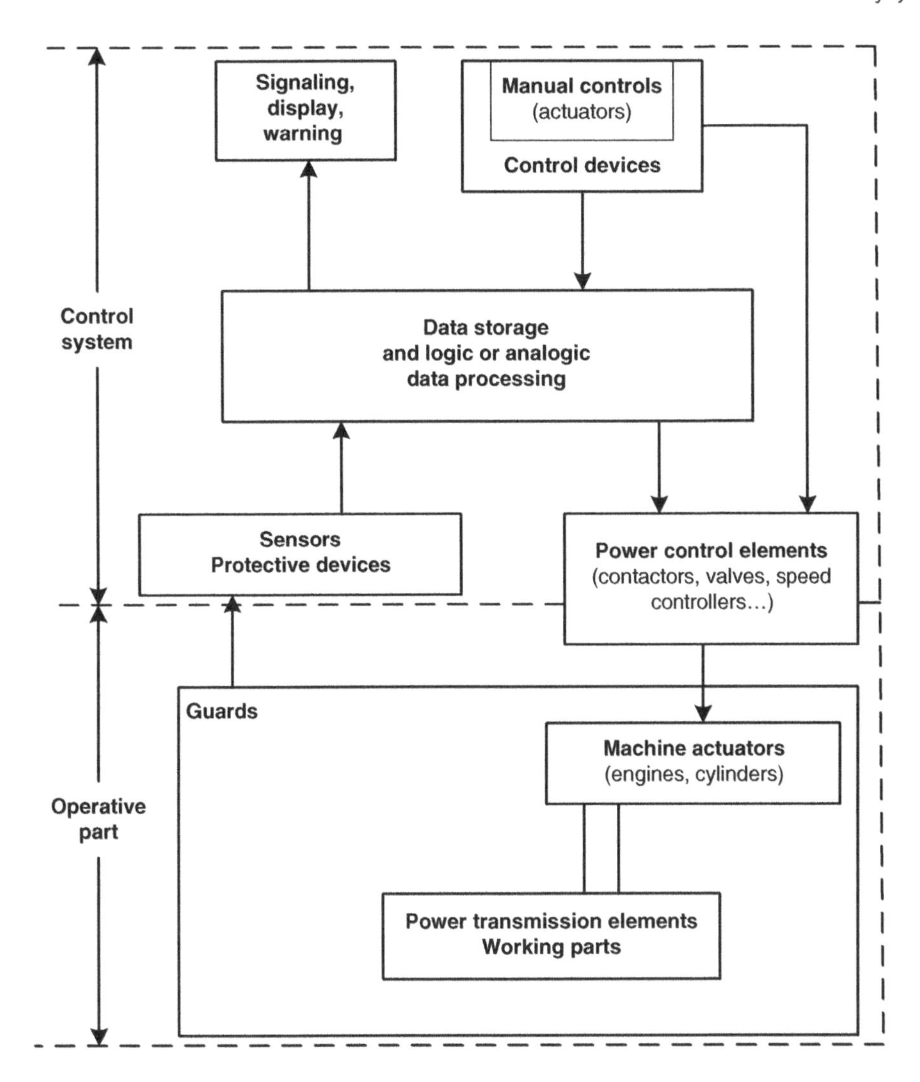

Figure 2.18 Schematic representation of a machine.

2.3.3 Implication of Implementing an Emergency Start Function

When implementing an emergency stop function, the "safe state" is a logical "0." That means any safety logic must drive its output to "0" or "no Energy" in order to bring the system to a safe state.

Consider for example a movement given by the rotation of a motor. If the motor stops, the movement stops as well. The majority of Safety systems reasons that way. That is also called the use of de-energization and is one of the Basic Safety Principle according to ISO 13849-2, table D.1.

> ***[ISO 13849-2] Table D.1 – Basic Safety Principle***
> ***Use of De-energization.*** *A safe state is obtained by de-energizing all relevant devices, e.g. by use of normally closed (NC) contact for inputs (push-buttons and position switches) and normally open (NO) contact for relays (see also ISO 12100:2010, 6.2.11.3).*

Exceptions can exist in some applications, e.g. where the loss of the electrical supply will create an additional hazard. Time-delay functions may be necessary to achieve a system safe state (see IEC 60204-1:2005, 9.2.2).

In the example given at § 1.13.2, a solenoid valve must be energized in order to close the steel flow from the ladle. This is an example of a safety system where the "safe state" is a logical "1." Attention has to be paid to its implementation since, in case of discrepancies or faults detected by the logic, **the output of the safety system has to be placed to a logical "1" and not to a "0."**

2.4 CE Marking, OSHA Compliance, and Functional Safety

2.4.1 CE Marking

The CE Marking, represented graphically in Figure 2.19, is the European approach to free movements of goods within the European Economic Area (EEA). Here is what Article 6 of the Machinery directive states:

[Machinery Directive 2006/42/EC] Article 6. Freedom of movement.

1. *Member States shall not prohibit, restrict or impede the placing on the market and/or putting into service in their territory of machinery which complies with this Directive.*

However, goods have to be safe! For that reason, the "New Approach" was developed in 1985, which restricts the content of legislation to "essential requirements," leaving the technical details to European harmonized standards.

A safe product can have CE Marking, but not all goods require a CE marking to be considered safe. There are 27 European Directives that require a CE Mark. For example, metal-enclosed switchgear or controlgears for rated voltages above 1 kV do not require a CE mark to be sold within the EEA. The reason is that they do not have a European Directive that gives Essential Safety Requirements for them.

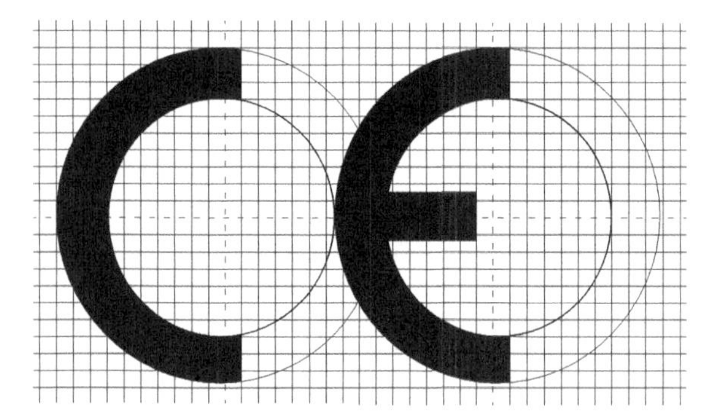

Figure 2.19 The CE marking.

So why a toy has a European Directive, the Toy Safety Directive 2009/48/EC, while a high voltage switchgear does not? Actually, the latter can be more dangerous than a toy. The reason is that toys can be bought by "untrained" people who cannot judge if they are buying a safe product. **The presence of the CE mark on a product means that it is safe to use.**

High voltage switchgears must comply with IEC 62271-200; however, depending upon the application, they may need to comply with other standards. You need to know that technical background in order to know which standards a component must comply with in order to be safe. The power of the CE marking is precisely that: by placing the CE mark on a product, the manufacturer is stating that the product complies with all the applicable EHSRs of all the applicable European Directives. That is a stronger guarantee of safety and easier to understand for a consumer than the conformity to specific technical standards.

The **conformité européenne** (CE) Mark is therefore the European Union's (EU) mandatory conformity marking, regulating the goods sold within the EEA since 1985. The CE marking represents a manufacturer's declaration that products comply with the EU's New Approach Directives. These directives not only apply to products within the EU but also to products that are manufactured outside Europe but will be sold in the EEA.

Most Products that require a CE Mark do not need third-party testing. For example, a circuit breaker can be sold in the European Market without having had any third-party testing nor certification, since the Low Voltage Directive (2014/35/EU) does not require the involvement of notified bodies or third-party laboratories. Here is what Article 12 states:

> ***[Low Voltage Directive 2014/35/EU] Article 12. Presumption of conformity on the basis of harmonised standards***
> *Electrical equipment which is in conformity with harmonised standards or parts thereof the references of which have been published in the Official Journal of the European Union shall be presumed to be in conformity with the safety objectives referred to in Article 3 and set out in annex I covered by those standards or parts thereof.*

That is valid for several products bearing the CE mark, and that is **one of the main differences between the approach to product safety in Europe compared to North America**. In most of the cases, the CE Mark is the manufacturer self-declaration that the product complies with all relevant essential requirements (e.g. health and safety requirements) of all the applicable directive(s). For more information about CE Marking, please refer to the Blue Guide [49].

2.4.2 The European Standardization Organizations (ESOs)

The need for safety and, at the same time, of free movement led to the creation of a European standardization policy, whose three main institutions are:

- **CEN:** European Committee for Standardization. CEN is a French acronym and stands for "Comité Européen de Normalisation".
- **CENELEC:** European Committee for Electrotechnical Standardization. CENELEC is also a French acronym and stands for "Comité Européen de Normalisation Électrotechnique".
- **ETSI:** European Telecommunications Standards Institute.

Before that happened, each country had its own standardization bodies, like BSI in the United Kingdom, DKE in Germany, or CEI in Italy, that could decide what characteristics a product needed to be considered safe. For example, the PE cable (protection earth or grounding cable) was green in some European countries and yellow in others. CENELEC harmonized standards (for example, EN 60204-1) states that the PE cable has to be GREEN-AND-YELLOW. That is an example of a compromise European countries had to go through, when discussing common Harmonized Standards.

Technical standards are more and more discussed and approved at international level:

- **ISO:** International Organization for Standardization prepares and publishes various types of "non-electrical" standards.
- **IEC:** International Electrotechnical Commission prepares and publishes international standards for all electrical, electronic, and related technologies.

Some ISO and IEC standards can be adopted, with or without modifications, by the ESOs, however, in order to become a Harmonized one, a standard has to be recognized as providing a "presumption of conformity." In other terms, if a manufacturer applies a specific harmonized standard, the machine can be considered safe for all the aspects dealt by that standard. It is said that a Harmonized Standard provides a "presumption of conformity." Here is what Article 7 of the machinery Directive states:

> ***[Machinery Directive 2006/42/EC] Article 7. Presumption of conformity and harmonised standards***
>
> 1. *Member States shall regard machinery bearing the CE marking and accompanied by the EC declaration of conformity, the content of which is set out in annex II, part 1, section A, as complying with the provisions of this Directive.*
> 2. ***Machinery manufactured in conformity with a harmonised standard**, the references to which have been published in the Official Journal of the European Union, **shall be presumed to comply with the essential health and safety requirements covered by such a harmonised standard**.*
> 3. *The Commission shall publish in the Official Journal of the European Union the references of the harmonised standards.*
> 4. *Member States shall take the appropriate measures to enable the social partners to have an influence at national level on the process of preparing and monitoring the harmonised standards.*

2.4.3 Harmonized Standards

There are more than 1000 harmonized standards to the Machinery Directive (MD). ISO classifies standards in three groups:

a. **Type-A standards** or basis standards. They give the basic safety concepts, principles for design, and general aspects that can be applied to machinery. ISO 12100 is a Type A standard.
b. **Type-B standards** or generic safety standards. They deal with one or more safety aspects or with one or more types of safeguards that can be used across a wide range of machinery:
   - **Type-B1** standards deal with particular safety aspects: for example safety distances (ISO 13855), surface temperatures (ISO 13732-x), and Functional Safety (ISO 13849-1): they give indications, in the form of data or methodology, how those particular safety aspects can be addressed.

- **Type-B2** standards provide the performance requirements for the design and construction of particular safeguards: for example two-hands controls (ISO 13851), interlocking devices (ISO 14119), pressure-sensitive devices (ISO 13856-x), and guards (ISO 14120).

c. **Type-C standards** They deal with detailed safety requirements for a particular machine or group of machines: for example Stationary Grinding Machines (ISO 16089).

IEC classifies standards again in three groups but with a bit different approach:

- **Basic Safety Publication** (BSP). Publication on a specific safety-related matter. A BSP is primarily intended for use by Technical Committees (TCs) in the preparation of standards, in accordance with the principles laid down in IEC Guide 104 and ISO/IEC Guide 51. An Example is IEC 60446 (*Basic and safety principles for man-machine interface, marking and identification – Identification of conductors by colors or alphanumerics*) that, being a BSP is not applicable to specific products and therefore will never be a harmonized standard. IEC 61508-1 [5] to -4 [8] are also BSP.
- **Group Safety Publication** (GSP). Publications covering all safety aspects of a specific group of products within the scope of two or more product TCs. Group safety publications are primarily intended to be stand-alone product safety publications, but they may also be used by TCs as source material in the preparation of their publications.
- **Product Safety Standard**. Standard covering all safety aspects of one or more products within the scope of a single product TC. Product includes equipment, software, installation, and service. IEC 62061 is an example of a product safety standard.

When it comes to publication in the **Official Journal of the European Union, IEC standards are classified as B-types.** Here is the definition given by the Official Journal:

> ***B-type standards:*** *B-type standards deal with specific aspects of machinery safety or specific types of safeguard that can be used across a wide range of categories of machinery. Application of the specifications of B-type standards confers a presumption of conformity with the essential health and safety requirements of the Machinery Directive that they cover when a C-type standard or the manufacturer's risk assessment shows that a technical solution specified by the B-type standard is adequate for the particular category or model of machinery concerned.*
> *Application of B-type standards that give specifications for safety components that are independently placed on the market confers a presumption of conformity for the safety components concerned and for the essential health and safety requirements covered by the standards (OJ C 348, 28.11.2013).*

Both EN ISO 13849-1 and EN IEC 62061 standards are harmonized to the Machinery Directive, and therefore **they both give the presumption of conformity** for the aspects they deal with. For example, compliance with one of the two standards gives a presumption of conformity to the EHSR (Essential Health and Safety Requirement) 1.2.1 stated by the Machinery Directive 2006/42/EC in annex I:

> ***[Machinery Directive 2006/42/EC] Annex I.*** *Essential health and safety requirements relating to the design and construction of machinery*
> ***1.2.1. Safety and Reliability of control systems***

Control systems must be designed and constructed in such a way as to prevent hazardous situations from arising. Above all, they must be designed and constructed in such a way that:

- *they can withstand the intended operating stresses and external influences,*
- *a fault in the hardware or the software of the control system does not lead to hazardous situations,*
- *errors in the control system logic do not lead to hazardous situations,*
- *reasonably foreseeable human error during operation does not lead to hazardous situations.*

  [...]

That is the reason why Functional Safety is important for CE Marking. Complying with either EN ISO 13849-1 or EN IEC 62061 gives the presumption of conformity for all aspects related to the Reliability of the SCS, like the EHSR 1.2.1.

2.4.4 Functional Safety in North America

In the United States, one of the main drivers of industrial safety is the Occupational Safety and Health Administration (OSHA). The Occupational Safety and Health Act of 1970 was signed into law on December 29, 1970, which technically "established" OSHA, a new federal regulatory agency within the Department of Labor. However, the OSH Act became "effective" on April 28, 1971. The purpose of this act was to provide safe and healthy working conditions.

The OSH Act places the responsibility on both the employer and the employee. This is quite different from the Machinery Directive, which requires suppliers to place machines on the market that are free from hazards. In the United States, a supplier may, in principle, sell a machine without any safeguarding; in this case, the employer needs to take all actions in order to make the machine safe for the employees. However, it should be noticed that, for new machines, the US legal system (non-regulatory, products liability) provides a strong incentive for machine builders to provide safeguarding.

OSHA publishes regulations in Title 29 of the Code of Federal Regulations (29 CFR). Regulatory standards pertaining to industrial machinery are published by OSHA in Part 1910 of 29 CFR. Despite the fact the OSHA standards still contain some technical details, **manufacturers mainly refer to Technical Standards in order to design safe machinery, because the requirements are kept up to date**. There are several ANSI Accredited Standard Developing Organizations (ASDs or SDOs), among which the ANSI B11 series of machinery safety standards and technical reports are the predominant and most important for machinery.

The approach to risk assessment and risk reduction [45] is fairly similar to what is described in ISO 12100, even if a different terminology is used. The three steps risk reduction method is translated in what is called the **Hazard Control Hierarchy**. The reason for the term is that when selecting the most appropriate risk reduction measures, they **should be applied in the order of hierarchy as they appear below**.

- **Inherently safe by design.** Like in ISO 12100, inherently safe design measures eliminate hazards or reduce the associated risks by a suitable choice of design features of the machine itself and/or interaction between the exposed persons and the machine.
- **Engineering controls.** They include guards, interlocking devices, presence sensing devices, and two-hand control devices. **SCS** plays a role at this level.
- **Administrative controls, including** Awareness means, Information for use (training and procedures), Supervision and Personal Protective Equipment (PPE).

2.4.4.1 The Concept of Control Reliable

The concept that the higher the risk, the higher the reliability of a Control System has to be, is present in the United States since a long time. We take, as an example, the ANSI/RIA R15.06 [79]. Table 2.10 was used to indicate the reliability of a Safety System.

Table 2.10 Safeguard selection matrix according to ANSI/RIA R15.06: 1999.

Risk level	Circuit performance
High	Control reliable
Medium	Single channel with monitoring
Low	Single channel
Negligible	Simple

Some definitions of terminology mentioned in the table are hereafter.

[ANSI/RIA R15.06:1999(R2009)] 4.5 Safety circuit performance

4.5.2 Single Channel. *Single channel safety circuits shall be hardware based or comply with 6.4, include components which* ***should be safety rated****, be used in compliance with manufacturers' recommendations and proven circuit designs (e.g. a single channel electromechanical positive break device which signals a stop in a de-energized state.)*

4.5.4 Control reliable. *Control reliable safety circuitry shall be designed, constructed and applied such that* ***any single component failure shall not prevent the stopping action of the robot****.*

These circuits shall be hardware based or comply with 6.4, and include automatic monitoring at the system level.

a) *The monitoring shall generate a stop signal if a fault is detected. A warning shall be provided if a hazard remains after cessation of motion;*
b) *Following detection of a fault, a safe state shall be maintained until the fault is cleared.*
c) *Common mode failures shall be taken into account when the probability of such a failure occurring is significant.*
d) *The single fault should be detected at time of failure. If not practicable, the failure shall be detected at the next demand upon the safety function.*

The issue with the concepts described, like Control Reliable, is that there is no specific standard (Type B, in ISO language) that defines them: in other words, considering we are in the 1990s and early 2000s, there is nothing equivalent to EN 954-1 in US voluntary or regulatory standards. Those concepts are simply mentioned and defined in Type C standards.

For that reason, as familiarity and understanding of ISO 13849-1 and IEC 62061 grows in North America, their voluntary consensus standards are beginning to informatively, and sometimes even normatively, reference those standards, not only in Canada but even in the United States.

[CSA Z432 – 2016] [48] 8.2 Safety control system performance criteria

8.2.1 General. *Safety-related parts of control systems (mechanical, electrical, electronic, fluidic etc.) for newly manufactured machinery shall provide functional safety performance as determined by ISO 13849-1 or IEC 62061. Existing, or rebuilt or redeployed machinery, should meet ISO 13849-1 or IEC 62061.*

2.4.4.2 Functional Safety in the United States

The B11 Functional Safety standard is ANSI B11.26 [47]. It is a Guideline for the Design of SCS **using ISO 13849-1**. It provides both requirements and guidance for the implementation of safety-related control functions as they relate to electrical, electronic, pneumatic, hydraulic, and mechanical components of control systems. Hereafter its definition of Control Reliability.

> ***[ANSI B11.26 – 2018] 5.4.3 Control reliable methodology (ANSI B11)***
> *Designers shall confirm that the control reliability of the control system meets the applicable requirements and is sufficient for the required design specification.*
>
> ***Informative Note 1:*** *Control reliability refers to the capability of the machine control system, the engineering control – devices, other control components and related interfacing* ***to achieve a safe state in the event of a failure*** *within the SRP/CS. This method has been used historically in the B11 series of standards. See also annex I.*
>
> ***Informative Note 2:*** *Because some failures cannot be detected until the completion of a cycle or a portion of the cycle, loss of safety functions can occur for a portion of the machine cycle. Other failures cannot be detected until a demand is made on the safety-related function. An example of such a safety-related function may be the use of an engineering control – device that is actuated infrequently (e.g., an interlocked guard for maintenance) that must be cycled (guard opened) to detect a failure. When a failure is detected, the safety-related function should meet the requirements of this subclause.*
>
> ***Informative Note 3:*** *Control reliability:*
>
> - *is one of the design strategies that may be used to meet these requirements;*
> - *cannot prevent a repeat cycle in the event of a major mechanical failure or in the presence of multiple simultaneous component failures;*
> - *is not provided by simple redundancy;*
> - *must have monitoring to confirm that redundancy is maintained.*
>
> ***Informative Note 4:*** *While the requirements of control reliability are not directly comparable to the requirements of ISO 13849-1 (1999) or ISO 13849-1 (2015), for the purposes of this standard, complying with Category 3 or 4 and/or Performance Level "d" or "e", at a minimum, will satisfy the requirements of control reliability.*

The primary objectives of the 2018 edition of ANSI B11.26 standard are the detailed schematic diagrams, the "Circuit Analysis Tables," and the detailed annexes for understanding performance levels and category block diagrams **as outlined in ISO 13849-1.** The intent is to clarify and provide direction for functional safety applications in current and future equipment installations.

Since ANSI B11.26 is a fairly new standard, several other standards in other machinery sectors (plastics, woodworking, etc.) have not caught up to it in their own revision cycles. Some mention Control Reliable generally, some mention Categories, and some even mention PLs (or both or all three). The ANSI B11 series of (Type-C) standards will reference ANSI B11.26, as they get revised.

It is also worth mentioning how OSHA is embracing ISO 13849-1 and ISO 13849-2 as the functional safety standard for Robots, as indicated in section IV, chapter 4 of its Technical Manual (OTM).

3

Main Parameters

3.1 Failure Rate (λ)

3.1.1 Definition

The instantaneous Failure Rate is a commonly used measure of reliability: we already discussed it in Chapter 1.

A failure rate is typically represented by the lower case Greek letter lambda (λ), and it has units of inverse time. It is a common practice to use units of "failures per billion (10^9) hours": this unit is known as **FIT** or Failures in Time. For example, a particular integrated circuit will experience seven failures per billion operating hours at 25 °C and, thus, it has a failure rate of 7 FITs [67]. Please consider that if a component has a FIT of 1, that is not to say that the device has a lifetime of 1 billion years but rather that, if you have 1 million components running for 1000 hours, you can expect one failure due to random hardware failure issues.

According to IEC 61508, there are four types of Failures:

- Safe failures;
- Dangerous failures;
- No Effect failures; and
- No Part failures.

Hereafter their definitions:

> ***[IEC 61508-4] 3.6 Fault, failure and error***
> ***3.6.7 Dangerous Failure.*** *Failure of an element and/or subsystem and/or system that plays a part in implementing the safety function that:*
>
> *a) Prevents a safety function from operating when required (demand mode) or causes a safety function to fail (continuous mode) such that the EUC is put into a hazardous or potentially hazardous state; or*
> *b) Decreases the probability that the safety function operates correctly when required*
>
> ***3.6.8 Safe Failure.*** *Failure of an element and/or subsystem and/or system that plays a part in implementing the Safety function that:*
>
> *a) Results in the spurious operation of the safety function to put the EUC (or part thereof) into a safe state or maintain a safe state; or*
> *b) Increases the probability of the spurious operation of the safety function to put the EUC (or part thereof) into a safe state or maintain a safe state*

Functional Safety of Machinery: How to Apply ISO 13849-1 and IEC 62061, First Edition. Marco Tacchini.

> ***3.6.13 No Part Failure.*** *Failure of a component that plays no part in implementing the safety function.*
>
> ***Note:*** *The no part failure is not used for SFF calculations.*
>
> ***3.6.14 No Effect Failure.*** *Failure of an element that plays a part in implementing the safety function but has no direct effect on the safety function.*

Both no effect failures and no part failures were added in the 2010 version of IEC 61508 **to prevent being able to influence the SFF calculation by considering circuitry not relevant for the reliability of the safety function:** both types of failures should not be used for the SFF calculation. **No effect failures** were not mentioned in the previous edition of IEC 62061, and **they are now important** to understand some new aspects related to the failure of electromechanical components.

3.1.2 Detected and Undetected Failures

Besides being Safe or Dangerous, each failure can also be classified as Detected or Undetected:

> ***[IEC 61508-4] 3.8 Confirmation of safety measures***
> ***3.8.8 Detected*** *(revealed – overt). In relation to hardware, detected by the diagnostic tests, Proof Tests, operator intervention (for example physical inspection and manual tests), or through normal operation.*

A dangerous failure detected by a diagnostic test is a revealed failure and can be considered as a safe failure only if effective measures, automatic or manual, are taken.

> ***[IEC 61508-4] 3.8 Confirmation of safety measures***
> ***3.8.9 Undetected*** *(unrevealed – covert). In relation to hardware, undetected by the diagnostic tests, Proof Tests, operator intervention (for example physical inspection and manual tests), or through normal operation.*

Therefore, the Failure Rate is the sum of five elements:

$$\lambda_T = \lambda_{SD} + \lambda_{SU} + \lambda_{DD} + \lambda_{DU} + \lambda_{NE}$$

- λ_{SD}: Safe detected failure rate
- λ_{SU}: Safe undetected failure rate
- λ_{DD}: Dangerous detected failure rate
- λ_{DU}: Dangerous undetected failure rate
- λ_{NE}: No effect failure rate
- λ_T: Total failure rate

Please consider that, in recent years, there is a tendency to clarify that **detected failures** are the ones revealed by online testing, automatic diagnostic testing or, with the new edition of IEC 62061 and ISO 13849-1, as "Faults detected by the process." Those failures are the ones included in:

- $\boldsymbol{\lambda_{DD}}$ in low demand mode or
- **DC** $\boldsymbol{\lambda_D}$ in high demand mode.

Undetected failures are instead the ones that can only be detected by "off-line testing" like the Proof Test. Undetected failures are indicated as:

- λ_{DU} in low demand mode.
- In high demand mode, undetected failures can be calculated as $(\mathbf{1}-\boldsymbol{DC})\cdot\boldsymbol{\lambda_D}$.

3.1.3 Failure Rate for Electromechanical Components

3.1.3.1 Input Subsystem: Interlocking Device

Let's consider a failure in an electromechanical component like a position switch (interlocking device). **The safety function** is based upon the detection that its normally closed free voltage contact (defined by ISO 14119 as "output system") is open. If an operator enters a safeguarded area by opening an interlocked movable guard, the output system normally de-energizes in order to perform the safety function. If it doesn't, it is a dangerous failure and the system goes into a **dangerous state**.

When the operator goes out of the safeguarded area and closes the gate, the contact is supposed to close (energization). If it doesn't, it is a safe failure and the system remains in a **safe state**.

The possible failures of the position switch (input subsystem) with a normally closed contact are:

- (C ⇒ O) The contact will open while the gate stays closed: safe failure. It can be considered very unlikely, and therefore **the failure rate related to this failure is considered** $\boldsymbol{\lambda_S \approx 0}$.
- (C ⇒ C) The contact will not open when the gate is opened: **dangerous failure** $\boldsymbol{\lambda_D}$.
- (O ⇒ C) The contact will close "by itself," while the door stay opened: it's a **dangerous failure** $\boldsymbol{\lambda_D}$.
- (O ⇒ O) The contact will not close once the gate is closed. This failure is not relevant for the opening of the guard door safety function, and it **has an influence only on the Availability of the machine**. That means **it is a no effect failure** $\boldsymbol{\lambda_{NE}}$, as previously defined, and not a safe failure.

That reasoning is described in the new edition of IEC 62061, and it can have quite an impact on the failure rate analysis that will be done from now onwards on electromechanical components. The reasoning means, when evaluating the SFF of electromechanical Input Subsystems, that normally $\lambda_S \approx 0$ and therefore:

$$SFF = \frac{\lambda_S + \lambda_{DD}}{\lambda_S + \lambda_D} = \frac{\lambda_{DD}}{\lambda_D} = DC$$

3.1.3.2 Input Subsystem: Pressure Switch

A similar example can be done for a **pressure switch without electronics on board**. Possible failures are:

- (C ⇒ O) The high-pressure switch will "open" while the pressure remains within the safe range. It can be considered very unlikely, and therefore **the failure rate related to this failure is considered** $\boldsymbol{\lambda_S \approx 0}$.
- (C ⇒ C) The high-pressure switch will not "open" in case of high pressure is detected: **dangerous failure** $\boldsymbol{\lambda_D}$.
- (O ⇒ C) The high-pressure switch will close "by itself," while the high pressure persists: it's a **dangerous failure** $\boldsymbol{\lambda_D}$.

- (O ⇒ O) The high-pressure switch will not "close" once the pressure returns within the safe range. This failure is not relevant for the high pressure safety function and **has an influence only on the Availability of the machine/process**. That means it is a no effect failure λ_{NE} and not a safe failure.

In the case of analog electronic transmitters, the situation is different since electronics itself may be subject to safe detected failures.

3.1.3.3 Output Subsystem: Solenoid Valve

Let's now consider, as an output subsystem, a solenoid valve that prevents methane gas from entering a furnace. As an example, the safety sub-function is the following: the valve shall close in case the automatic burner control system (ABCS) does not detect the burner flame. Possible failures of the gas solenoid valve are:

- The valve will close while the furnace is normally working, despite a flame is present and the valve coil is energized: safe failure. We suppose the closure is not due to the lack of the electrical signal; the signal is present (1 ⇒ 1) and, despite that, the valve suddenly closes. It can be considered very unlikely, and therefore **the failure rate related to this failure is considered $\lambda_S \approx 0$.**
- The valve will not close when "no flame" is detected; in other terms, its coil is de-energized (1 ⇒ 0), but the valve gets stuck and does not close: **that is a dangerous failure λ_D**.
- The valve will open "by itself" despite having the coil de-energized (0 ⇒ 0): it's a **dangerous failure λ_D**.
- The valve will not open when the ABCS is ready to start the furnace, despite the presence of the electrical signal from the ABCS to the valve (0 ⇒ 1). This failure is not relevant for our safety function, and it **has an influence only on the Availability of the furnace**. That means it is a no effect failure λ_{NE} and not a safe failure.

That means, also in this case, $\lambda_S \approx 0$ and therefore $SFF = DC$.

3.1.3.4 Output Subsystem: Power Contactor

Let's now consider, as an output subsystem, a contactor that prevents a saw to turn. The safety sub-function is the following: when the contactor coil is de-energized, the power contacts open. Possible contactor failures are the following:

- The power contacts will open while the saw is normally working, despite nobody for example, entered the safeguarded area: safe failure.
  We suppose the opening is not due to the lack of the electrical signal; the signal is present, the coil is energized (1 ⇒ 1) and, despite that, the power contacts suddenly open. It can be considered very unlikely, and therefore **the failure rate related to this failure is considered $\lambda_S \approx 0$.**
- The power contacts will not open when, for example, a person enters the safeguarded area; in other terms, its coil is de-energized (1 ⇒ 0), but the contacts get stuck and do not open: **that is a dangerous failure λ_D**.
- The power contacts close "by themselves" despite having the coil de-energized (0 ⇒ 0): it's a **dangerous failure λ_D**.
- The power contacts will not close once the safety function is reset, and the saw start button is activated, despite the contactor coil is energized (0 ⇒ 1). This failure is not relevant for our

safety function and **has an influence only on the Saw availability.** That means it is a no effect failure λ_{NE} and not a safe failure.

That means, also in this case, $\lambda_S \approx 0$ and therefore $SFF = DC$.

That reasoning is valid for the majority of electromechanical components. Those components would be defined as Type A, according to IEC 61508-2 (§ 3.4.7.1). The reasoning is not applicable to Type B components.

3.2 Safe Failure Fraction

The Safe Failure Fraction (SFF) was introduced in IEC 61508 as a measure used to determine the minimum Hardware Fault Tolerance (HFT) of a safety subsystem.

The **SFF** of a component is defined as:

[IEC 61508-4] 3.6 Fault, failure and error

3.6.15 Safe Failure Fraction SFF. *Property of a safety related element that is defined by the ratio of the average failure rates of safe plus dangerous detected failures and safe plus dangerous failures. This ratio is represented by the following equation:*

$$SFF = \frac{\sum \lambda_{S\,avg} + \sum \lambda_{DD\,avg}}{\sum \lambda_{S\,avg} + \sum \lambda_{DD\,avg} + \sum \lambda_{DU\,avg}}$$

when the failure rates are based on constant failure rates the equation can be simplified to:

$$SFF = \frac{\sum \lambda_S + \sum \lambda_{DD}}{\sum \lambda_S + \sum \lambda_{DD} + \sum \lambda_{DU}}$$

In IEC 62061, a different definition is given.

[IEC 62061] 3.2 Terms and definitions

3.2.54 Safe Failure Fraction (SFF). *Fraction of the overall failure rate of a subsystem that does not result in a dangerous failure.*

Note 1 to entry: *The diagnostic coverage (if any) of each subsystem in SCS is taken into account in the calculation of the probability of random hardware failures. The safe failure fraction is taken into account when determining the architectural constraints on hardware safety integrity (see § 7.4).*

Note 2 to entry: *"No effect failures" and "no part failures" (see IEC 61508-4) is not used for SFF calculations.*

$$SFF = \frac{\lambda_S + \lambda_{DD}}{\lambda_S + \lambda_D}$$

At first sight, the definition of IEC 62061 does not seem right; but you need to consider that dangerous detected failures, since they are detectable, will not generate any dangerous situation: that is the reasoning behind the definition. **SFF is used in both high and low demand mode** with the same meaning.

In order to estimate the different failure rates, an analysis (e.g. fault tree analysis, failure mode and effects analysis) of each subsystem shall be performed. You have to consider that whether a failure is safe or dangerous it depends on the SCS and the intended safety function. The probability of each failure mode shall be determined based on the probability of the associated fault(s), taking into account the intended use. It may be derived from sources such as failure rate data collected from field, or component failure data from a recognized industry source or failure rate data derived from the results of testing and analysis.

Information on component failure rates can be found in several sources, including: *MIL-HDBK 217 F, MIL-HDBK 217 F, SN 29500 Parts 7 and 11, IEC 61709, FMD-2016, OREDA Handbook, EXIDA Safety Equipment Reliability Handbook and EXIDA Component Reliability Database (CRD) Handbook.*

The SFF is the proportion of "safe" failures among all failures. A "safe" failure is either a failure that is safe by design, or a dangerous failure that is immediately detected and corrected. IEC standards define a safe failure as a failure that does not have the potential to put the SIS in a hazardous or fail-to-function state. A dangerous detected failure is a failure that can prevent the SIS from performing a specific SIF, but when detected soon after its occurrence, for example by online diagnostics, the failure is considered to be "safe" since the Diagnostics can bring the system to a safe state. In some cases, the SIS can automatically respond to a dangerous detected failure as if it were a true demand, for example, causing the shutdown of the process [68].

Many electronic safety devices have built-in diagnostics such that most dangerous failures become Dangerous Detected failures, and they will therefore have a high SFF, often greater than 90%. Mechanical safety devices, for which internal diagnostics is not feasible, will have, in general, a low SFF.

3.2.1 SFF in Low Demand Mode: Pneumatic Solenoid Valve

SFF is related to the capability of a component, or a channel, to have automatic self-diagnosis.

Let's consider a pneumatic **Solenoid Valve** in the process industry. It is important to understand how that component is normally used. As shown in drawing Figure 3.1, the valve is normally part of a three-components output:

- The Solenoid Valve itself
- A Pneumatic Actuator
- An on/off Process Safety Valve

That can be considered part of a channel, or it can be considered a safety subsystem.

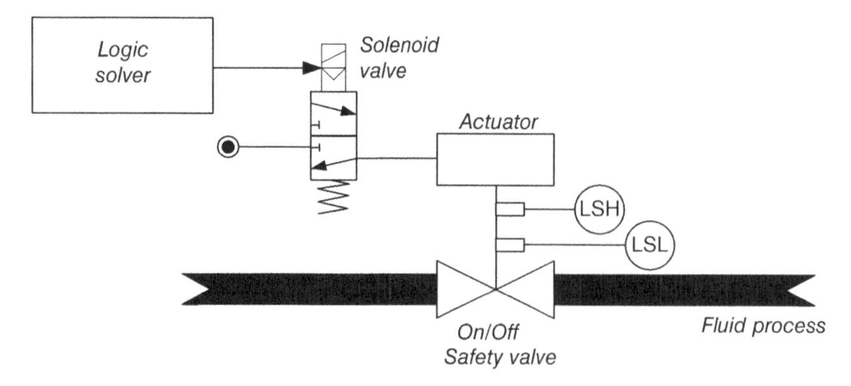

Figure 3.1 Use of a 3/2 solenoid valve in a process application.

The way the channel works is the following: as soon as a dangerous condition is detected (a high pressure, for example), the solenoid valve is de-energized, air to the actuator is cut, and the on/off safety process valve closes in few seconds.

3.2.1.1 Example

Looking at the component certificate, the failure rates are shown in Table 3.1.

It is important to highlight that those values have been estimated with **the valve working in low demand mode.**

Table 3.1 Solenoid valve reliability data.

Solenoid valve failure rate data (Source: Exida SERH 2015 – 03 final elements – item 3.1.23)		Normal	PVST
		Per 10^9 hours (FITs)	
Fail dangerous detected	λ_{DD}	0	267
Fail dangerous undetected	λ_{DU}	324	32
Fail safe detected	λ_{SD}	0	156
Fail safe undetected	λ_{SU}	126	15
No effect failure	λ_{NE}	221	311

Being the solenoid valve an electromechanical component, considering the reasoning the new edition of IEC 62061 makes and described in § 3.1, in the future, you may argue if Fail Safe undetected are in reality No effect failures.

P.V.S.T. means **Partial Valve Stroke Test**. During this test, the solenoid valve is de-energized for a very short time, and then it is re-energized. During this time, the on/off process valve has just the time to start closing and then to go back to a fully open position. The process valve is equipped with an open (LSH) and a closed (LSL) position switch. During this partial closure, the Safety System that performs the PVST is able to detect that the solenoid valve has toggled, since the process valve started closing and came back to full opening. The LSH limit switch opened and then closed again, indicating the slight valve movement.

Let's calculate the SFF and DC in both cases: with and without PVST.

Normal Situation: No partial stroke test is done; that means the solenoid valve is kept energized for more than one year.

$$SFF = \frac{\lambda_S + \lambda_{DD}}{\lambda_S + \lambda_D} = \frac{126}{126 + 324} = 28\%$$

$$DC = \frac{\lambda_{DD}}{\lambda_D} = \frac{0}{324} = 0$$

It is clear that, in case no partial stroke test is done, there is no way to detect any failure in the solenoid valve: it is an electromechanical component with no self-diagnostic capabilities.

Partial Valve Stroke Test: the test should be performed **at least an order of magnitude more frequently than the Proof Test**. The solenoid valve is still operated (completely switched off and on again) less than once per year. We can use the failure rate data in the PVST column.

$$SFF = \frac{\lambda_S + \lambda_{DD}}{\lambda_S + \lambda_D} = \frac{156 + 15 + 267}{156 + 15 + 267 + 32} = \frac{438}{470} = 93.2\%$$

$$DC = \frac{\lambda_{DD}}{\lambda_D} = \frac{267}{267 + 32} = \frac{267}{299} = 89.3\%$$

In this case, the fact of being able to automatically detect if the solenoid valve has toggled, the channel or subsystem has a Medium Level of Diagnostic Coverage and an SFF between 90 and 99%.

Table 3.2 Architectural constraints for Type A components.

	Hardware fault tolerance		
Safe failure fraction of an element	**0**	**1**	**2**
$SFF < 60\%$	SIL 1	SIL 2	SIL 3
$60\% \leq SFF < 90\%$	SIL 2	SIL 3	SIL 4
$90\% \leq SFF < 99\%$	SIL 3	SIL 4	SIL 4
$SFF \geq 99\%$	SIL 3	SIL 4	SIL 4

As shown in Table 3.2, in a low demand mode application, the subsystem **could** reach SIL 3 ($HFT = 0$ and $SFF \geq 90\%$). However, in order to confirm the safety level, the Reliability of the whole sub-function has to be calculated: that means not only the Solenoid Valve but also the Actuator and the ON/OFF Safety Valve.

3.2.2 SFF in High Demand Mode: Pneumatic Solenoid Valve

In case the same valve is used in high demand mode, the data from the previous example cannot be used. A pneumatic valve (Figure 3.2) used in high demand mode does not have a constant failure rate, and therefore a B_{10D} value has to be indicated by the valve manufacturer and used in the calculations.

3.2.2.1 Example for a 1oo1 Architecture

As an example, we take the value recommended by IEC 62061 in table C.1: $\mathbf{B_{10D} = 20\,000\,000}$.

Being an Electromechanical component used in high demand mode $SFF = DC$.

In order to make a straight comparison with the low demand mode example, let's suppose the Safety System simply de-energizes the 3/2 valve, which is directly monitored 2920 times in a year (once every hour for eight hours a day).

This can be a Basic Subsystem Architecture C (1oo1D) according to IEC 62061 (please refer to § 7.2.5).

Considering the architectural constraints, being $HFT = 0$, up to SIL 3 could be reached. However, it is important to verify that one of the following conditions is satisfied in Architecture C (please refer to § 7.2.1).

- **the sum of the diagnostic test interval and the time to perform the specified fault reaction** function to achieve or maintain a safe state shall be shorter than the process safety time (e.g. see ISO 13855);

- or, when operating in high demand mode of operation, **the ratio of the diagnostic test rate to the demand rate shall be equal to or greater than 100.** This last condition aligns Architecture 1oo1D with Category 2 of ISO 13849-1.

Neither of these two conditions is met:

- The test rate is equal to the demand rate.
- In case a fault is detected (the valve did not close), there is no possibility to cut the air to the cylinder, being a single channel architecture.

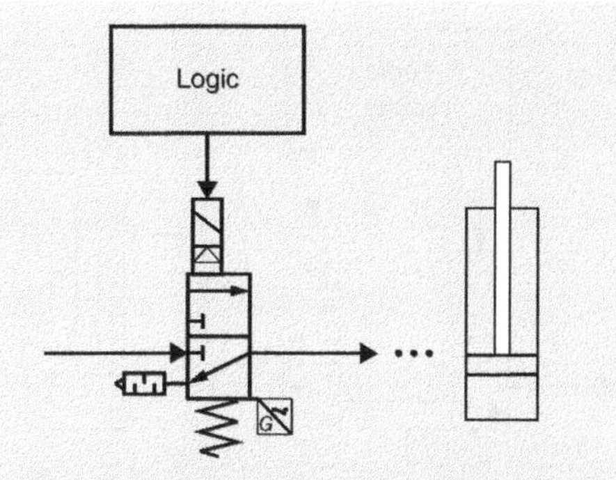

Figure 3.2 Use of a 3/2 solenoid valve in machinery.

Therefore, the subsystem has to be considered a Basic Subsystem Architecture A (1oo1) consisting of an electromechanical component (solenoid valve). This situation limits the maximum achievable SIL to **SIL 1**.

The PFH_D value of the subsystem can be calculated using the equation for Architecture A:

$$PFH_D = \lambda_D$$

$$\lambda_D = \frac{0.1 \cdot n_{op}\ (1/\text{years})}{B_{10D} \cdot 8760\ \text{hours/years}} = 1.67 \cdot 10^{-9}\ \text{h}^{-1} = PFH_D$$

Conclusion: the subsystem can only reach **SIL 1** with a PFH_D equal to $1.67 \cdot 10^{-9}$.

3.2.2.2 Example for a 1oo2D Architecture

Sometimes, the safety system de-energizes both the 3/2 valve (directly monitored with a limit switch mounted on the spool) and the 5/2 valve "monitored by the process," like shown in Figures 3.3 and 3.4 (please refer to § 3.3.5). We assume, for this example, that the former is de-energized once every hour and the latter once every 30 seconds; we assume the machine works 8 hours a day, 365 days/year.

In this case, the output subsystem is an Architecture 1oo2D. Supposing $\beta = 2\%$, the calculation is the following:

$$PFH_D = (1-\beta)^2 \cdot \left[\lambda_{De_1} \cdot \lambda_{De_2} \cdot (DC_1 + DC_2) \cdot \frac{T_2}{2} + \lambda_{De_1} \cdot \lambda_{De_2} \cdot (2 - DC_1 - DC_2) \cdot \frac{T_1}{2}\right] + \beta \cdot \frac{(\lambda_{De_1} \cdot \lambda_{De_2})}{2}$$

$\boldsymbol{T_1}$ = min (Useful Lifetime; Proof Test);
Useful Lifetime = min (Mission Time; T_{10D});
Mission Time = 20 years;

$$T_{10D} = \frac{B_{10D-SV1}}{n_{op}} = \frac{2 \cdot 10^7}{8 \cdot 365} = \frac{2 \cdot 10^7}{2920} = 6849\ \text{years} > 20\ \text{years}$$

Useful lifetime = 20 years.

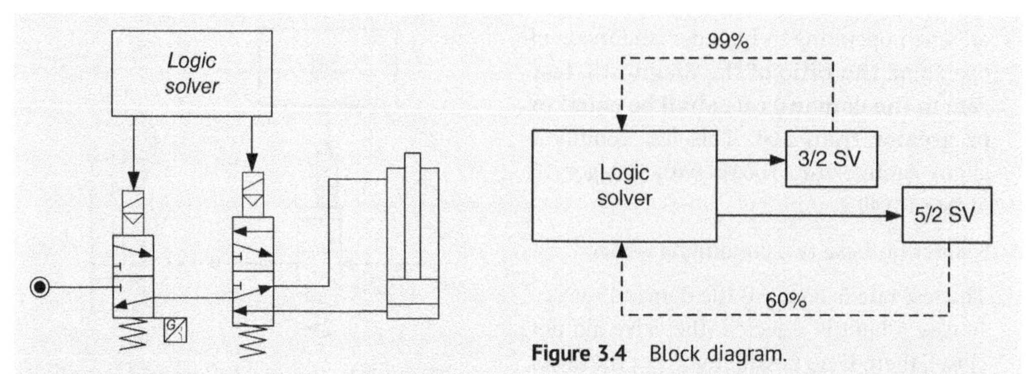

Figure 3.4 Block diagram.

Figure 3.3 Pneumatic diagram.

Proof Test = 20 years

T_1 = 20 years = 175 200 hours

T_2 (the longer diagnostic test interval of the two channels) $= \dfrac{1}{(n_{op}/\text{year})/((\text{working hours})/\text{year})}$

$= \dfrac{1}{2920/2920} = 1\ \text{hour}$

$$PFH_D = (1-0.02)^2 \cdot \left[1.67\cdot10^{-9}\cdot1.67\cdot10^{-9}\cdot(0.99+0.60)\cdot\frac{1}{2}+1.67\cdot10^{-9}\cdot1.67\cdot10^{-9} \cdot(2-0.99-0.60)\cdot\frac{175\,200}{2}\right]+0.02\cdot\frac{(1.67\cdot10^{-9}\cdot1.67\cdot10^{-9})}{2}=9.6\cdot10^{-14}\,h^{-1}$$

Being Electromechanical components used in high demand mode $SFF = DC$.

$$\boldsymbol{DC_{TOT} = SFF_{TOT}} = \frac{DC_1\cdot\lambda_{D1}+DC_2\cdot\lambda_{D2}}{2\cdot\lambda_D} = \frac{0.99\cdot1.67\cdot10^{-9}+0.60\cdot1.67\cdot10^{-9}}{2\cdot1.67\cdot10^{-9}} = 0.795$$

Considering the architectural constraints for $HFT = 1$, maximum achievable SIL with $60\% \le \boldsymbol{SFF} < 90\ \%$ is SIL 2.

Conclusion: the subsystem can reach **SIL 2** with a $PFH_D = 9.6\cdot10^{-14}\ h^{-1}$.

3.2.3 SFF and Electromechanical Components

In Machinery, and in general for components used in high demand mode, like Interlocking devices, Solenoid Valves, or Contactors, **the component manufacturer does not** normally **provide the different types of failure rates**.

The FMEDA analysis is typically done for electronic components used in Low demand mode. FMEDAs are sometimes done for Solenoid valves, but there is rarely one done for interlocking devices or for Contactors. IEC 61511-1 requires FMEDA analysis for each component used in a safety system, while ISO 13849-1 and IEC 62061 don't. In high demand mode, the value of the Diagnostic Coverage of the subsystem is needed in order to be able to calculate λ_{DD} as DC λ_D and λ_{DU} as (1-DC)$\cdot\lambda_D$.

IEC 61508 was introduced to deal with the complexity of electronic components used in safety systems. In machinery, up to 2006, electronic components could not be used in safety systems; please refer to § 2.1.

3.2.3.1 The Advantage of Electronic Sensors

A lot has been done since that time and Electronic components are now widely used in safety systems. They are designed to have very low Systematic Failures and very good Reliability values. Moreover, **they have the advantage of being able of self-diagnostic** and often, especially in the process industry, sensors provide an analog signal (usually a 4–20 mA).

In many applications, analog input signals are better than digital ones. A **Pressure Switch** has no internal diagnostics and can only provide a 0 (open contact) or a 1 (closed contact) to a Programmable Logic Controller (PLC).

A **Pressure Transmitter** can have internal diagnostics and, outside the 4–20 mA range, it can communicate its status to a safety system. For example, when the current drops below 3.6 mA, it may mean that the component is not working properly anymore; please refer to the NAMUR standard, explained in the next paragraph.

That is the reason why pressure transmitters, or other electronic signal transmitters used in the process industry, are provided, by their manufacturer, with several Reliability data like λ_S, λ_{DD}, λ_{DU}, etc. In such electronic components, **the diagnostic is mainly made inside the component itself**. That is the basis upon which the process industry works: **safety components are electronic components, are "intelligent" and are provided with all relevant Reliability and Diagnostic data**; moreover, they communicate when they do not work properly.

3.2.3.2 SFF and DC for Electromechanical Components

When the machinery industry was confronted with the IEC 61508 probabilistic theory, this was one of the issues it faced: **components used were electromechanical, with digital output and no possibility for internal diagnostic**. That is the reason why, if you work in Machinery safety, you are used to the fact that the diagnostic coverage depends upon the way the component is cabled to the Logic Solver and the way the logic solver is programmed. In other words, the DC (ISO 13849-1) or the SFF (IEC 62061) are **not linked to the component itself!**

Let's take, as an example, a power contactor, used like in Figure 3.5. The status of the power contacts is monitored by the Mirror Contacts (§ 4.12.2.1). They provide direct monitoring of the power contacts. That means, in principle, $a > 99\%$ diagnostics.

However, since the component is used in a single channel application, in high demand mode, that subsystem is a Category 1 according to ISO 13849-1, or Architecture A according to IEC 62061. For both standards, the DC cannot reach 60%.

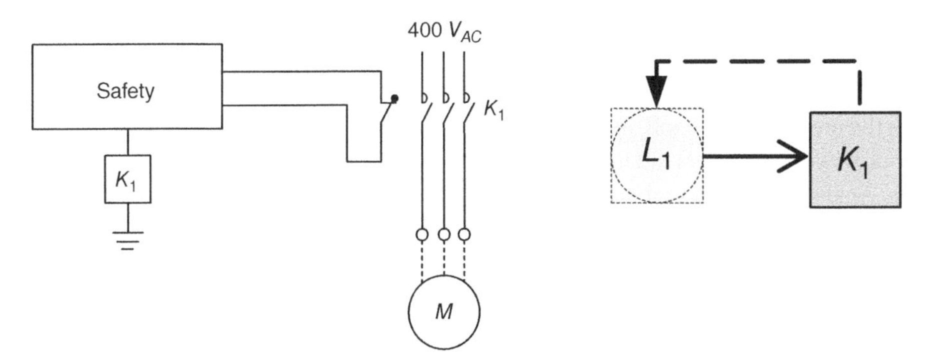

Figure 3.5 Single channel output subsystem and its reliability block diagram.

If two contactors in series are used, like in Figure 3.6, since each component is used in a two-channel application, in high demand mode, the output subsystem is a Category 4, according to ISO 13849-1, or Architecture D according to IEC 62061. For both standards, the DC can be assumed ≥99%.

Again, in high demand mode, the component Diagnostic Coverage is not really linked to the component itself but to the way it is connected inside its subsystem.

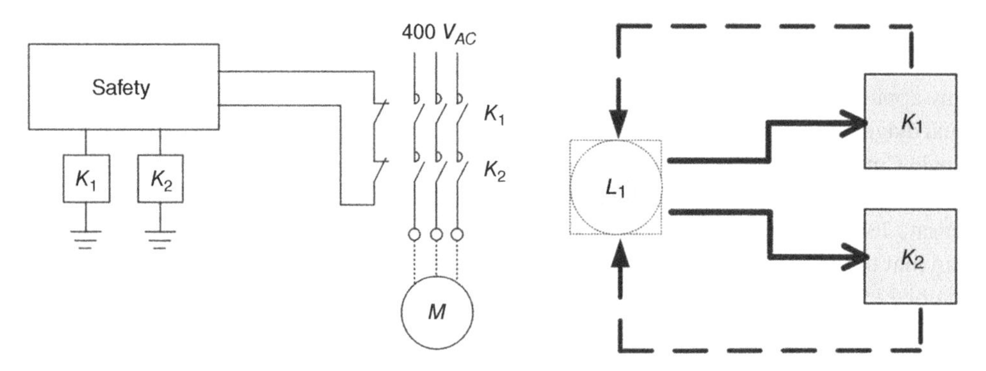

Figure 3.6 Two-channel output subsystem and its reliability block diagram.

Annex D "*Failure modes of electrical/electronic components*" of the previous edition of IEC 62061 was written to help in this aspect. The new edition of the standard has removed that approach and **replaced it with the more pragmatic approach of ISO 13849-1**, detailed in its annex E. The new annex D in IEC 62061 is titled "*Examples for diagnostic coverage (DC).*"

Another important clarification contained in the new edition of IEC 62061, is that, in case of electromechanical components used in high demand mode, normally, $\lambda_S = 0$ (please refer to § 3.1.3) and therefore

$$SFF = DC$$

That means: **for electromechanical components used in single channel (high demand mode), it is not possible to reach more than SIL 1.**

In the past, several people reached the conclusion that, by using IEC 62061 and a single electromechanical component ($HFT = 0$), referring its table 6 (Table 7.1 in this book), it would be possible to reach SIL 3, provided a very high level of SFF could be declared. However, in case of a single electromechanical component, that has no electronics and therefore no possibility of self-diagnostics, that is not possible, due to the lack of diagnostic coverage.

3.2.4 SFF in Low Demand Mode: Analog Input

As explained in the previous paragraph, in low demand mode applications there is a predominance of analog sensors, while in high demand mode of operation there are mainly digital ones, like interlocking devices. Analog sensors have a predominance of electronics inside, which means the possibility of self-diagnostics that, combined with an analog signal, allow a high level of diagnostics to be transmitted to the Safety logic.

When we analyze the Reliability of an input signal, we should look at two possible faults:

- **the Sensor;** and
- **the connection** between the Sensor and the Logic Solver.

Moreover, Systematic failures should be considered for the connection between the sensor and the process.

For Analog sensors, the component failure rates are provided by the manufacturer. Moreover, the use of 4–20 mA signal allows a certain amount of diagnostics, like component failure, to be communicated by driving the signal to out of range, as indicated in Figure 3.7. A German Standard called **Namur NE43** recommends that values ≤3.6 mA or ≥21 mA be used to communicate a component failure.

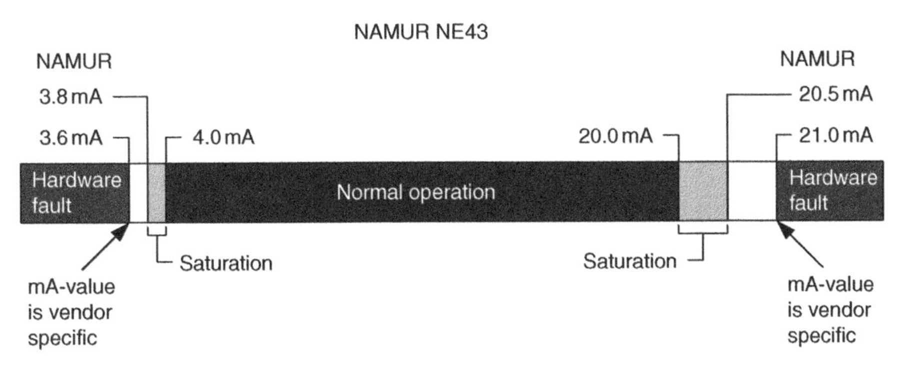

Figure 3.7 NAMUR recommendation NE43.

Table 3.3 shows an example of Reliability data for a Pressure Transmitter (Type B component).

Table 3.3 Pressure transmitter reliability data.

Honeywell ST 3000 with HART 6 (Source: Exida SERH 2015 - 01 sensors - item 1.6.78)		HART protocol Per 10^9 hours (FITs)
Fail low		89
Fail high		20
Fail detected		268
Fail dangerous undetected	λ_{DU}	40
No effect failure	λ_{NE}	64
Route 1_H SFF (%)		90.5
Useful life (years)		50

Here is the definition of some of the items in the table given by Exida:

- ***Fail Low:*** *a failure that will result in an output current that is lower than 4 mA or under range per NAMUR recommendation 43.*
- ***Fail High:*** *a failure that will result in an output current that is higher than 20 mA or over range per NAMUR recommendation 43.*

- ***Fail Detected:*** *a failure that is detected by internal diagnostic, whose external effect depends on equipment item settings. In a transmitter, for example, a detected failure could result in over range or under range output depending on a jumper setting.*

Referring to Table 3.4, being a Type B component with *SFF* > 90%, it can be used in SIL 2 safety functions.

Table 3.4 Architectural constraints for Type B components.

Safe failure fraction of an element	Hardware fault tolerance		
	0	1	2
SFF < 60%	Not allowed	SIL 1	SIL 2
60% ≤ *SFF* < 90%	SIL 1	SIL 2	SIL 3
90% ≤ *SFF* < 99%	SIL 2	SIL 3	SIL 4
SFF ≥ 99%	SIL 3	SIL 4	SIL 4

Regarding the connection between the sensor and the logic, failures due to a circuit interruption or a short circuit can be easily detected by an out-of-range setting in the Logic Solver.

Bottom line: in case of a low demand mode of operations, with a single sensor, a SIL 2 level of Reliability could be reached by the safety function.

3.2.5 SFF and DC in High Demand Mode: The Dynamic Test and Namur Circuits

In case of an interlocking device, the same aspects seen in the previous paragraph should be taken into account:

- Component failure.
- Failure in the connection between the component and the Logic Solver.

Its failure rate data are normally expressed as

- $\boldsymbol{B_{10D}}$ in case of electromechanical components. The signal is normally a Voltage Free Contacts (VFC).
- $\mathbf{PFH_D}$ in case of electronic components. The signal is normally an OSSD one (please refer to § 4.12).

The issue with a VFC is that, in case of short circuit, the Logic solver still reads a digital one, despite the interlocked movable guard is, for example, open. A useful function, now available in most Logic Solver, is the **Dynamic test: i**t detects a possible short circuit, and therefore it improves the subsystem diagnostic coverage.

> ***[ISO 13849-1] 3.1 Terms and definitions***
> ***3.1.51 dynamic test.*** *Executing either software or operating hardware, or both, in a controlled and systematic way, so as to demonstrate the presence of the required behaviour and the absence of unwanted behaviour.*
>
> ***Note 1 to entry:*** *The test fails if monitoring (3.1.29) did not detect the change as expected.*
>
> ***Note 2 to entry:*** *The use of test pulses is a common technology of dynamic testing and is widely used to detect short circuits or interruptions in signal paths or malfunctions.*

Usually, a logic solver has one or two sources of trigger. The source has the behavior shown in Figures 3.8 and 3.9.

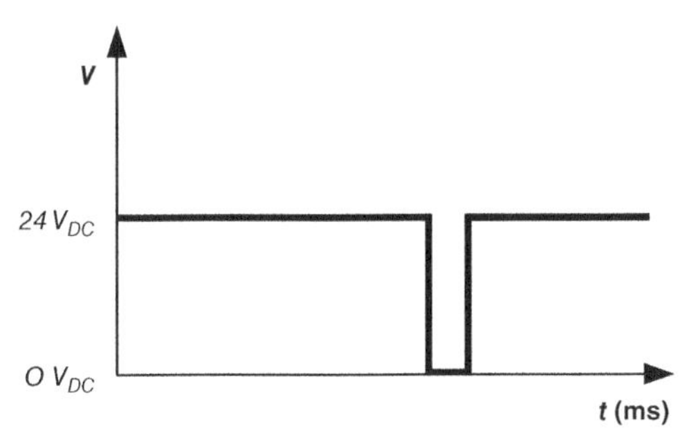

Figure 3.8 Logic solver with one trigger source.

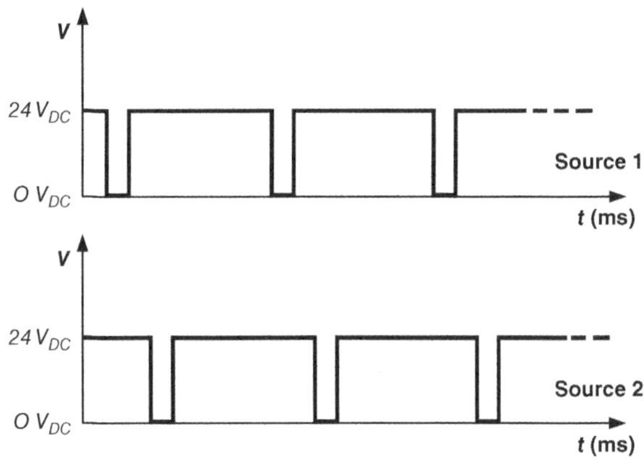

Figure 3.9 Logic solver with two trigger sources.

In case of an interlocking device with two Interlocking Monitoring Contacts, each VFC can be powered by one of the two triggering sources, labeled TR1 and TR2 in Figure 3.10.

Input labeled "IN1" expects a short "logic zero" when TR1 generates a short "logic zero." The same for the input labeled "IN2."

In that way, if a short circuit happens between the two circuits, the logic solver detects it.

The trigger function can also be available for the outputs of a logic solver. **The function can be used to detect short circuits on the output wiring**.

Both ISO 13849-1 and IEC 62061 standards give an indication of DC values. The trigger function is mentioned as one way to improve the diagnostic of the subsystem.

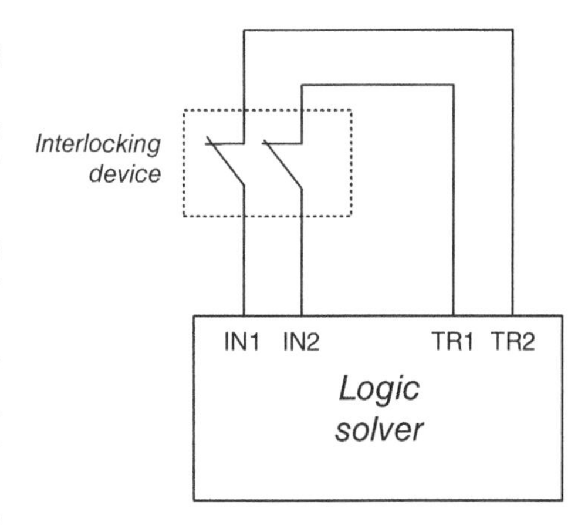

Figure 3.10 An interlocking device safety subsystem.

3.2.5.1 Namur Type Circuits

An issue remains in case of short circuit between wires of the same circuit: in our example, it would be a short circuit between "TR1" and "IN1" that would "bypass" the VFC: that fault would not be detected.

A way to detect that type of short circuit is to use the VFC in a "Namur type circuit," shown in Figure 3.11.

- When the contact is closed, the logic solver reads $24V_{DC}/1.2\,\Omega = 20$ mA.
- When the contact is open, the logic solver reads $24V_{DC}/(1.2\,\Omega{+}3.9\,\Omega) = 4.7$ mA.

In case of a short circuit in the field between the two wires, the logic solver would detect an out of range error. The logic solver could also detect short circuits between different input circuits.

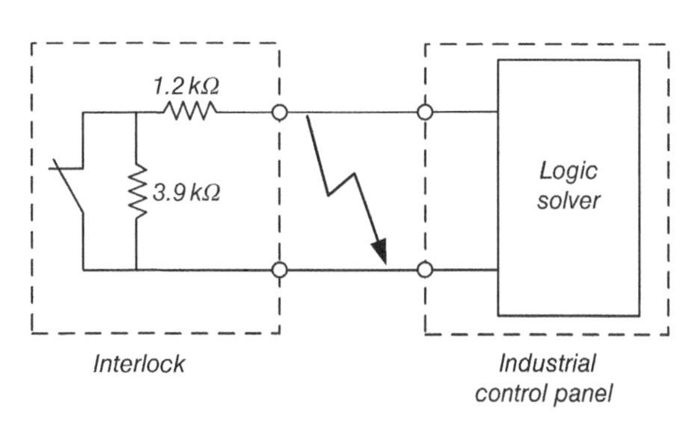

Figure 3.11 NAMUR type digital input.

The values are just examples. Different values can be used, provided they are coordinated with the logic solver.

3.2.5.2 Three Wire Digital Input

A third way to detect a short circuit is to have the connections shown in Figure 3.12. It is an example of a High-pressure switch:

- As pressure rises: 1 NC opens, 2 NO closes
- As pressure falls: 1 NC closes, 2 NO opens

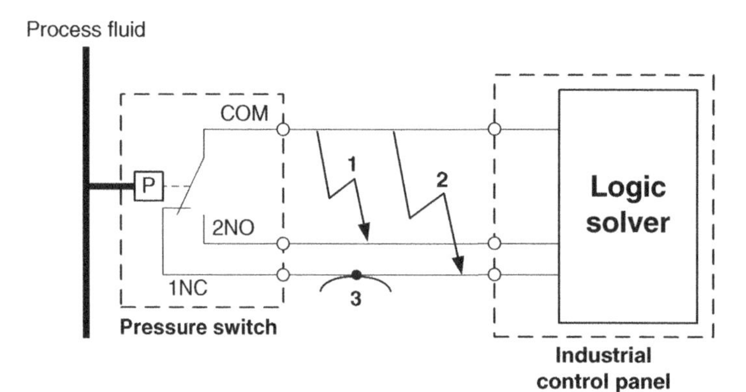

Figure 3.12 Pressure switch 3 wire connection.

The connection can be considered as having $HFT = 1$, provided a fault exclusion on the sensing element is done:

- Short circuit 1 is recognized immediately.
- Short circuit 2 is recognised as soon as there is a demand upon the safety function.
- Short Circuit 3, with wiring from other sensors, is recognised either immediately or as soon as there is a demand upon the safety function.

3.2.6 Limits of the SFF Parameter

Considering the definition of SFF, the safety of a component can be enhanced by making the dangerous-failure rate lower, and the safe-failure rate higher, assuming the total failure rate of the component does not change. Therefore, the following situation can occur: a component manufacturer has designed and developed a product with an estimated dangerous failure rate of $2{\cdot}10^{-8}$; however, the SFF is estimated at 50%. The company modifies the design in order to increase the Safe failures and/or the Dangerous Detected failures; however, the modified component will not assure more safety and will cause economical losses for the user, since its process will be subject to more spurious trips.

Therefore [77], does a high SFF indicate a safer design? Reliability experts, system integrators, and end users have questioned the suitability of SFF as an indicator of a safe design. The reason is that Safe failures are not always positive for safety, since spurious trips may create other hazardous situations, during the shut-down clearance and the process restart. Moreover, the SFF may credit unneeded hardware, since the SFF gives credit to high rate of "safe" failures, and for producers, it is a business advantage to claim a high SFF. With a high SFF, components may be used in configurations with low HFT, which means more business for the component manufacturer.

3.2.6.1 Example

Consider the following two components used in $HFT = 0$ subsystem in high demand mode (IEC 62061).

Component 1:

- $\lambda_{DU} = 50$ FIT
- $\lambda_{DD} = 0$ FIT
- $\lambda_S = 0$ FIT
- $SFF = 0$
- $PFH_D = 50$ FIT
- Max SIL reachable: SIL 1

Component 2:

- $\lambda_{DU} = 50$ FIT
- $\lambda_{DD} = 0$ FIT
- $\lambda_S = 4950$ FIT
- $SFF = 99\%$
- $PFH_D = 50$ FIT
- Max SIL reachable: SIL 3

In other terms, both components have the same PFH_D; one is "intrinsically safe" and has no safe failures. The second has a much higher total failure rate but the majority are safe failures.

The second component, despite having the same Average probability of dangerous failure per hour as the first one, can be used up to SIL 3, only because it has a lot of "Safe Failures".

Again, that is the reason why, in the second edition of IEC 61508, the concept of **No Effect Failures** was introduced: to avoid the overestimation of Safe Failures.

3.3 Diagnostic Coverage (DC)

We already discussed about the importance of testing the channels of the Safety Function. Diagnostic Coverage represents the automatic testing of a component or of a channel. Let's start with the definition:

> ***[ISO 13849-1] 3.1 Terms and definitions***
> ***3.1.35 Diagnostic Coverage*** *(DC). Measure of the effectiveness of diagnostics, which may be determined as the ratio between the failure rate of detected dangerous failures and the failure rate of total dangerous failures.*

It is calculated using the following formula,

$$DC = \frac{\sum \lambda_{DD}}{\sum(\lambda_{DD} + \lambda_{DU})}$$

where λ_{DD} is the rate of detected dangerous hardware failures and λ_D ($\lambda_D = \lambda_{DD} + \lambda_{DU}$) is the total rate of dangerous hardware failures.

DC is mainly used in high demand mode, where λ_{DD} and λ_{DU} are not common parameters. However, those are hidden in the formulas used in high demand mode (IEC 62061), given the fact:

- $\lambda_{DD} = DC \cdot \lambda_D$
- $\lambda_{DU} = (1 - DC) \cdot \lambda_D$

Component redundancy can be used to improve the Reliability of a Safety Control System; however that has to be matched with **monitoring** of the so-called **Functional channel.**

Often in machinery, the logic unit takes care of the diagnostic function of both input and output devices. That is different from the process industry, whereby the sensor often contains electronics that allows self-diagnostic (please refer to § 3.2).

Let's consider a Safety Control System that stops a motor using two contactors in series. In case the status of two contactors is monitored (Figure 3.13), there is a diagnostic function (automatic testing) that can detect dangerous failures, like the fact that the contactor coils are de-energized, but the power contacts did not open. This diagnostic function is external to the component.

This type of monitoring is defined in ISO 13849-1 annex E as "*Redundant shut-off path with monitoring of the outputs by logic and test equipment*" and a value of $DC = 99\%$ is given to the output subsystem.

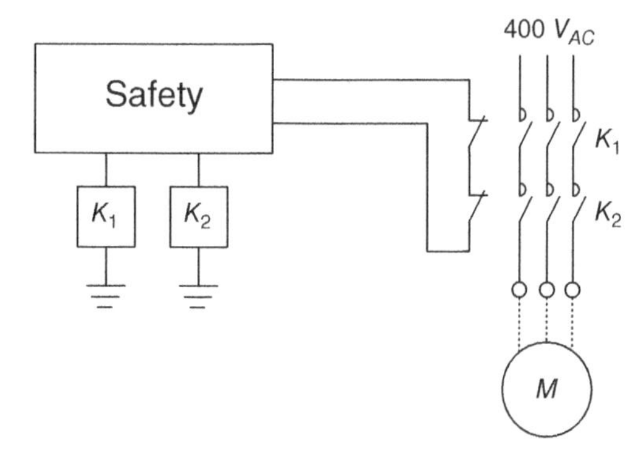

Figure 3.13 Direct on Line motor under two contactors.

Therefore, **the presence of a Diagnostic function improves the system Reliability.** For example, an operator enters twice a day into a safeguarded area. Every time he opens the interlocked movable guard, the contactor coils are de-energized and the motor stops. In that way, twice a day, the two contactors are tested to see if they are still properly working. The situation changes if the contactors open twice a year: accumulation of faults may result in both contactors being faulty: when a demand is finally placed on the safety function, the motor will not stop. That is the reason why, in low demand mode, it is important to have electronics inside components (typically sensors) or to perform some sort of either "automatic partial test" on the component (§ 3.7.3.3) or "off line" testing (§ 3.7).

In principle, for the estimation of the DC of a component, a Failure Mode Effects and Diagnostic Analysis (FMEDA) should be used. However, both ISO 13849-1 and IEC 62061 help the machine manufacturer with indication of possible levels of DC, based upon the type of interconnection: It is a diagnostic external to the component. In the past, that was only available in ISO 13849-1 (annex E); with the new edition, it is now available in IEC 62061 as well (annex D).

In recent years, more and more interlocking devices or, in general, sensors like SPEs (Sensitive Protective Equipment) [31] contain electronic components, and therefore they can detect internal failures. However, for those devices that are normally used in high demand mode applications, no failure rates are given: the manufacturer simply provides the component PFH_D value.

One final remark is the comparison between SFF and DC. They are strongly linked to the percentage of Dangerous non-Detected Failures. However, IEC 62061 has the advantage of an explicit link to the fact that a component can have a very high Reliability index (PFH_D) but, if the percentage of λ_{DU} vis a vis the other failure rates is high, that component cannot be used in a high Reliability control functions: that is the essence of the concept of **Architectural Constraint** (§ 3.4). A sort of Architectural constraint is present in ISO 13849-1 as well since, normally, each Safety Subsystems has to be associated to a Category.

3.3.1 Levels of Diagnostic

According to ISO 13849-1, there are four levels of Diagnostic Coverage; they are indicated in Table 3.5.

Table 3.5 Diagnostic coverage.

Level	Range
None	$DC < 60\%$
Low	$60\% \leq DC < 90\%$
Medium	$90\% \leq DC < 99\%$
High	$DC \geq 99\%$

You may wonder why the range does not have equal intervals as shown in Figure 3.14.

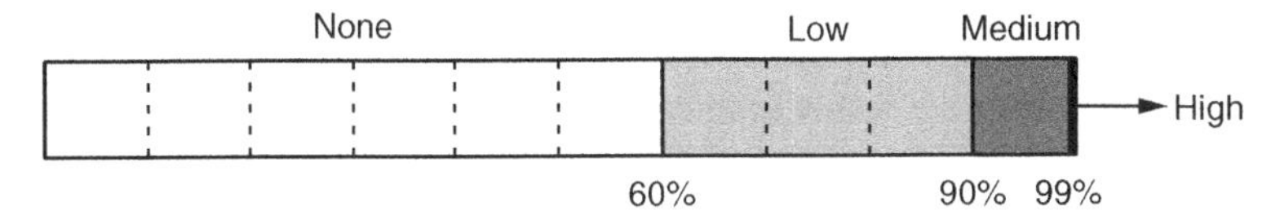

Figure 3.14 A graphical representation of the diagnostic coverage range.

The reason is that $(1 - DC)$, rather than DC, is a characteristic measure for the effectiveness of the test. $(1 - DC)$ forms a kind of logarithmic scale fitting to the logarithmic PL-scale. A DC value less than 60% has only slight effect on the Reliability of the tested system, and it is therefore called "none." A DC value greater than 99% for complex systems is very hard to achieve. To be practicable, the number of ranges was restricted in the standard to four. The indicated values of 60, 90, and 99% are assumed within an accuracy of 5%.

3.3.2 How to Estimate the DC Value

The method favored by ISO 13849-1 is based upon a reasoned conservative estimate of the DC, based upon how the component is connected to the rest of the components in the subsystem and if they are in a single or in a redundant channel. The values obtained are followed by the calculation of the DC_{avg} from the individual DC values by means of an averaging formula. Many types of connections and related possible tests can be classified as typical standard measures for which estimated DC values are listed in annex E of ISO 13849-1 (Tables 3.6 and 3.7).

Detection of a dangerous failure is only the beginning. **In order for the test to be passed, for PL d and PL e, a safe state that presents no further hazard must be reached in time**. This includes an effective shut-off path, which, for example in the case of single-channel tested systems (Category 2), entails a requirement for a second shut-off element. This is required in order to initiate and maintain the safe state when the test has detected failure of the normal shut-off element (block "O" on the safety-related block diagram).

That remark is valid for subsystems working in low demand mode as well: in general, according to IEC 61508-2, a Dangerous failure can be considered Detectable only if it is possible to bring the system to a safe state:

> ***[IEC 61508-2] C.2 Determination of diagnostic coverage factors***
> *[...] The calculations to obtain the diagnostic coverage, and the ways it is used,* ***assume that the EUC can operate safely in the presence of an otherwise dangerous fault that is detected by the diagnostic tests****. If this assumption is not correct then the E/E/PE safety-related system shall be treated as operating in a high demand or a continuous mode of operation (see § 7.4.8.3, 7.4.5.3 and 7.4.5.4).*

The meaning of "*shall be treated as operating in high demand mode*" is that only two options can be chosen, still remaining in the low demand mode domain:

- The sum of the diagnostic test interval and the time to perform the specified **action to achieve or maintain a safe state** is less than the process safety time; or
- The ratio of the diagnostic test rate to the demand rate equals or exceeds 100.

To express the above concept is a different way: regardless if we are in high or low demand mode, dangerous detected failures are considered safe, because the assumption is that a safe state can be achieved once the failure is detected. A dangerous detected failure cannot be considered as such if either it is not possible to bring the subsystem to a safe state or if the diagnostic test is performed less than 100 times the expected demand rate.

In high demand mode, where the risk is low (up to $PL_r = c$) and when the initiation of a safe state is not possible (for example, owing to welding of the contacts of the final switching device) it may be sufficient, in Category 2, for the output of the test equipment (OTE), to provide a warning (ISO 13849-1).

3.3.3 Frequency of the Test

In high demand mode, usually, the Diagnostic test is done at the moment when there is a request upon the safety function. For example, when entering a safeguarded area, the motors (generating the hazards) are stopped, and the relative contactor status is tested (verification that they have opened).

The mathematical modeling of High demand mode of operations works well in practice when the test is frequent (like a few times per hour). The longer the period without testing, the "less reliable" the mathematical results are; we mention the issue in Chapter 5. In essence, if the demand upon the safety system is few times per year, the machine or the user should initiate the test. The test should preferably be performed automatically; alternatively, the test interval may be triggered automatically. Only in exceptional cases should it be assured by organizational measures.

3.3.4 Direct and Indirect Testing

In order to estimate the DC, all possible dangerous failures should be considered. Being able to understand which failures can be detected helps the calculation of the DC value.

In general, a test on a component can be direct or indirect. For example, on a pneumatic valve, a direct test could be in case the spool position is monitored: that is the case for the valve in Figure 3.15. However, also in that case, it could be an indirect monitoring! In general, the valve manufacturer knows if that is a direct or an indirect monitoring of the spool. An example of the latter is when a REED sensor is used to monitor the spool.

To a **direct monitoring,** a 99% DC is normally given; please refer to § 3.3.6.

A valve can also be monitored with a pressure switch, mounted on the output ports as shown in Figure 3.16. In principle, this is an **indirect monitoring**, but that does not mean it cannot reach 99%: in principle, indirect monitoring is not worse that the direct one.

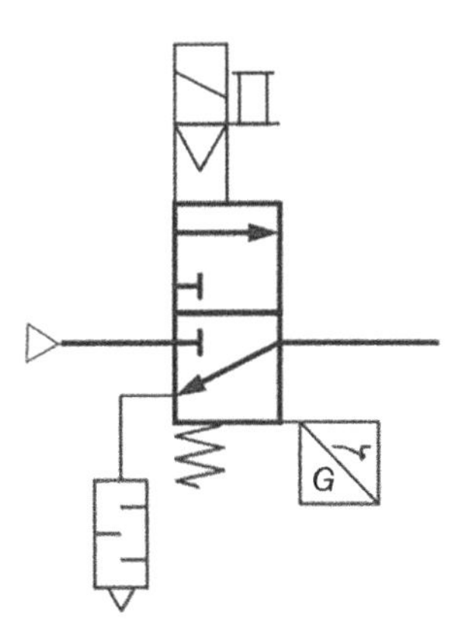

Figure 3.15 Solenoid valve with monitored of the spool.

3.3.4.1 DC for the Component and for the Channel

To determine the DC, the focus is **how many of the dangerous failures of the component can be detected**. Of course, the detection has to lead the system into a safe state, before a hazard can arise. In Category 3 and 4, there is still the second channel executing the safety function; in Category 2, the testing has to be performed 100 times more often than the demand on the safety function or the reaction has to be fast enough to reach the safe state before the hazard can arise.

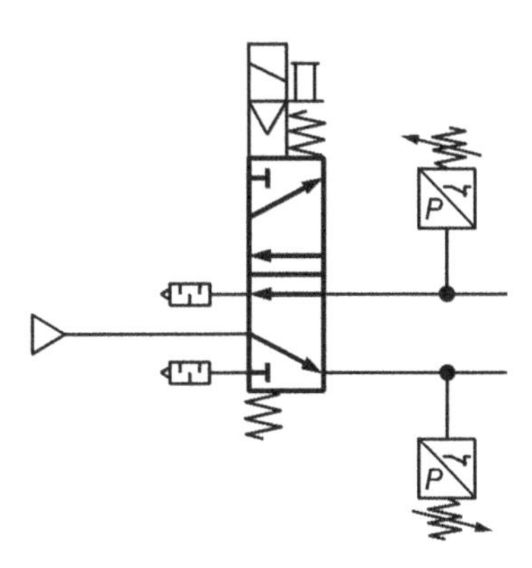

Figure 3.16 Monitored solenoid valve with the use of pressure switches.

For this decision, **it is not important whether the diagnostics can locate the failure to a specific component, as long as the dangerous malfunction in the channel is detected**. The task to locate the failure and to replace the failed component can be performed offline by the maintenance team while the Safety Control System remains in a safe lockout state. Therefore, this "repair task" is no longer a "safety task" in this perspective.

That is the reason why **the monitoring** of the safety function made of a 5/2 Solenoid Valve **can be done either on the spool, with pressure switches, or even on the cylinder,** as shown in Figure 3.17.

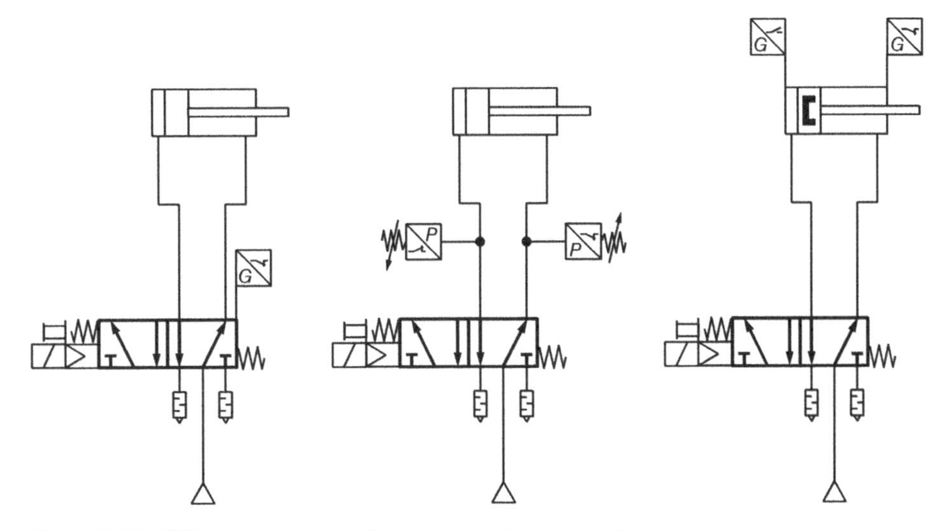

Figure 3.17 Different ways to monitor a pneumatic output subsystem.

3.3.5 Testing by the Process

Normally, the Diagnostic Testing is done automatically by a Safety Control system. In Category 2, the Markov model calculations are based upon monitoring by an external Control System, not necessarily a safety one, but one having MTTF values.

Both IEC 62061 and ISO 13849-1 accept a **testing made by an operator**. That is what is defined as "Fault detection by the process," and the DC value can go from 0 to 99%. **It is a concept introduced in the new edition of both standards.**

The example made in both ISO 13849-1 annex E and IEC 62061 annex D, is related to a printing process. In such a process, there is one of the rollers whose rotation speed is monitored by a rotary encoder. In case of a fault of the encoder, the operator would see the printing defects and stop the line. The DC for the sensor used to monitor the rotation speed can be estimated equal to 90% and up to 99%. The monitoring is done by the automation control system, with supervision by the operator. That is a good example of what "testing by the process" means.

In general, the fault has to be visible to the operator. Therefore, fault detection by the process can only be applied if the safety-related component is involved in the production process. Moreover, and that is another condition stated by both standards, the monitoring has to be one **part of two redundant functional channels executing the safety function.**

Two redundant channels are in Category 3, Category 4, and Basic Subsystem Architecture D: those are the only configurations where testing by the process can be used **in one of the two channels, but not in both**! Monitoring by the process cannot be used in Architecture C, nor in Category 2.

Let's consider a cylinder having a dangerous movement (Figure 3.18). In case both valves are controlled by the Safety System, that could be a 1oo2D architecture according to IEC 62061 (Category 3 or 4 of ISO 13849-1). The question is the DC level, since only the cut-off valve is tested automatically. Provided that a failure in the movement of the cylinder results in an obvious fault on the production line and therefore an operator would notice and stop the machine, a $DC > 60\%$ can be the result of the assessment. That means the output subsystem may be a Category 3 or even 4, or a Basic Subsystem Architecture D.

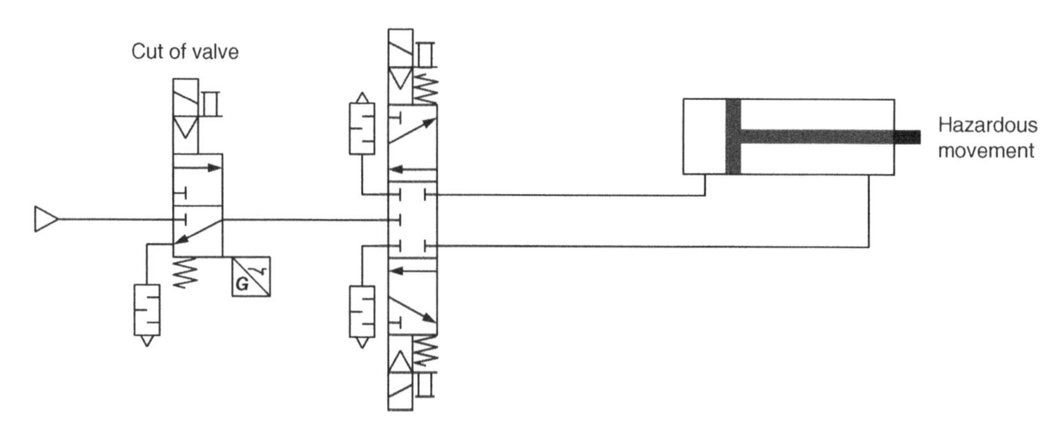

Figure 3.18 A possible Category 3 or 4 architecture according to ISO 13849-1 or 1oo2D according to IEC 62061.

Again, that is possible **only if one of the two channels is monitored automatically**. That means, in the example of Figure 3.18, **in case the 3/2 solenoid valve would not be monitored,** the output subsystem could not be considered neither a 1oo2D nor a Category 3 or 4 system, and therefore it would not be possible to attribute any DC to neither the 3/2 nor the 5/3 Valve.

3.3.6 Examples of DC Values

Tables 3.6 and 3.7, give estimates of the diagnostic coverage, depending upon the type of monitoring performed by the Safety Function.

Table 3.6 Diagnostic coverage estimates for the input subsystem.

Measure	Diagnostic coverage (*DC*)	Examples
Input device		
Cyclic test stimulus by dynamic change of the input signals	90%	Automatically changing an output to check whether the input connected with this output will change state
Plausibility check, e.g. use of normally open and normally closed mechanically linked contacts	99%	Compare automatically a normally closed contact with a normally open contact off a single sensor
Cross monitoring of inputs without dynamic test	percentage to be defined depending on the specific application (0% to 99%), depending on how often a signal change is done by the application (NOTE 1)	
Cross monitoring of input signals with dynamic test if short circuits are not detectable (for multiple I/O)	90%	Comparing the signals of two position monitoring devices by realizing the dynamic test through automatically moving the devices between the two positions and thus the expected position can be compared with effective position (without short circuit detection)
Cross monitoring of input signals and intermediate results within the logic (L) and temporal and logical software monitor of the program flow and detection of static faults and short circuits (for multiple I/O)	99%	Electronic devices continuously checking its functioning. Typically this measure is used in complex electronics.
Indirect monitoring (e.g. monitoring by a pressure switch, electrical position monitoring of actuators)	90–99%, depending on the application (NOTE 2)	Monitoring a cylinder is in its end position and remains in this end position. IEC 62061 note: Care should be taken the system can detect a failure before a dangerous situation can exist (e.g. if the cylinder leaves its position, it should be possible to place the system automatically in a safe state)
Direct monitoring (e.g. electrical position monitoring of control valves, monitoring of electromechanical devices by mechanically linked contact elements)	99%	Monitoring the functioning of a contactor by a mechanically linked NC contact
Fault detection by the process	percentage to be defined depending on the specific application (0%–99%), depending on the application; this measure alone is not sufficient for SIL 3. (NOTE 1, NOTE 2, NOTE 3)	• Degradation of the outcome of the production process indicates a probable future loss of the safety function • A measured value (e.g. level) does not correspond with the expected value
Monitoring some characteristics of the sensor (response time, range of analog signals, e.g. electrical resistance, capacitance)	60%	Checking an analog value remains within predefined borders (e.g. 12–17 mA)

Table 3.7 Diagnostic coverage estimates for the output subsystem.

Measure	Diagnostic coverage (*DC*)	Examples
Output device		
Monitoring of outputs by one channel without dynamic test	Percentage to be defined depending on the specific application, e.g. depending on how often a signal change is done by the application (NOTE 1)	
Cross monitoring of outputs without dynamic test	percentage to be defined depending on the specific application (0–99%); depending on how often a signal change is done by the application (NOTE 1)	
Cross monitoring of output signals with dynamic test without detection of short circuits (for multiple I/O)	90%	Check if the two 3/2 exhaust valves have switched off by making use of a pressure switch and switching on the valves one by one to see if a difference in pressure occurs
Cross monitoring of output signals and intermediate results within the logic (L) and temporal and logical software monitor of the program flow and detection of static faults and short circuits (for multiple I/O)	99%	Measuring the speed after activation of Safely Limited Speed which is compared with the expected program values
Redundant shut-off path with monitoring of the actuators by logic and test equipment	99%	Monitoring both NC mechanically linked contacts (placed in series or in parallel on the logic) of a redundant contactor arrangement
Indirect monitoring (e.g. monitoring by pressure switch, electrical position monitoring of actuators)	90–99%, depending on the application (NOTE 2)	Monitoring a cylinder is in its end position and remains in this end position. IEC 62061 note: Care should be taken the system can detect a failure before a dangerous situation can exist (e.g. if the cylinder leaves its position it should be possible to place the system automatically in a safe state)
Fault detection by the process	Percentage to be defined (0–99%) depending on the application; this measure alone is not sufficient for the required PL e or SIL 3 (NOTE 1, NOTE 2, NOTE 3)	• Degradation of the outcome of the production process indicates a probable future loss of the safety function • A measured value (e.g. level) does not correspond with the expected value
Direct monitoring (e.g. electrical position monitoring of control valves, monitoring of electromechanical devices by mechanically linked contact elements)	99%	Monitoring the functioning of a contactor by a mechanically linked NC contact

Note 1 (Tables 3.6 and 3.7): regarding the DC measure "Cross monitoring of inputs without dynamic test", an example are two pressure switches on the same piping. Each one can be used as diagnostic of the other. However, that works only if the system works in high demand mode of operation (for example, several times each month). The less the demand upon the safety function, the less effective such a monitoring is. That is the reason why ISO 13849-1 gives the following indication for Category 3 and 4 (the same approach is valid when applying IEC 62061),

1) $r_t < 1/\text{year}$ — *DC* is 0%
2) $r_t \geq 1/\text{year}$ — *DC* is limited to 90%
3) $r_t \geq 1/\text{month}$ — *DC* is limited to 99%

Note 2 (Tables 3.6 and 3.7): regarding the DC measure related to "Fault detection by the process", it is important to compare:

- the **Demand rate** of the safety function (r_d);
- with respect to the **Process diagnostic (test) rate** (r_t).

The idea is that the more often the operator observes the correct functioning of the safety system, the higher is the level of Diagnostic applied to the Functional Channel. ISO 13849-1 gives the following indication:

1) $r_t/r_d > 1$ — *DC* is limited to 60%
2) $r_t/r_d > 10$ — *DC* is limited to 90%
3) $r_t/r_d > 100$ — *DC* is limited to 99%

Note 3 (Tables 3.6 and 3.7): When the DC measure "fault detection by the process" is combined with other DC measures as listed in the above tables, this measure can still be included in the DC estimation of the block, even for PLr e or for SIL 3.

ISO 13849-2:2012, annex E, presents a complete worked example for the validation of fault behaviour and diagnostic means on an automatic assembly machine.

Please refer to § 4.14 for the evaluation of diagnostic coverage for series connected interlocking devices with potential free contacts.

The DC measure "fault detection by the process" can only to be applied if the safety-related component is involved in the production process, e.g. a PLC or sensors are used for workpiece processing and as part of one of two redundant channels executing the safety function. The appropriate DC level depends on the overlap of the commonly used resources (logic, inputs/outputs etc.). For example, when all faults of a rotary encoder on a printing machine lead to highly visible interruption of the printing process, the DC for this sensor used to monitor a safely limited speed is estimated as 90 up to 99%.

3.3.7 Estimation of the Average DC

In case a channel is made of more than one subsystem element (example in Figure 3.19), or a subsystem is made of more than one subsystem element (example in Figure 3.20), one average DC for the whole SRP/CS performing the safety function can be calculated (example in Figure 3.21), in order to estimate the Reliability of the subsystem.

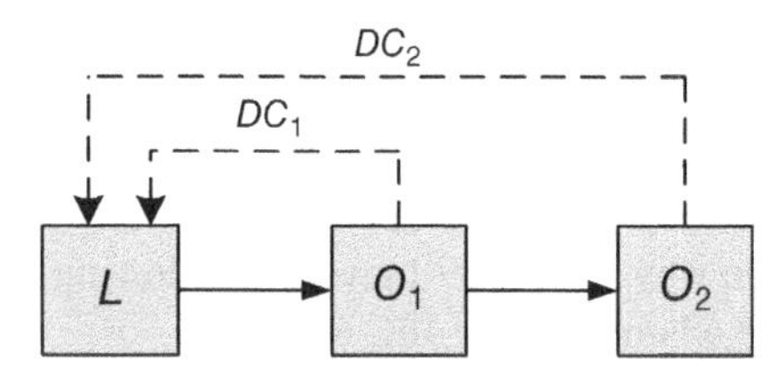

Figure 3.19 Example of a channel made of two subsystems (O_1 and O_2).

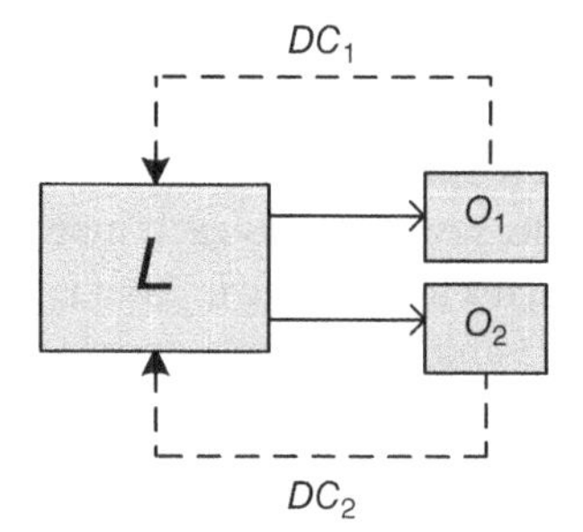

Figure 3.20 Example of a subsystem made of two subsystem elements (O_1 and O_2).

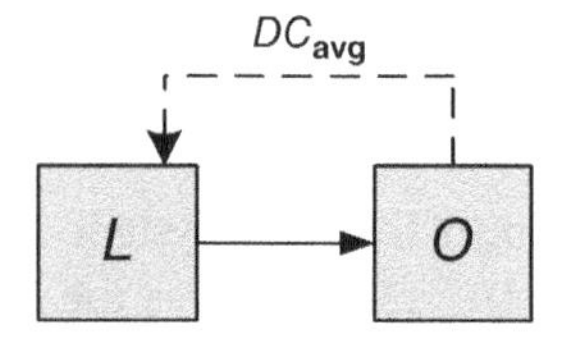

Figure 3.21 DC average.

The formula to be used to calculate the Average Diagnostic Coverage, in both cases, is the following:

$$DC_{avg} = \frac{(DC_{O1}/MTTF_{D-O1}) + (DC_{O2}/MTTF_{D-O2})}{(1/MTTF_{D-O1}) + (1/MTTF_{D-O2})}$$

The general formula is the following:

$$DC_{avg} = \frac{(DC_1/MTTF_{D1}) + (DC_2/MTTF_{D2}) + \dots + (DC_n/MTTF_{Dn})}{(1/MTTF_{D1}) + (1/MTTF_{D2}) + \dots + (1/MTTF_{Dn})}$$

All components of the Safety Control System, not subject to fault exclusion, have to be considered. Please notice that each DC refers to the tested part and not to the testing device. Components without failure detection (e.g. which are not tested) have $DC = 0$ and contribute only to the denominator of DC_{avg}.

3.4 Safety Integrity and Architectural Constraints

3.4.1 The Starting Point

Safety integrity means both **Hardware safety integrity** and **Systematic safety integrity**. What does that mean?

Let's look at the definitions:

> ***[IEC 62061] 3.2 Terms and definitions***
> ***3.2.23 Systematic Safety Integrity***. *Part of the safety integrity of an SCS or its subsystems relating to its resistance to systematic failures in a dangerous mode.*
>
> ***[IEC 62061] 3.2 Terms and definitions***
> ***3.2.22 Hardware Safety Integrity***. *Part of the safety integrity of an SCS or its subsystems relating to random hardware failures in a dangerous mode of failure.*
>
> ***[IEC 61508-4] 3.5 Safety functions and safety integrity***
> ***3.5.10 Software Safety Integrity Level.*** *Systematic capability of a software element that forms part of a subsystem of a safety-related system.*

Systematic failures (hardware or software), and consequently Systematic safety integrity, cannot be quantified. On the other hand, **Random** Hardware Failures usually can.

A Safety Control System needs a certain level of Safety Integrity in order to be "reliable" or, in other terms, to be "immune" from both **Systematic and Random failures**.

Hardware random failures are quantifiable and are taken into considerations thanks to values given by the component manufacturer, like Failure Rates, $MTTF_D$, and PFH_D. The issue is how to tackle the Systematic Failures. That is done by guaranteeing a certain level of **Systematic Capability**, that is the terminology used in IEC 61508. The Systematic Capability applies to a safety component with respect to its confidence that the **Systematic Safety Integrity** meets the requirements of the specified **Safety Integrity Level (SIL)**. IEC 62061 uses instead the term **Systematic Integrity**.

3.4.2 The Systematic Capability

Here the definition from IEC 61508-4 [8]:

> ***[IEC 61508-4] 3.5 Safety functions and safety integrity***
> ***3.5.9 Systematic Capability.*** *Measure (expressed on a scale of SC 1 to SC 4) of the confidence that the systematic safety integrity of an element meets the requirements of the specified SIL, in respect of the specified element safety function, when the element is applied in accordance with the instructions specified in the compliant item safety manual for the element.*

For a Safety PLC, its **systematic capability** will include considerations on both hardware and software failures. For an electro-mechanical interlocking device, used in a safety system, only considerations on systematic hardware failure should be taken into account. In other terms, the systematic capability applies to an element with respect to its confidence that the systematic safety integrity meets the requirements of the specified safety integrity level.

The issue is that Systematic Failures cannot be quantified. Therefore, IEC 61508 gives requirements, based upon experience and judgment gained in the industry, so that the target failure measure, associated with the safety function, can be considered to be achievable.

In Functional Safety, people tend to focus their attention to formulas and block diagrams. However, probably, **the most important aspect is the process followed to design, build, validate and maintain a safety control system.** Compliance with that process has the purpose of guaranteeing the lack of Systematic Failures (due to, for example, mistakes in the engineering phase) and therefore confirming the Systematic Capability of the safety Control or Instrumented System.

The concept of **systematic capability** was introduced in the second edition of IEC 61508. Since the term was not present in the first edition, terminology like "SIL n capable" component, or "SIL n compliant" component appeared in datasheets and documents: which generated confusion!

3.4.2.1 Systematic Safety Integrity

With the new edition of IEC 61508 series, that was clarified: a device, with a systematic capability of SIL 2 (SC 2), for example, meets the **systematic safety integrity** of SIL 2 when applied in accordance with the instructions stated by its manufacturer. That means, even if its failure rates and SFF allow the component to reach SIL 3, it can only be used in a SIL 2 safety subsystem. That also means, even if it is used in redundancy with another component and, together, their subsystem can reach SIL3, the safety system can only reach SIL 2.

When a component manufacturer considers achieving IEC 61508 compliance for a particular product, according to Exida [69], there are about 340 unique requirements that need to be met. Of these 340 requirements, about 90 are product specific and 250 are related to the development process used by the manufacturer. The majority of the development process requirements focus on the development and testing of the software portions of an equipment item.

Once again, the overall objective of the Systematic Safety Integrity is that a product is well made, without design errors: from requirements through manufacturing, validation, and testing.

Pls notice that if an end user claims **Prior Use** on a component, the end user basically takes responsibility for the quality of the development process.

There are three possible Routes to compliance:

- **Route 1s,** requirements for the avoidance (prevention) and requirements for the control of systematic failures.
- **Route 2s,** evidence that the equipment is "**proven in use.**"
- **Route 3s,** for pre-existing software elements only.

For compliance with IEC 61508-2, it is necessary to meet the requirements of Route 1s or Route 2s, and for pre-existing software elements, Route 3s.

3.4.3 Confusion Generated by the Concept of Systematic Capability

The confusion comes from the fact that, thanks to the concept of Systematic Safety Integrity, **it is possible to give a SIL level to a component**. That is unusual since the SIL concept belongs to a safety function and not to a component.

Let's consider an electronic sensor, for example a pressure transmitter, that has the following data:

- **Systematic Capability:** SC 3 (SIL 3 Capable, in the old language).
- **Random Capability:** Type B Element, SIL 1 with $HFT = 0$.

From the above data, we understand that the component was tested following the principle of IEC 61508, and its main application is in low demand mode. If the component has no significant electromechanical parts, that does not make any difference, and you can use those data for high demand mode applications as well.

3.4.3.1 Random Capability

Let's now understand what data mean: let's start from the Random Capability. Table 3.8 summarizes the Failure rate data given in the product certificate.

Table 3.8 Pressure transmitter reliability data.

Rosemount 2051C (Source: Exida SERH 2015 - 01 sensors - item 1.6.91)		Coplanar Per 10^9 hours (FITs)
Fail low		23
Fail high		31
Fail detected		174
Fail dangerous undetected	λ_{DU}	46
No effect failure	λ_{NE}	63
Route 1_H SFF (%)		83.5
Useful life (years)		50

Since the SFF is between 60 and 90%, from Table 3.10, we know that, in $HFT = 0$, the component can be used in a SIL 1 Safety System, not higher. That does not mean the component is SIL 1: the component is capable of being used in a SIL 1 safety system, if used as a single component. However, for example, if it is used in a 1oo2 architecture ($HFT = 1$), the safety system can reach SIL 2.

Bottom line: SIL "n" referred to an $HFT =$ "m" is about the **component capability** and it is not correct to state that the component is SIL "n": **SIL is a characteristic of the safety system and not of the component.**

3.4.3.2 Systematic Capability

We clarified the importance of the systematic capability of a component; we can state that the Rosemount 2051C is SIL 3 Capable from a systematic safety integrity point of view, but again, we cannot state that the component is SIL 3. Therefore, please do not get confused by the "SIL n CAPABLE" logo that is shown on some component third-party certificates.

3.4.3.3 ISO 13849-1

The standard recognises the importance of random failures, but it does not use the FMEDA approach: the component reliability data can be given in one of the following three forms:

- $\boldsymbol{B_{10D}}$, normally used for Electromechanical Components;
- **MTTF$_D$**; or
- **PFH$_D$** in case of complex electronic components.

The architectural constraints are taken into consideration in the Categories and the limitation in the PFH$_D$ that is possible to achieve in each of them.

The standard also recognizes the importance of Systematic Failures, but it did not develop a "PL n Capable" concept.

For that reason, it is incorrect to state that a solenoid valve is, for example, PL c. A solenoid valve, suitable for use in a safety control system, can have a B_{10D} value if pneumatic or a MTTF$_D$ value if hydraulic, but it cannot have neither a PL level, nor a Category!

3.4.4 The Safety Lifecycle

The product Safety Lifecycle is the most important aspect of IEC 61508. In essence, a component can claim a certain level of Systematic Capability only if its overall safety lifecycle complies with the following recommendations.

The requirements start when the first **concept** of the product is defined; it goes through a **Hazard and Risk Analysis** till the **Safety Requirements Specification** are defined. At this point the new product is defined on paper.

The next phase is the **realization phase** which should be under the control of well-defined **planning**.

Once the product is manufactured, it has to go through a **Safety Validation** phase. It can then be sold to a machine or process manufacturer who integrates the product in a Safety Control System. However, in the product manual, details for the **Installation**, **Maintenance** and possible **Repairs** need to be specified. Figure 3.22 details all the steps.

The safety lifecycle is critical for the development of components to be used in a Safety Control System (SCS). For example, if a manufacturer developed a product several years ago and he did not follow this process, he cannot simply put the product under test, estimate the Failure Rates and then claim it can be used in an SCS or in an SIS.

The safety lifecycle is probably less critical in an SCS used in high demand mode, but I imagine that statement would not find everybody in agreement. In any case, that is probably the reason why IEC 62061 only covers those aspects of the safety lifecycle that are related **to safety requirements allocation** (step 5 of the Safety Lifecycle) through to **safety validation** (Step 13). Requirements for maintenance, repair, and possibly decommissioning are provided in the Information for use.

3.4.5 The Software Safety Lifecycle

On the other hand, the Lifecycle is fully used in software development. IEC 62061 uses a V-model: a way to structure the software design into small parts called **Modules**. The model does not introduce any sequence for the creation of specifications or for the implementation.

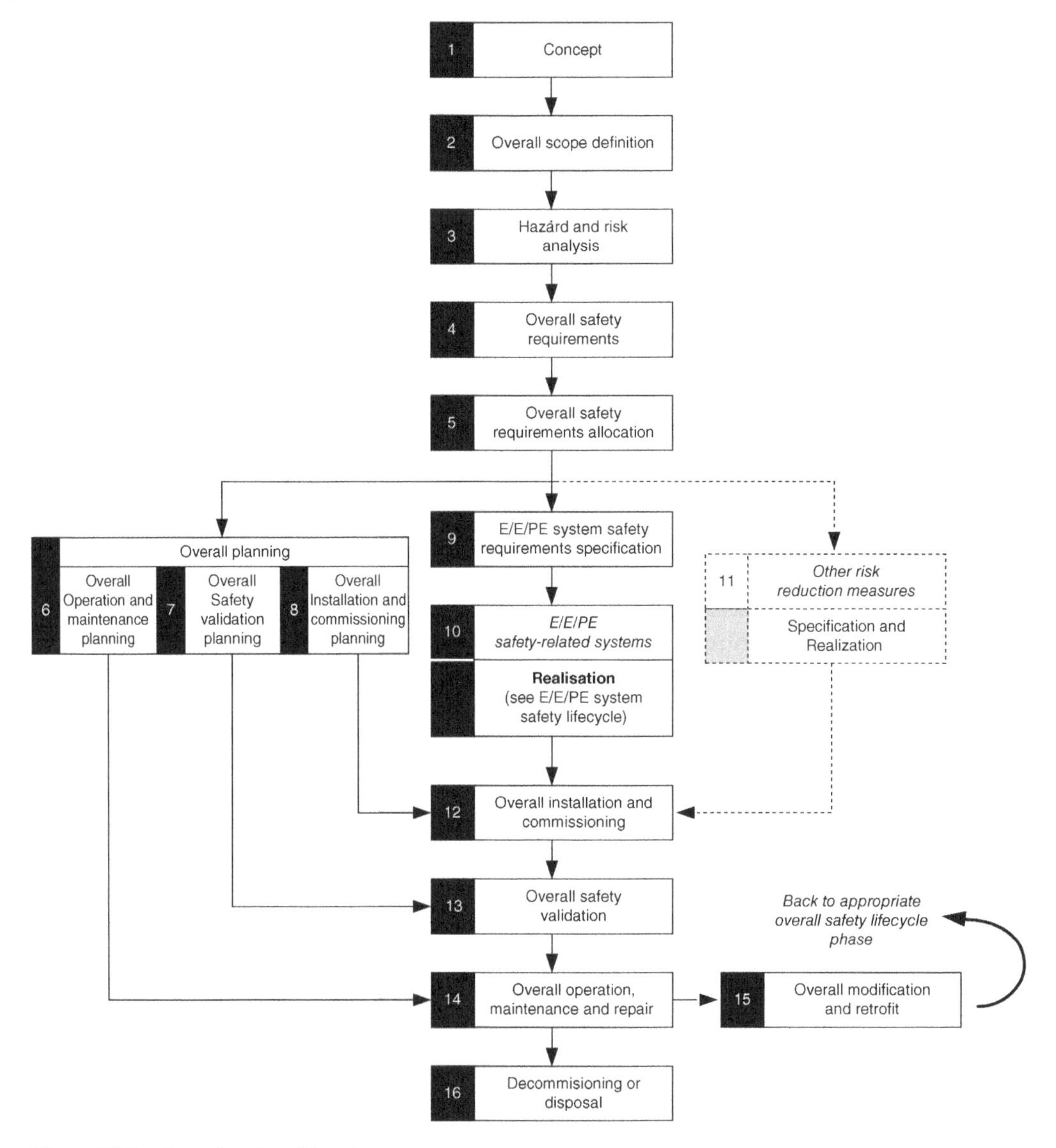

Figure 3.22 Overall safety lifecycle.

IEC 62061 focuses on the so-called Software Level 1, which is an application software that uses a so-called Limited Variable Language.

In that case, the V-Model for a module of the software is shown in Figure 3.23.

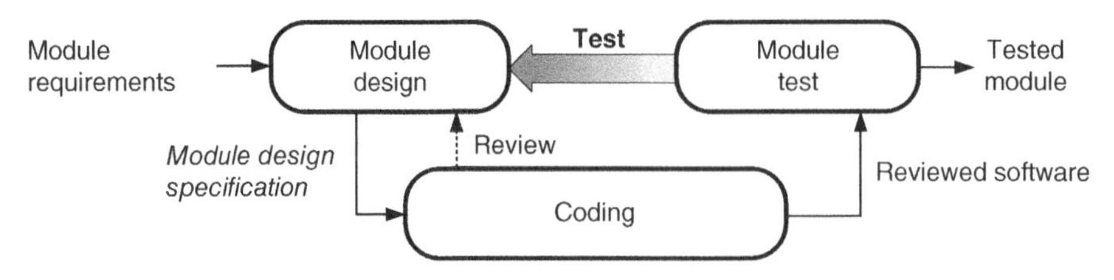

Figure 3.23 V-Model for a part (module) of a Software Level 1.

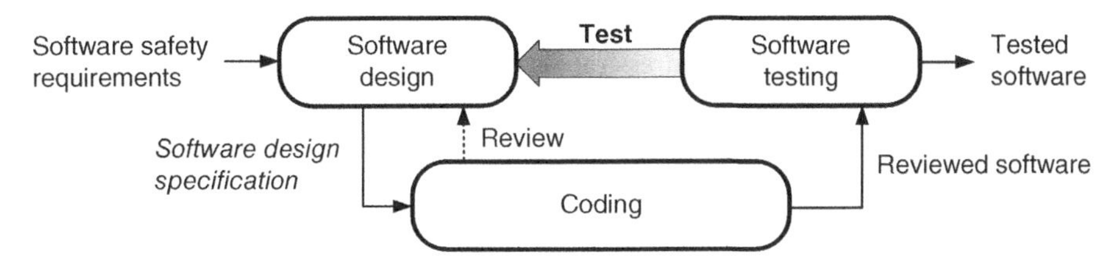

Figure 3.24 V-Model for a Software Level 1.

On the left side of the V-model, the output of each phase is reviewed. Review means to check the output of a phase in the V-model against the requirements of the input of the same phase.

The arrow "Review" represents the first step of the software verification.

The arrow "Test" represents the results of test cases according to the specification and, in addition, the need for more precise test case requirements and specifications. The results of Figure 3.23, done for each module, is the input to the coding of Figure 3.24.

3.4.6 Hardware Fault Tolerance

The concept of HFT is used in IEC 61508 to indicate the ability of a hardware subsystem to continue performing a required function in the presence of faults or errors. The HFT is given as a digit, where $HFT = 0$ means that, in case of one fault, the function (e.g., a pressure measurement) is lost. $HFT = 1$ means that if a channel fails, there is another one that is able to perform the same function: in other terms, the subsystem can tolerate one failure and still be able to function. A subsystem of three channels that are voted 2oo3 is functioning as long as two of its three channels are functioning. This means the subsystem can tolerate one channel failure and still function normally. The HFT of the 2oo3 voted group is, therefore, $HFT = 1$.

Let's now look at its definition:

> ***[IEC 62061] 3.2 Terms and definitions***
> ***3.2.35 Hardware Fault Tolerance*** *(HFT). Property of a subsystem to potentially lose the safety function upon at least $N + 1$ faults.*

Fault tolerance is referred to as **HFT,** and it is used from a hardware perspective where additional sensors, logic solver, or final elements are added to reduce the probability of random failures

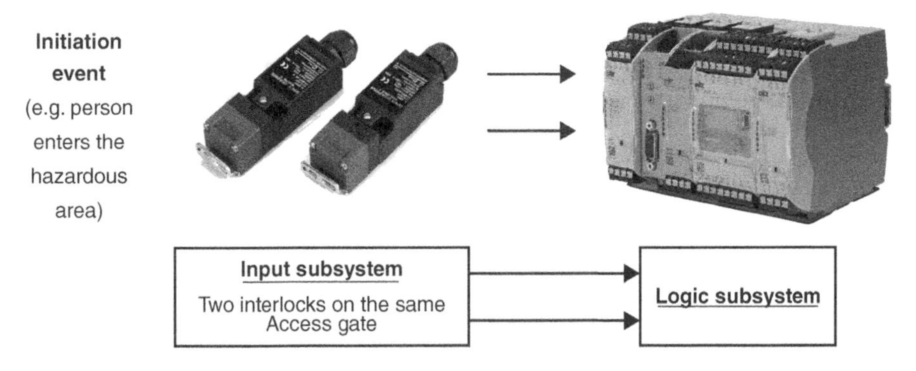

Figure 3.25 $HFT = 1$ input safety subsystem.

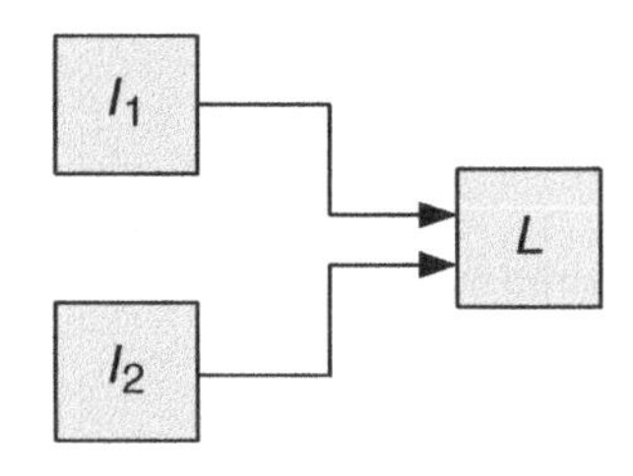

Figure 3.26 *HFT* = 1 input safety sub-function.

causing the failure of these safety functions. However, HFT is not very useful in addressing systematic failures particularly when identical technologies are used.

Once again, an HFT of N means that $N + 1$ faults could cause a loss of the safety function. In determining the HFT of a subsystem, measures taken to improve the Reliability of a safety function, like diagnostics, should not be considered. Moreover, where one fault directly leads to the occurrence of one or more subsequent faults, these are considered as a single fault.

Finally, when determining the HFT achieved, certain faults may be excluded, provided that the likelihood of them occurring is very low in relation to the safety integrity requirements of the subsystem. However, **any such fault exclusions shall be justified and documented**.

Figures 3.25 and 3.26 show an input safety subsystem with two interlocking devices on the same access gate: it has an $HFT = 1$.

We will see in § 3.4.7 that the concept of HFT is important for the determination of the Maximum SIL that can be claimed by a Safety-related Control System. The concept comes from IEC 61508-2, and it is used in case Route 1_{H} is used.

3.4.7 The Hardware Safety Integrity

If a machinery manufacturer buys a component that is certified following Route 1_{H}, what does it mean and how should he use it? Can the component be used in a SIL 2 safety function?

Let's start from the second question. In general, the highest safety integrity level (SIL) that can be claimed for a safety function is limited by the **hardware safety integrity constraints** achieved following one of two possible routes (applicable at system or subsystem level):

- **Route 1_{H}** is based on HFT and SFF concepts; or
- **Route 2_{H}** based on component Reliability data from field feedback, increased confidence levels and HFT for specified safety integrity levels.

Please consider that which route to choose is application and sector dependent.

3.4.7.1 Type A and Type B Components

Components used in a Safety Function can be classified as Type A or Type B.

According to IEC 61508-2, § 7.4.4.1.2 a component **can be regarded as Type A** if

a) *the failure modes of all constituent components are well defined; and*
b) *the behavior of the element under fault conditions can be completely determined; and*
c) *there is sufficient dependable failure data to show that the claimed rates of failure for detected and undetected dangerous failures are met (see § 7.4.9.3 to 7.4.9.5).*

Electromechanical interlocking devices are examples of Type A component.

A component can be regarded as **Type B** if [IEC 61508-2: § 7.4.4.1.3]:

a) *the failure mode of at least one constituent component is not well defined; or*
b) *the behavior of the element under fault conditions cannot be completely determined; or*
c) *there is insufficient dependable failure data to support claims for rates of failure for detected and undetected dangerous failures (see § 7.4.9.3 to 7.4.9.5).*

A 4–20 mA transmitter is normally a Type B component.

3.4.8 Route 1_H

Historically, this was the only way to determine the maximum SIL that can be claimed by a Safety Function. Here are the steps to be followed: IEC 61508-2, § 7.4.4.2

1) Divide the Safety-related system in **subsystems**.
2) **For each subsystem** calculate the SFF for all elements in the subsystem separately. In case of redundant element configurations, the SFF may be calculated by taking into consideration the additional diagnostics that may be available (e.g. by comparison of redundant elements).
3) **For each element,** use the achieved SFF and HFT of 0 to determine the maximum safety integrity level that can be claimed from column 2 of Table 3.9 (same as table 2 of IEC 61502-2) for **Type A elements;** in case of **Type B elements**, Table 3.10 should be used (same as table 3 of IEC 61502-2).
4) The maximum safety integrity level that can be claimed for an E/E/PE safety-related system shall be determined by the subsystem that has achieved the lowest safety integrity level.

For Route 1_H, each safety component must have **all the failure rates coming from an FMEDA Analysis**.

Table 3.9 Maximum allowable safety integrity level for a safety function carried out by a Type A safety-related element or subsystem.

	Hardware fault tolerance		
Safe failure fraction of an element	**0**	**1**	**2**
$SFF < 60\%$	SIL 1	SIL 2	SIL 3
$60\% \leq SFF < 90\%$	SIL 2	SIL 3	SIL 4
$90\% \leq SFF < 99\%$	SIL 3	SIL 4	SIL 4
$SFF \geq 99\%$	SIL 3	SIL 4	SIL 4

Table 3.10 Maximum allowable safety integrity level for a safety function carried out by a Type B safety-related element or subsystem.

	Hardware fault tolerance		
Safe failure fraction of an element	**0**	**1**	**2**
$SFF < 60\%$	Not allowed	SIL 1	SIL 2
$60\% \leq SFF < 90\%$	SIL 1	SIL 2	SIL 3
$90\% \leq SFF < 99\%$	SIL 2	SIL 3	SIL 4
$SFF \geq 99\%$	SIL 3	SIL 4	SIL 4

3.4.8.1 Route 1_H and Type A Component: Example

Table 3.11 shows an example of Reliability data for a Pressure Switch defined as a Type A component.

Being a Type A component with $SFF < 90\%$, its use in low demand mode is limited to SIL 2 safety functions.

Table 3.11 Pressure switch reliability data.

FEMA DCM series (Source: Exida SERH 2015 – 01 sensors – item 1.6.73)		Stainless still membrane	6002, NBR membrane
		Per 10^9 hours (FITs)	
Fail safe undetected	λ_{SU}	143	141
Fail dangerous undetected	λ_{DU}	79	89
No effect failure	λ_{NE}	81	118
Route 1_H *SFF* (%)		64.4	61.3
Useful life (years)		10	10

3.4.8.2 Route 1_H and Type B Component: Example

Table 3.12 shows an example of Reliability data for a Pressure Transmitter defined as a Type B component.

Table 3.12 Pressure transmitter reliability data.

Honeywell ST 3000 (Source: Exida SERH 2015 – 01 sensors – item 1.6.77)		HART protocol	DE protocol
		Per 10^9 hours (FITs)	
Fail low		89	85
Fail high		48	48
Fail detected		207	207
Fail dangerous undetected	λ_{DU}	146	140
No effect failure	λ_{NE}	227	223
Route 1_H *SFF* (%)		70.2	70.8
Useful life (years)		50	50

Here is the definition of some of the items in the table given by Exida:

- ***Fail Low:*** *a failure that will result in an output current that is lower than 4 mA or under range per NAMUR recommendation 43.*
- ***Fail High:*** *a failure that will result in an output current that is higher than 20 mA or over range per NAMUR recommendation 43.*
- ***Fail Detected:*** *a failure that is detected by internal diagnostic, whose external effect depends on equipment item settings. In a transmitter, for example, a detected failure could result in over range or under range output depending on a jumper setting.*

Being a Type B component with $SFF < 90\%$, its use in low demand mode is limited to a SIL 1 safety functions

3.4.9 High Demand Mode Safety-Related Control Systems

When using IEC 62061, we need to apply the concept of Architectural Constraints. Moreover, **Route 1_H is the only one that can be followed**.

Finally, the highest safety integrity level that can be claimed for an SCS is limited by the HFT and SFF of the subsystems that carry out that safety function.

In this case, for the Architectural Constraint, there is only one table that can be followed, and it is similar to the one for Type B components. Table 3.13 (same as table 6 of IEC 62061) specifies the highest safety integrity level that can be claimed by a Safety-related Control Subsystem in high demand mode, taking into account the HFT and SFF of that subsystem.

Table 3.13 Maximum SIL that can be claimed by a safety control subsystem.

	Hardware fault tolerance (*HFT*)		
Safe failure fraction (*SFF*)	**0**	**1**	**2**
$SFF < 60\%$	Not allowed (Note 1)	SIL 1	SIL 2
$60\% \leq SFF < 90\%$	SIL 1	SIL 2	SIL 3
$90\% \leq SFF < 99\%$	SIL 2	SIL 3	SIL 3
$SFF \geq 99\%$	SIL 3	SIL 3	SIL 3

Note 1: For subsystems which have a $SFF < 60\%$, $HFT = 0$ and that use well-tried components, SIL 1 can be achieved.

Comparing Table 3.13 with Table 3.10, it is clear that SIL 4 is not considered in the machinery sector. In the previous edition of IEC 62061, the maximum SIL that could be claimed was defined as **SIL Claim** (SILCL).

3.4.9.1 Example

In this we consider a pressure switch from a different manufacturer. Looking at Table 3.14, the component is suitable for use on both air and gas piping, in Furnace applications. The Reliability data are given in case of usage of less than once every 10 hours. **Data is valid for high demand mode of operations only**.

Being a Type A component with $SFF > 90\%$, its use in high demand mode can be in SCS up to SIL 2. That is now in contrast with the reasoning of IEC 62061: please refer to § 3.1.

Table 3.14 Pressure switch reliability data.

Source : Certificate dated July 2014	
HFT	0
CCF	70
SFF	>90%
λ_D	3 FIT
$MTTF_D$	36 317 years
T_{10D}	3 632 years

3.4.10 Route 2_H

The concept of Route 2_H was introduced, for the first time, in the 2010 edition of IEC 61508. Its use is linked to the concept of Proven in Use (IEC 61508) or Prior Use (IEC 61511-1).

Basically, it is possible to reach a Reliability level without information on the SFF of the component, provided the Failure rate field data are both available and reliable. Therefore, the Reliability data used when quantifying the effect of random hardware failures shall be:

a) based on field feedback for elements in use in a similar application and environment;
b) based on data collected in accordance with international standards (e.g., IEC 60300-3-2 or ISO 14224:); and
c) evaluated according to:
    i. the amount of field feedbacks;
    ii. the exercise of expert judgment; and, where needed,
    iii. the undertaking of specific tests.

That is needed in order to estimate the average and the uncertainty level, for example, the 90% confidence interval or the probability distribution of each Reliability parameter (e.g., failure rate) used in the calculations.

The 90% confidence interval of a failure rate λ is the interval $[\lambda_{5\%}, \lambda_{95\%}]$ in which its actual value has a probability of 90% to belong to: in this case, λ has a probability of 5% to be better than $\lambda_{5\%}$ and worse than $\lambda_{95\%}$. On a pure statistical basis, the average of the failure rate may be estimated by using the "maximum likelihood estimate" and the confidence bounds $(\lambda_{5\%}, \lambda_{95\%})$ may be calculated by using the χ^2 function. The accuracy depends on the cumulated observation time and the number of failures observed. The Bayesian approach may be used to handle statistical observations, expert judgment and specific test results. This can be used to fit relevant probabilistic distribution functions for further use in Monte Carlo simulation.

If route 2_H is selected, then the Reliability data uncertainties shall be taken into account when calculating the target failure measure (i.e. PFD_{avg}), and the system shall be improved until there is a confidence greater than 90% that the target failure measure is achieved.

With Route 2_H, the maximum SIL achievable based upon the HFT of a subsystem is indicated in Table 3.15:

Table 3.15 Minimum HFT required in Route 2_H.

	HFT = 0	*HFT* = 1	*HFT* = 2
High or continuous demand mode	**SIL 1** (§ 7.4.4.3.1 e)	**SIL 2** (§ 7.4.4.3.1 c) **SIL 3** (§ 7.4.4.3.1 b)	Not applicable
Low demand mode	**SIL 2** (§ 7.4.4.3.1 d)	**SIL 3** (§ 7.4.4.3.1 b)	**SIL 4** (§ 7.4.4.3.1 a)

The paragraphs indicated in the table are from IEC 61508-2 [6]

Despite § 7.4.4.3.1a also includes high demand mode of operation, SIL 4 is not applicable in Machinery. Morevore, despite being defined for both high and low demand mode of operation, nether ISO 13849-1 (§ 6.1.2) nor IEC 62061 (§ 7.2) allow the use of subsystems designed according to **Route 2_H in high demand mode**.

3.5 Mean Time to Failure (MTTF)

It is a measure of a component Reliability. Actually, there are two parameters normally used:

MTBF: Mean Time Between Failures. From its name, it is clear that a component can fail more than once. Indeed the MTBF is used for components that can be repaired. It is not used in high demand mode Functional Safety Standards.

MTTF: Mean Time to Failure. It is used in case of components that cannot be repaired or are not considered to be repaired.

> ***[ISO/TR 12489] 3 Terms and definitions***
> ***3.1.29 Mean Time to Failure – MTTF:*** *expected time before the item fails.*
>
> ***Note 1 to entry:*** *The MTTF is classically used to describe the time to failure for a non-repairable item or* ***to the first failure for a repairable item.*** *When the item is as good as new after a repair, it is also valid for the further failures.*

Typically in high demand mode, when a component fails, it is replaced, given its limited cost: this is one of the few advantages of using electromechanical components! Back to the MTTF, the one that refers to Dangerous Faults ($MTTF_D$) is the one used in Functional Safety.

As discussed in Chapter 1, its value corresponds to the mean time in years when 63.2% of components has failed to the dangerous side.

The main source of the $MTTF_D$ should be the component manufacturer. In case the component manufacturer does not give a value, ISO 13849-1 and, with the new edition, IEC 62061, both provide conservative values for commonly used components. In case the component is missing from that list, a value of 10 years can, in general, be used.

3.5.1 Examples of MTTF Values

In case the manufacturer does not provide the value of the $MTTF_D$ of its component, the values in Table 3.16 can be used, **provided all the following conditions are respected**:

a) The manufacturer of the component confirms that he used basic and well-tried safety principles (please refer to Chapter 4), or he built the components following the relevant standards mentioned in Table 3.16, for example, by stating it in the component datasheet.
b) The manufacturer of the component specifies how to install the product and what are the operating conditions that need to be respected.
c) The machine manufacturer designs the safety system using basic and well-tried safety principles for the implementation and operation of the component.

In case of Hydraulic components, $MTTF_D$ values are normally given, while for Pneumatic components, international technical standards, like ISO 19973-1 [21], recommend the use the parameter B_{10D}. Even if the $MTTF_D$ values for hydraulic components depend on the number of operations, the wear effects are different than those for pneumatic, mechanical, or electromechanical components. The reason for the different approach is that, despite being Electromechanical, Hydraulic Valves are "naturally" continuously lubricated: for that reason **they are not so much influenced by the number of operations** like Pneumatic valves are.

I remind that, in case of a constant failure rate, the $MTTF_D$ is equal to the reciprocal of the dangerous failure rate.

$$MTTF_D = \frac{1}{\lambda_D}$$

Table 3.16 Typical values of $MTTF_D$ or B_{10D} for common components.

	Basic and well-tried safety principles (ISO 13849-2)	Other relevant standards	Typical $MTTF_D$ (years) or B_{10D} (cycle) values
Mechanical components	Tables A.1 and A.2		$MTTF_D = 150$
Hydraulic components with $n_{op} \geq 1\,000\,000$ cycles/year[a]	Tables C.1 and C.2	ISO 4413	$MTTF_D = 150$
Hydraulic components $1\,000\,000 > n_{op} \geq 500\,000$[a]	Tables C.1 and C.2	ISO 4413	$MTTF_D = 300$
Hydraulic components $500\,000 > n_{op} \geq 250\,000$[a]	Tables C.1 and C.2	ISO 4413	$MTTF_D = 600$
Hydraulic components $n_{op} \leq$ 250 000 cycles/year[a]	Tables C.1 and C.2	ISO 4413	$MTTF_D = 1\,200$
Pneumatic components	Tables B.1 and B.2	ISO 4414	$B_{10D} = 20\,000\,000$[c]
Relays and contactor relays with small load (mechanical load)	Tables D.1 and D.2	IEC 61810-3, IEC 60947	$B_{10D} = 20\,000\,000$
Relays and contactor relays with maximum load	Tables D.1 and D.2	IEC 61810-3, IEC 60947	$B_{10D} = 400\,000$
Proximity switches with small load (mechanical load)	Tables D.1 and D.2	IEC 60947, ISO 14119	$B_{10D} = 20\,000\,000$
Proximity switches with nominal load	Tables D.1 and D.2	IEC 60947, ISO 14119	$B_{10D} = 400\,000$
Contactors with small load (mechanical load)	Tables D.1 and D.2	IEC 60947	$B_{10D} = 20\,000\,000$
Contactors with nominal load	Tables D.1 and D.2	IEC 60947	$B_{10D} = 1\,300\,000$ (see Note 1)
Position switches[b]	Tables D.1 and D.2	IEC 60947, ISO 14119	$B_{10D} = 20\,000\,000$
Position switches (with separate actuator, guard-locking)[b]	Tables D.1 and D.2	IEC 60947, ISO 14119	$B_{10D} = 2\,000\,000$
Emergency stop devices[b]	Tables D.1 and D.2	IEC 60947, ISO 13850	$B_{10D} = 100\,000$
Push buttons (e.g. enabling switches)[b]	Tables D.1 and D.2	IEC 60947	$B_{10D} = 100\,000$

Note 1: B_{10D} is estimated as two times B_{10} (50% dangerous failure) if no other information (e.g. product standard or results of analysis) is available.
"Nominal load" or "small load" should take into account safety principles described in ISO 13849-2:2012, like over-dimensioning of the rated current value. "Small load" means, for example, 20%.
Note 2: Emergency stop devices according to IEC 60947-5-5 and ISO 13850 and enabling switches according to IEC 60947-5-8 can be considered as an $HFT = 0$ or $HFT = 1$ subsystem depending on the number of electrical output contacts and on the fault detection in the subsequent SCS.
Each contact element (including the mechanical actuation) can be considered as one channel with a respective B_{10D} value. For enabling switches according to IEC 60947-5-8, this implies the opening function by pushing through or by releasing. In some cases, it can be possible that the machine builder can apply fault exclusion according to table A–D of ISO 13849-2:2012, considering the specific application and environmental conditions of the device.
Note 3: The $MTTF_D$ for mechanical components refers exclusively to mechanically moving components/parts (not to housing).

[a] Mechanical components refer to components being part of a Safety Function, like springs. There are other static mechanical components, like housing, that are important for the systematic part of the safety component but, in general, they do not have random failure data.
[b] B_{10D} calculation for hydraulic components is not permitted as a reverse calculation from the indicated $MTTF_D$ values.
[c] If fault exclusion for direct opening action is possible.

3.5.2 Calculation of MTTF$_D$ and λ_D for Components from B_{10D}

As described in Chapter 1, MTTF$_D$ and B_{10D} are linked by the following equation:

$$MTTF_D = \frac{B_{10D}}{0.1 \cdot n_{op}} \qquad \text{[Equation 3.5.2]}$$

where:

$$n_{op} = \frac{d_{op} \cdot h_{op} \cdot (3600\,\text{s})/(\text{h})}{t_{cycle}}$$

and with the following assumptions having been made on the application of the component:

- $\boldsymbol{h_{op}}$ is the mean operation, in hours per day;
- $\boldsymbol{d_{op}}$ is the mean operation, in days per year;
- $\boldsymbol{t_{cycle}}$ is the mean time between the beginning of two successive cycles of the component (e.g. shifting of a valve) in seconds per cycle.

The maximum time the component can be used (constant failure rate) is

$$T_{10D} = \frac{B_{10D}}{n_{op}}$$

That is the reason why, when the T_{10D} value for a component is less than the Mission Time (20 years), the machinery manufacturer shall inform the user to replace the component at or before the T_{10D} period ends (§ 1.12). Component replacement based on the T_{10D} value will help maintaining the expected performance level of the safety function.

If the ratio of dangerous failure (RDF) given by the component manufacturer is estimated at less than 50%, than T_{10D} value is limited to $2{\cdot}T_{10}$.

In case no B_{10D} is given by the component manufacturer, it is permitted to determine the B_{10D} by the following formula:

$$B_{10D} = \frac{B_{10}}{RDF}$$

λ_D and B_{10D} are linked by the following equation:

$$\lambda_D = \frac{0.1 \cdot C}{B_{10D}} = \frac{0.1 \cdot n_{op}}{B_{10D} \cdot (8760\,\text{hours})/(\text{a})}$$

where C is the duty cycle or mean operation per hour:

$$C = \frac{n_{op}}{8760}$$

3.5.3 Estimation of MTTF$_D$ for a Combination of Systems

Having a n-series of components, the MTTF$_D$ of the whole channel can be calculated with the following formula:

$$\frac{1}{MTTF_D} = \sum_{i=1}^{N} \frac{1}{MTTF_{Di}}$$

where MTTF$_{Di}$ is the MTTF$_D$ of each component which contributes to the safety function.

If two redundant components are present (two functional channels), the MTTF can be calculated with the following formula:

$$MTTF_D = \frac{2}{3}\left[MTTF_{DCh1} + MTTF_{DCh2} - \frac{1}{1/(MFFT_{DCh1}) + 1/(MFFT_{DCh2})}\right]$$

where MTTF$_{DChi}$ is the MTTF$_D$ of each channel which contributes to the safety function.

3.5.3.1 Example for Channels in Series

Let's consider the Safety Function represented in Figure 3.27.

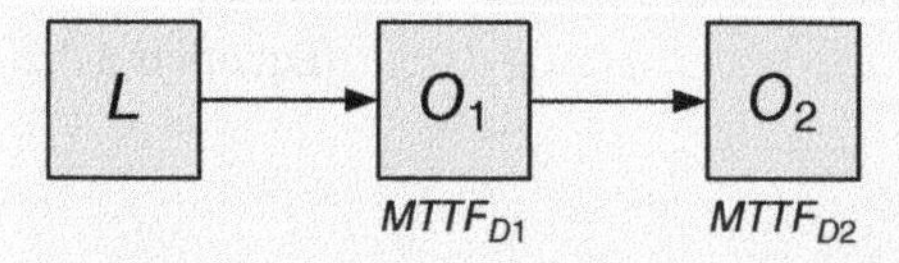

Figure 3.27 Single channel output subsystem with two elements.

The $MTTF_D$ of the two element output subsystem can be calculated using the following formula:

$$MTTF_D = \left(\frac{1}{MTTF_{O_1}} + \frac{1}{MTTF_{O_2}}\right)^{-1}$$

If the $MTTF_D$ of O_1 is = 150 years and the one of O_2 is 100 years, the MTTF of the output system in Figure 3.28 is equal to:

$$MTTF_D = \left(\frac{1}{150} + \frac{1}{100}\right)^{-1} = 60 \text{ years}$$

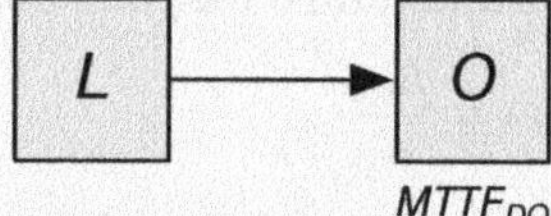

Figure 3.28 Single channel output subsystem with one element.

3.5.3.2 Example for Redundant Channels

Considering the subsystem in Figure 3.29.

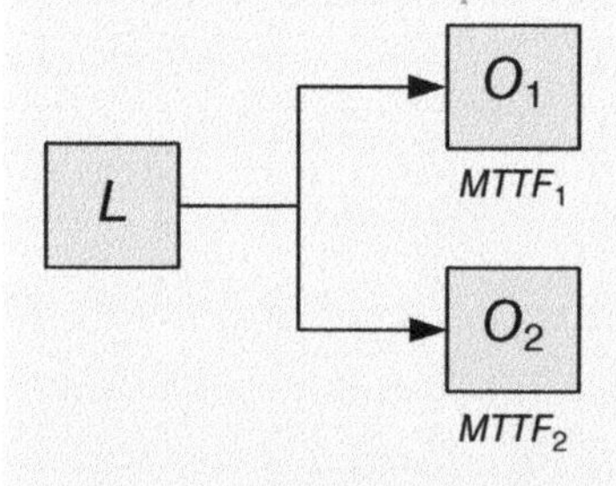

Figure 3.29 Two channels output subsystem.

The $MTTF_D$ of the two redundant elements can be calculated using the following formula:

$$MTTF_D = \frac{2}{3}\left[MTTF_{DO1} + MTTF_{DO2} - \frac{1}{1/(MFFT_{DO1}) + 1/(MFFT_{DO2})}\right]$$

If the $MTTF_D$ of O_1 is =150 years and the one of O_2 is 100 years, the MTTF of the output subsystem (Figure 3.30) is equal to:

$$MTTF_D = \frac{2}{3}\left[150 + 100 - \frac{1}{\frac{1}{150} + \frac{1}{100}}\right] = 126 \text{ years}$$

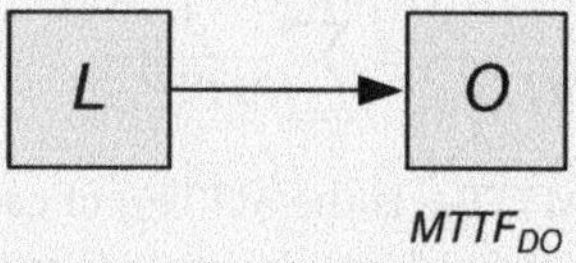

L → O

$MTTF_{DO}$

Figure 3.30 Single channel output subsystem with one element.

3.6 Common Cause Failure (CCF)

3.6.1 Introduction to CCF and the Beta-Factor

Redundancy increases reliability, however two identical components used in a redundant channel may fail at the same time due to a common cause. Therefore, CCF has to be considered when redundant channels or components are used. Hereafter the definition from IEC 62061, very similar to the one given in ISO 13849-1

> ***[IEC 62061] 3.2 Terms and definitions***
> ***3.2.56 Common Cause Failure (CCF).*** *Failure, that is the result of one or more events, causing concurrent failures of two or more separate channels in a multiple channel subsystem, leading to failure of a safety function.*

From an academic point of view, if the probability of failure of a single component is $\boldsymbol{p}$, the probability of failure of two independent redundant similar components (1oo2) is $\boldsymbol{p}^2$ and of three independent redundant similar components (1oo3) is $\boldsymbol{p}^3$.

Despite it seems in theory possible to reduce the probability of failure to any level just by implementing the right level of redundancy, some factors exist that limit the potential probability reduction expected from the redundancy. Those factors are linked to the potential for common causes (CCF), and systematic failures to cause multiple components to fail. Figure 3.31 shows how common cause failures can be modeled as a single fault.

A simple model, called the **beta-factor model**, was introduced by Fleming (1975) to incorporate CCFs into Reliability models. The idea of the beta-factor model is to split the failure rate λ of an item into two parts:

- λ_i: the rate of individual failures, i.e., failures that affect only the specific item and
- λ_c: the rate of failures that affect all the items in a voted group, that is, the rate of CCFs, such that

$$\lambda = \lambda_i + \lambda_c$$

The parameter β was introduced as the fraction of all failures of an item that are CCFs.

$$\beta = \frac{\lambda_c}{\lambda}$$

That means, we can express λ_i and λ_c as

- $\lambda_i = (1 - \beta)\cdot\lambda$
- $\lambda_c = \beta\cdot\lambda$

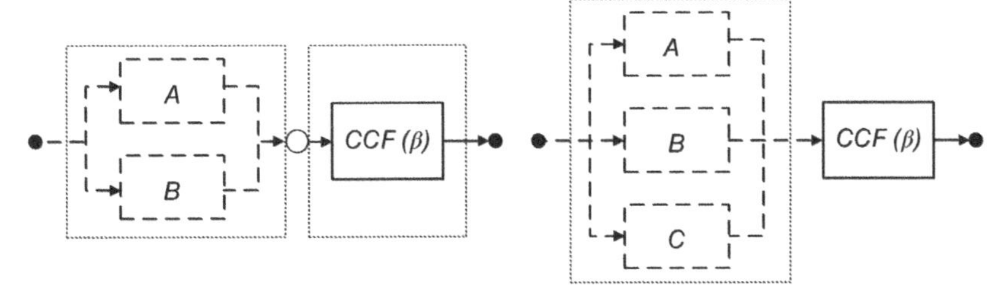

Figure 3.31 The beta-factor model.

The beta-factor β, can also be interpreted as a conditional probability: when a failure of the item is observed, β is the probability that this failure, in fact, is a CCF. A consequence of the beta-factor model is that any failure is either an individual failure affecting a single item, or a CCF affecting all the items of the voted group.

Common Cause Initiators (CCI) are for example the effects of electric fields and electromagnetic waves that can commonly influence the channels. Also, the ambient temperature can be an initiator, as well as the degradation of components and parts due to mechanical forces or vibrations. Finally, humidity, pollution, and the impacts of commonly supplied circuitry, by issues related to the power source (like power outage, over- and undervoltage, or transient effects) are common cause initiators.

Therefore β is strongly influenced by local, plant-specific conditions, and it is not likely that a relevant estimate for β in any generic data source can be found. However, that is probably more critical in low demand mode than in high demand mode.

3.6.2 How IEC 62061 Handles the CCF

The standard has a qualitative approach. Table 3.17 contains criteria that reduces the probability of Common Cause Failures. For each listed measure, only the full score or nothing can be claimed. If a measure is only partly fulfilled, the score according to this measure is zero.

Table 3.17 Criteria for an estimation of CCF according to IEC 62061.

Item	Reference	Score
Separation/segregation		
Are SCS signal cables for the individual channels routed separately from other channels at all positions? For example: - Signal cables for the individual channels separate from other channels at all positions or sufficiently shielded (connected to protective earth) - Short circuit detection provided - Sufficient clearances and creepage distances on printed-circuit boards	1a	5
Where information encoding/decoding is used, is it sufficient for the detection of signal transmission errors?	1b	10
Are SCS signals and power cables/sources separate at all positions or sufficiently shielded (no interference from any other electrical system to the SCS signals, see IEC 60204-1:2016, annex H)?	2	5
If subsystem elements can contribute to a CCF, are they provided as physically separate devices in their local enclosures?	3	5
Diversity/redundancy		
Does the subsystem employ different technologies for example, one electronic or programmable electronic and the other an electromechanical relay or a hydraulic valve?	4	8
Does the subsystem employ elements that use different physical principles (e.g. sensing elements at a guard door that use mechanical and magnetic sensing techniques)?	5	10
Does the subsystem employ elements with temporal differences in functional operation and/or failure modes?	6	10
Do the subsystem elements have a diagnostic test interval of $\leq$1 min?	7	10

Table 3.17 (Continued)

Item	Reference	Score
Complexity/design/application		
Is cross-connection between channels of the subsystem prevented with the exception of that used for diagnostic testing purposes?	8	2
Assessment/analysis		
Has an analysis been conducted to establish sources of common cause failure and have predetermined sources of common cause failure been eliminated by design?	9	9
For example, over voltage, over temperature, over pressure, etc. (see § 7.3.2.3)		
Are field failures analyzed with feedback into the design?	10	9
Competence/training		
Do subsystem designers understand the causes and consequences of common cause failures?	11	4
Environmental control		
Are the subsystem elements likely to operate always within the range of temperature, humidity, corrosion, dust, vibration, etc., over which it has been tested, without the use of external environmental control? (See IEC 60068 (all parts)	12	9
Is the subsystem immune to adverse influences from electromagnetic interference (See IEC 61326-3-1 or IEC 61000-6-7)	13	9

Note 1: An alternative item (e.g. references 1a and 1b) is given in table F.1 where it is intended that a claim can be made for a contribution towards avoidance of CCF from only the most relevant item.
Note 2: Similar criteria can be derived for other technologies based on the same principles.

This overall score can be used to determine a common cause failure factor (β) using Table 3.18.

Table 3.18 Criteria for an estimation of CCF according to IEC 62061.

Overall score	Common cause failure factor (β)
$\leq$35	10% (0.1)
36–65	5% (0.05)
66–85	2% (0.02)
86–100	1% (0.01)

As it will be shown in Chapter 7, CCF plays a role only if

1. $HFT = 0$ and $SFF \geq 60\%$ or
2. $HFT = 1$

3.6.3 How ISO 13849-1 Handles the CCF

The ISO standard has a similar qualitative approach, and CCF is not relevant for Category 1. There is no explicit β factor calculation since the standard assumes a $\beta = 2\%$. For that reason, the machinery manufacturer has to collect at least 65 points, using a similar table to IEC 62061 and shown in Table 3.19.

Table 3.19 Scoring process and quantification of measures against CCF.

No.	Measure against *CCF*	Score
1	**Separation/segregation**	15
2	**Diversity**	20
3	**Design/application/experience**	
3.1	Protection against over-voltage, over-pressure, over-current, over-temperature, etc.	15
3.2	Components used are well-tried	5
4	**Assessment/analysis**	5
5	**Competence/training**	5
6	**Environmental**	
6.1	Prevention of EMC or impurity of the pressure medium	25
6.2	Other influences	10
	Total	(max. achievable 100)
Total score[a]	**Measures for avoiding *CCF***	
65 or better	Meets the requirements	
Less than 65	Process failed ⇒ choose additional measures	

[a] Where technological measures are not relevant, points attached to this column can be considered in the comprehensive calculation.

For each listed measure, the full score can only be claimed if the measure is fully implemented. If a measure is only partly fulfilled, a score of zero must be assumed. The single items are similar to the ones detailed by IEC 62061. Please refer to annex F of ISO 13849-1 for further details.

3.7 Proof Test

The Diagnostic Coverage is done in an "automatic way." For example, every time a contactor coil is de-energized, the control system detects if the power contacts have opened; if they have not, the control system can take actions, like generating an alarm. The Diagnostic test is a **functional test** that allows only detectable failures to be identified.

The **Proof Test** is usually performed **offline** and its aim is to detect all possible failures, including the normally undetected ones; those are failures that cannot be detected by the Diagnostic Coverage (DC). Hereafter is the definition:

> ***[IEC 61508-4] 3.8 Confirmation of safety measures***
> ***3.8.5 Proof Test.*** *Periodic test performed to detect dangerous hidden failures in a safety-related system so that, if necessary, a repair can restore the system to an "as new" condition or as close as practical to this condition.*

Therefore, a Proof Test can reveal dangerous faults which are undetected by automatic diagnostic tests. That means, a component manufacturer should do an FMEDA and list all dangerous undetected failures of its component. Then, he needs to specify the **Proof Test procedure** to be followed by the end user in order to detect all those dangerous undetected failures during the Proof Test activity, that is also called "Periodical Test."

3.7.1 Proof Test Procedures

The Proof Test procedure **needs to be specified by the component manufacturer** and cannot be decided or defined by the user.

3.7.1.1 Example of a Proof Test Procedure for a Pressure Transmitter

In case of Analog Sensors, it is relatively easy to find the proof test procedures. Tables 3.20 and 3.21 show what Emerson's Rosemount states for its 3051 Pressure Transmitter with 4–20 mA HART.

Table 3.20 Example of a partial proof test procedure.

Steps	*Actions*
1	*Bypass the safety function and take appropriate action to avoid a false trip*
2	*Use HART communications to retrieve any diagnostics and take appropriate action.*
3	*Send a HART command to the transmitter to go to the high alarm current output and verify that the analogue current reaches that value [This tests for compliance voltage problems such as a low loop power supply voltage or increased wiring resistance. This also tests for other possible failures].*
4	*Send a HART command to the transmitter to go to the low alarm current output and verify that the analogue current reaches that value [This tests for possible quiescent current related failures].*
5	*Inspect the Transmitter for any leaks, visible damage or contamination.*
6	*Remove the bypass and otherwise restore normal operation*

Suggested Partial Proof Test. *The suggested proof test described in the table hereafter will detect 51% of possible DU failures in the Rosemount 3051 Coplanar Differential and Coplanar Gage and 41% of possible DU failures in the Rosemount 3051 Coplanar Absolute, In-Line Gage and Absolute.*

Suggested Comprehensive Proof Test. *The suggested proof test described in Table 3.21 hereafter will detect 90% of possible DU failures in both the Rosemount 3051 Coplanar Differential and Coplanar Gage and the Rosemount 3051 Coplanar Absolute, In-Line Gage and Absolute.*

Table 3.21 Example of a comprehensive proof test procedure.

Steps	*Actions*
1	*Bypass the safety function and take appropriate action to avoid a false trip*
2	*Use HART communications to retrieve any diagnostics and take appropriate action.*
3	*Send a HART command to the transmitter to go to the high alarm current output and verify that the analogue current reaches that value (This tests for compliance voltage problems such as a low loop power supply voltage or increased wiring resistance. This also tests for other possible failures).*
4	*Send a HART command to the transmitter to go to the low alarm current output and verify that the analogue current reaches that value (This tests for possible quiescent current related failures).*
5	*Inspect the Transmitter for any leaks, visible damage or contamination.*
6	***Perform a two-point calibration of the transmitter*** *over the full working range and verify the current output at each point.*
7	*Remove the bypass and otherwise restore normal operation*

Tables 3.22 and 3.23 show the Failure Rates stated in the Exida report:

Table 3.22 Failure rates.

Device	λ_{SD}	λ_{SU}	λ_{DD}	λ_{DU}	SFF
Emerson's Rosemount 3051 4–20 mA HART Pressure Transmitter: Coplanar Differential and Coplanar Gage	*0*	*84*	*258*	*32*	(342/374=) *91%*
Emerson's Rosemount 3051 4–20 mA HART Pressure Transmitter: Coplanar Absolute, In-Line Gage and Absolute	*0*	*94*	*279*	*41*	(373/414=) *90%*

Proof Test Coverage: the Proof Test Coverage for the various product configurations is given in Table 3.23.

Table 3.23 Proof test coverage.

Device	Coplanar differential and coplanar gage (%)	Coplanar absolute, in-line gage and absolute (%)
Rosemount 3051 – partial	*51*	*41*
Rosemount 3051 – comprehensive	*90*	*90*

3.7.1.2 Example of a Proof Test Procedure for a Solenoid Valve

In case of electromechanical components, it is harder to find the proof test procedure. In Table 3.24 what ASCO states in the Safety Manual of a **pneumatic solenoid valve** is shown.

The objective of Proof Testing is to detect failures within an ASCO Solenoid that are not detected by any automatic diagnostics of the system. Of main concern are undetected failures that prevent the safety instrumented function from performing its intended function.

The frequency of Proof Testing, or the Proof Test interval, is to be determined in Reliability calculations for the safety instrumented functions for which an ASCO Solenoid is applied. The Proof Tests must be performed more frequently than or as frequently as specified in the calculation in order to maintain the required safety integrity of the safety instrumented function.

The following Proof Test is recommended. Any failures that are detected and that compromise Functional Safety should be reported to ASCO Valve.

Table 3.24 Example of a proof test procedure for a pneumatic solenoid valve.

Steps	Actions
1	*Bypass the safety PLC or take other appropriate action to avoid a false trip, following company Management of Change (MOC) procedures*
2	*Inspect the external parts of the solenoid valve for dirty or clogged ports and other physical damage. Do not attempt disassembly of the valve.*
3	*De-energize the solenoid coil and observe that the actuator and valve move. Energize the solenoid after a small movement of the valve.*
4	*Inspect the solenoid for dirt, corrosion or excessive moisture. Clean if necessary and take corrective action to properly clean the air supply. This is done to avoid incipient failures due to dirty air.*
5	*Record any failures in your company's SIF inspection database. Restore the loop to full operation.*
6	*Remove the bypass from the safety PLC or otherwise restore normal operation*

This test will detect approximately 99% of possible DU failures in the solenoid (Proof Test Coverage).

The person(s) performing the Proof Test of an ASCO Solenoid should be trained in SIS operations, including bypass procedures, solenoid maintenance and company Management of Change procedures. No special tools are required.

As you can see, the procedure consists basically in:

- **De-energize** the component and verify that the process valve moves (it opens or it closes).
- Remove dirt, verify air filters and check for damages.

Being electromechanical, there is not much more that can be done. A similar procedure could be devised for contactors or other electromechanical components.

3.7.2 How the Proof Test Interval Affects the System Reliability

$\boldsymbol{T_i}$ indicates the Proof Test Interval. The value is used in the calculation of PFD_{avg} in low demand mode. For example, in case of a 1oo1 system:

$$PFD_{avg} = \frac{\lambda_{DU} \cdot T_i}{2}$$

Therefore, **the Proof Test has a direct influence on the Reliability of a safety function.**

Please consider that, in low demand mode of operation, the formulas are based upon the underlying assumption that *the expected interval between demands is at least an order of magnitude greater than the proof test interval* [10 – § B.3.1]. In the 1997 edition of IEC 61508-4, a low demand mode of operation was defined as "... *where the frequency of demands for operation made on a safety-related system is no greater than one per year and* ***no greater than twice the proof test frequency.***" That was changed in the second edition by changing the definition and adding the above language in IEC 61508-6.

3.7.2.1 Example

Let's consider a component with a $\lambda_{DU} = 4000$ FIT $[10^{-9}\ \text{h}^{-1}]$; That means $\lambda_{DU} = 4\ 10^{-6}\ \text{h}^{-1}$. If the component is subject to a Proof Test every four years:

$$PFD_{avg} = \frac{\lambda_{DU} \cdot T_i}{2} = \frac{4 \cdot 10^{-6} \cdot 4 \cdot 8760}{2} = 7 \cdot 10^{-2}\text{h}^{-1}$$

If the same component is subject to a Proof Test every year, its PFD becomes:

$$PFD_{avg} = \frac{\lambda_{DU} \cdot T_i}{2} = \frac{4 \cdot 10^{-6} \cdot 1 \cdot 8760}{2} = 1.75 \cdot 10^{-2}\,\text{h}^{-1}$$

The effectiveness of the Proof Test will depend on both the failure coverage and the repair effectiveness. In practice, detecting 100% of the hidden dangerous failures is, in general, not achievable: the manufacturer has to indicate that, by indicating the Proof Test Coverage (§ 3.7.3.1).

3.7.3 Proof Test in Low Demand Mode

In low demand mode applications, there are components that can be subject to either **a Partial Proof Test** or to a **Comprehensive or Full Proof Test.**

[63] Considering a 1oo1 system, the concept of Proof Testing is illustrated in Figure 3.32. The safety system unreliability, PFD(t), increases with time but, once the Proof Test is completed, then the PFD(t) returns to zero, meaning that the SIF returns to its designed status. That is based upon the fact that the Safety Instrumented System (SIS), with respect to the specified SIF, is now restored to the "as new" condition, after completion of the Proof Test. During this test, all the unrevealed dangerous failures have been detected and repaired. This is defined as a **Perfect Proof Test.**

Example 1

Let's consider a component with $\lambda_{DU} = 514\ \text{FIT} = 5.14{\cdot}10^{-7}\ \text{h}^{-1}$.

If the Proof test interval is every year ($T_i = 1$ year), the PFD_{avg} is shown in Figure 3.32.

$$PFD(t) = \lambda_{DU} \cdot t = 5.14 \cdot 10^{-7} \cdot t$$

$$PFD_{avg} = \frac{\lambda_{DU} \cdot T_i}{2} = \frac{5.14 \cdot 10^{-7} \cdot 1 \cdot 8760}{2} = 2.25 \cdot 10^{-3}$$

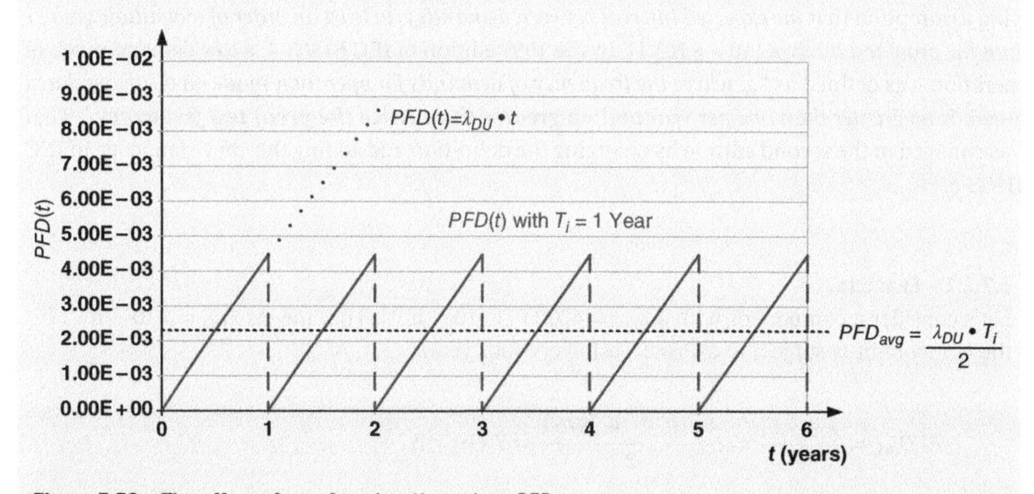

Figure 3.32 The effect of proof testing (1 year) on PFD_{avg}.

Example 2

If the Proof test interval increases to two years, as shown in Figure 3.33, the PFD_{avg} doubles. Therefore, the probability that the SIF will fail increases.

$$PFD(T_i) = \lambda_{DU} \cdot T_i = 5.14 \cdot 10^{-7} \cdot T_i$$

$$PFD_{avg} = \frac{\lambda_{DU} \cdot T_i}{2} = \frac{5.14 \cdot 10^{-7} \cdot 2 \cdot 8760}{2} = 4.5 \cdot 10^{-3}\,\text{h}^{-1}$$

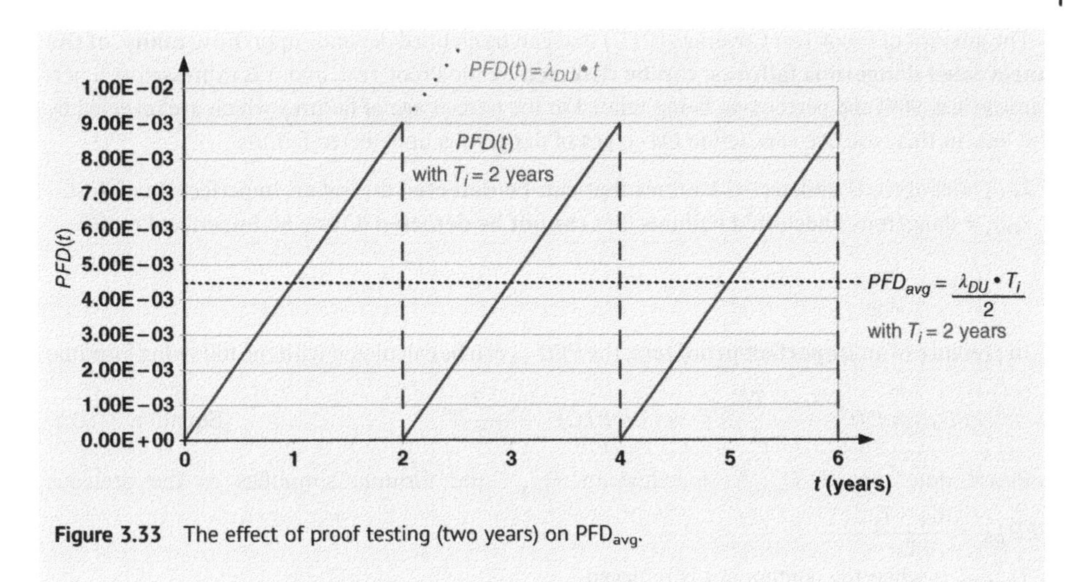

Figure 3.33 The effect of proof testing (two years) on PFD_{avg}.

3.7.3.1 Imperfect Proof Testing and the Proof Test Coverage (PTC)

A Perfect Proof Test detects 100% of the dangerous undetected failures. In practice, that is difficult to achieve; the dangerous failures that are not detected at each Proof Test will continue to be present and will increase the PFD(t), based upon Equation 1.10.1, seen in Chapter 1.

To understand the concept, we need to recognise that, especially for components like a process valve or a contactor or, in general, components that contain mechanical moving parts, we need to consider two types of hardware failures:

- Sudden failure
- Degraded failure

The **sudden failure** is typically a random hardware failure that may happen at any time. The sudden failure does not have any memory, so the rate of sudden failures is typically constant over time. A typical **degraded failure** is a failure that comes from **corrosion, erosion, deposition, vibration, high temperature** and other effects from the process. These effects accumulate slowly with time; therefore, the degraded failures have a memory. It also means that the rate of degraded failures changes over time. An imperfect proof test does not mean the proof test procedure is wrong; it means that there are degraded hardware failures that cannot be detected. Due to that, given enough time, the PFD_{avg} will exceed the target, which is necessary to maintain the required risk reduction within the overall system for the hazard being protected against.

The concept which defines the effectiveness of a Proof Test is referred to as **Proof Test Coverage** (PTC).

> ***[IEC 62061] 3.2 Terms and definitions***
> ***3.2.48 proof test coverage****. term given to the percentage of dangerous undetected failures that are detected by a defined proof test procedure.*
>
> ***Note 1 to entry:*** *It measures the effectiveness of a proof test and ranges from 0 % to 100 % (perfect proof-test).*
>
> ***Note 2 to entry:*** *For example, a PTC of 95 % states that 95 % of all possible undetected failures will be detected during the proof test. It doesn't include aging or degradation not directly related to the safety function failure.*
>
> ***Note 3 to entry:*** *The PTC can be estimated by the means of Failure Mode and Effects Analysis (FMEA) in conjunction with engineering judgement based on sound evidence.*

The amount of Proof Test Coverage (PTC) that can be claimed depends upon **how many, of the unrevealed dangerous failures, can be detected** by the Proof Test, and it is expressed as a percentage, e.g. 90%: the percentage being related to the percentage of failures which are revealed by the test. In this case, we can define two types of dangerous undetected failures:

$\boldsymbol{\lambda_{DU_d}}$ = dangerous undetected Failures that **can be detected** during an Imperfect Proof Test.
$\boldsymbol{\lambda_{DU_u}}$ = dangerous undetected Failures that **cannot be detected** during an Imperfect Proof Test.

$$PTC = \frac{\lambda_{DU_d}}{\lambda_{DU}}$$

In presence of an **Imperfect proof test**, the PFD_{avg} can be calculated with the following formula:

$$PFD_{avg} = PTC \cdot \frac{\lambda_{DU} \cdot T_{i_{imperfect}}}{2} + (1 - PTC) \cdot \frac{\lambda_{DU} \cdot T_{i_{perfect}}}{2} \qquad \text{[Equation 3.7.3.1]}$$

Please note that if $T_{i_{imperfect}}$ is equal to $T_{i_{perfect}}$ the formula simplifies to the previous $PFD_{avg} = \frac{\lambda_{DU} \cdot T_i}{2}$

$T_{i\ prefect}$ is when the component is replaced.

3.7.3.2 Partial Proof Test (PPT)

Let's consider the example from Section 3.2.1 and shown in Figure 3.34.

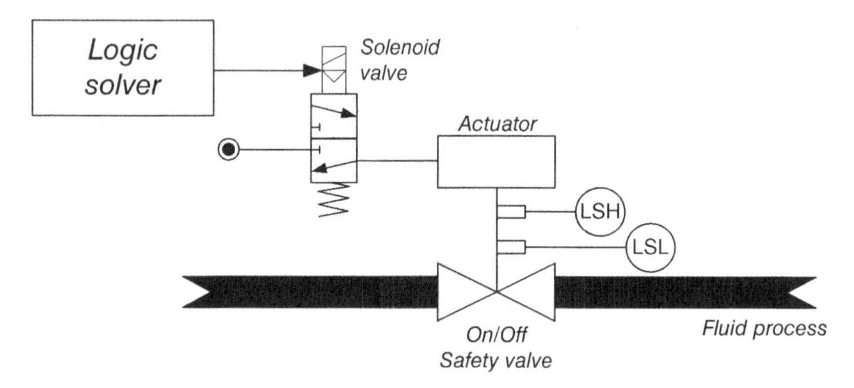

Figure 3.34 ON/OFF safety valve and its output subsystem.

The output subsystem is subject to an automatic testing called **Partial Valve Stroke Testing**, with a higher frequency than the proof test: we call it **Partial Proof Test**. The difference with an imperfect proof test is that the former is automatic while the latter is offline. The formula to be used to calculate the PFD_{avg} is however similar to Equation 3.7.3.1.

In this case, we can define two types of dangerous undetected failures:

λ_{DU_d} = dangerous undetected Failures that **can be detected** during a Partial Proof Test.
λ_{DU_u}= dangerous undetected Failures that **cannot be detected** during a Partial Proof Test.

We can also define the **Partial Proof Test Coverage** as:

$$PPTC = \frac{\lambda_{DU_d}}{\lambda_{DU}}$$

Please notice that the λ_{DU} used in this case is normally different from the λ_{DU} used in case of an imperfect proof test PFD_{avg} calculation (Example 3.2.1.1).

In presence of a Partial Valve Stroke Test (and a Perfect Proof Test), PFD_{avg} can be calculated with the following formula:

$$PFD_{avg} = PPTC \cdot \frac{\lambda_{DU} \cdot T_{i_{partial}}}{2} + (1 - PPTC) \cdot \frac{\lambda_{DU} \cdot T_{i_{perfect}}}{2} \qquad \text{[Equation 3.7.3.2]}$$

The PPTC has the same "nature" of a Diagnostic Coverage in the sense that it is a done automatically.

Still referring to the previous example, despite the process valve is monitored, in case no partial stroke test is done (we are in low demand mode!), the diagnostic coverage of the system is = 0 and the $SFF = 28\%$. Therefore, the architecture constraint is limited to SIL 1. With a partial stroke test, not only the PFD_{avg} would improve, but also the subsystem could reach SIL 2, in the given example.

3.7.3.3 Example for a Partial Valve Stroke Test

Let's consider a process valve, required to close, to achieve the safety function: the failure modes are as shown in Table 3.25 [64]:

Table 3.25 The effect of proof testing on PFD_{avg}.

Failure mode of output subsystem	Failure rates (per year)	Detected by partial stroke testing?
Solenoid fails to vent	0.005	Yes
Valve sticks open	0.004	Yes
Valve doesn't fully close	0.001	No
Other unknown failures	0.006	No
Total	**0.016**	

In this case, PPTC (Partial Proof Test Coverage) is the ratio between the failure rate of undetected dangerous failures "detected" by the partial stroke test λ_{DU_d} and the failure rate of all undetected failures λ_{DU}

$$PPTC = \frac{\lambda_{DU_d}}{\lambda_{DU}} = \frac{0.005 + 0.004}{0.016} = 0.5625$$

If we assume that the plant is only shut down every four years, without partial stroke testing the component is tested every four years with a Perfect Proof Test. The PFD_{avg} for this valve would be:

$$PFD_{avg} = \frac{\lambda_{DU} \cdot T_i}{2} = \frac{0.016 \cdot 4}{2} = 0.032 \quad [A]$$

Please notice that, in case the frequency of the Proof Test is different for each component in the safety function, the PFD_{avg} can be calculated with the formula:

$$PFD_{avg} = \sum_j \frac{\lambda_{D_j} \cdot T_{i_j}}{2}$$

The formula when a Partial Stroke Test is implemented, in addition to the Perfect Proof Test, is:

$$PFD_{avg} = PPTC \cdot \frac{\lambda_{DU} \cdot T_{i_{partial}}}{2} + (1 - PPTC) \cdot \frac{\lambda_{DU} \cdot T_{i_{perfect}}}{2} = [B] + [C] = [D]$$

We assume the Proof Test to be Perfect ($PTC = 1$). Please also note that if $T_{i_{partial}}$ is equal to $T_{i_{perfect}}$ the formula is reduced to the previous $PFD_{avg} = \frac{\lambda_{DU} \cdot T_i}{2}$

If we do a Partial Valve Stroke Test every year and a Perfect Full Test every four years, than the value becomes:

$$PFD_{avg} = 0.5625 \cdot \frac{0.016 \cdot 1}{2} + (1 - 0.5625) \cdot \frac{0.016 \cdot 4}{2} = 0.0045 + 0.014$$
$$= 0.0185 = [B] + [C] = [D]$$

In this example, the valve **PFD$_{avg}$** is about 50% smaller, compared with the case without Partial Valve Stroke Test.

The same value of PFD$_{avg}$ can be achieved with a Perfect Full Test performed every 2.3 years

$$T_i = \frac{PFD_{avg} \cdot 2}{\lambda_{DU}} = \frac{0.0185 \cdot 2}{0.016} = 2.3\,\text{years}$$

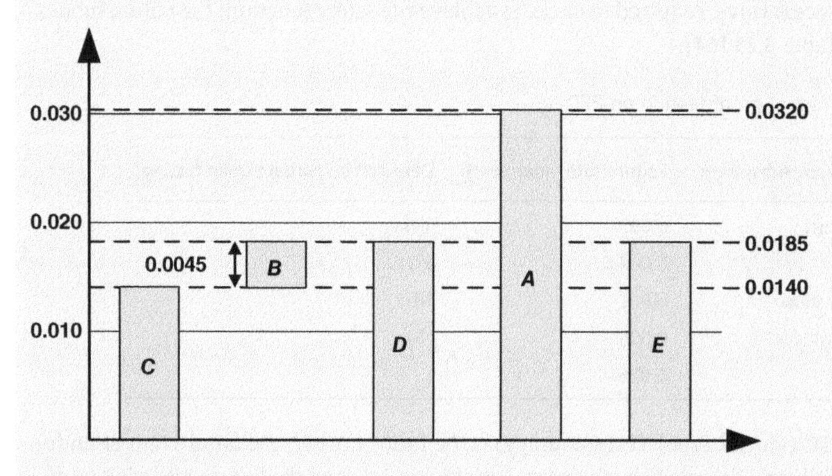

Figure 3.35 The effect of proof testing on PFD$_{avg}$.

In Figure 3.35:

- ***A*** represents the Valve PFD$_{avg}$ with a perfect proof test every four years.
- ***B* + *C* = *D*** represents the Valve PFD$_{avg}$ with a perfect proof test every four years and a Partial Valve Stroke Test every two years.
- ***E*** represents the Valve PFD$_{avg}$ with a perfect proof test every two years.

3.7.4 Proof Test in High Demand Mode

During the revision of IEC 62061, several discussions took place regarding the effectiveness of a Proof Test in high demand mode; especially for electromechanical components.

The first aspect to consider is that for Electromechanical components, manufacturers do not normally provide indications how a Proof Test on their components should be done. Moreover, a Proof Test on those components can only be partial due to the presence of **degraded failures**. In high demand mode applications for electromechanical components, it is recommended not to use the concept of Proof Test. That is, for example, clearly stated for Basic Subsystem Architecture A.

> ***[IEC 62061] 7.5.2.1 Basic Subsystem Architecture A: single channel without a diagnostic function***
> *[...] In high or continuous mode of operation, Architecture A shall not rely on a Proof Test.*

For the other architectures, the Proof Test interval is inside a parameter called T_1, that is the smaller between

- **Proof Test interval** of the perfect Proof Test **and**
- **The useful lifetime**, which is smaller between the component **Mission Time** and the parameter $\boldsymbol{T_{10D}}$.

In high demand mode, for electronic components, normally the PFH_D is provided by the manufacturer; therefore, the determination of the proof test interval is normally not required for the calculation of the reliability of the safety system.

3.8 Mission Time and Useful Lifetime

The Mission Time represents **the period within which the component failure rate is considered constant,** and it is a value defined by the component manufacturer; here is its definition.

> ***[ISO 13849-1] 3.1 Terms and definitions***
> ***3.1.36 Mission Time*** *T_M. Period of time covering the intended use of an SRP/CS.*

Both IEC 62061 and ISO 13849-1 assume a component constant failure rate period of 20 years, but its manufacturer may decide a longer time. However, based upon the usage frequency in the specific machinery, the period within which the failure rate is considered constant can be less than the 20 years indicated by the component manufacturer: for components with wear-out characteristics, that period is limited by the T_{10D}.

That is the reason why, in the new edition of IEC 62061 [12], it was decided to use the term **Useful Lifetime**; hereafter, its definition:

> ***[IEC 62061] 3.2 Terms and definitions***
> ***3.2.42 Useful Lifetime.*** *Minimum elapsed time between the installation of the SCS or subsystem or subsystem element and the point in time when component failure rates of the SCS or subsystem or subsystem element can no longer be predicted, with any accuracy.*

The Useful Lifetime can be defined as the minimum value between the Mission Time, indicated by the component manufacturer, and the T_{10D} calculated by the machinery manufacturer.

$$Useful\ \ Lifetime = \min\left(Mission\ Time; T_{10D}\right)$$

Please bear in mind that **the Mission Time $\boldsymbol{T_M}$ is a characteristic of the component not of the Safety Control System,** and it can only be specified by the component manufacturer. That is different from the **Proof Test** interval $\boldsymbol{T_i}$ or the $\boldsymbol{T_{10D}}$, which are determined by the machinery manufacturer.

Each component within the SCS can have a different Mission Time. Normally, at the end of the Mission Time, the component has to be replaced if the Safety Control System needs to continue its operations. In the simplified approach of ISO 13849-1, the Mission Time is assumed to be 20 years [ISO 13849-1; § 6.1.8]. IEC 62061 does not give an indication, since it focuses on the Useful Lifetime.

In high demand mode, if the useful lifetime of the subsystem is larger than or equal to the useful lifetime of the SCS, a Proof Test is not necessary, and in PFH_D estimations, T_1 is equal to the useful lifetime. Where it is smaller, the subsystem or subsystem element should be replaced during the useful lifetime of the SCS if a Proof Test is not possible. In IEC 62061, the parameter T_1 is defined as "*the proof test interval of the perfect proof test or useful lifetime whichever is the smaller.*"

The concept is also present in ISO 13849-1 and it is called "**operating life time**".

3.8.1 Mission Time Longer than 20 Years

Should the Mission Time of an SRP/CS exceed 20 years, the PFH_D values determined by means of the simplified method (annex K of the standard) are generally no longer valid. Under certain circumstances, this situation can, however, be addressed within the simplified approach, with a few improvements. The influence of the longer Mission Time can then be estimated, on the safe side from the Markov models upon which annex K of ISO 13849-1 is based, as follows: **for every five years' extension** of the Mission Time beyond 20 years, **a further 15% is added to the PFH_D for Categories 2, 3, and 4** (Categories B and 1 require no adjustment of the PFH_D).

However, the extension of the Mission Time is possible only when the manufacturer's information is available on the measures to be taken when the Mission Time is extended and only conditional upon these measures being implemented by the user.

4

Introduction to ISO 13849-1 and IEC 62061

4.1 Risk Assessment and Risk Reduction

Once a manufacturer told me: "*I have all my Safety Systems in PL e, so I do not need to do any risk assessment.*" That is not the right approach to Machinery safety!

As stated in ISO 12100, Machinery safety starts with a **Risk Assessment** (§9.3), followed by an often necessary **Risk Reduction**, as shown in Figure 4.1.

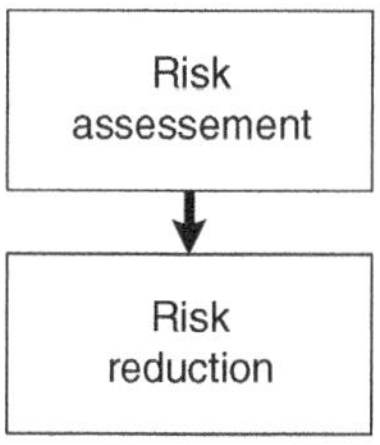

Figure 4.1 The risk analysis as required by ISO 12100.

Risk reduction can be achieved in three steps:

1. **Inherently safe design** measures. In other terms, the first step is to try to eliminate the risk.
2. **Safeguarding** and, if needed, **Complementary Protective Measures**, defined as **Engineering Controls** in B11.19 [46]. The risks that cannot be eliminated should therefore be reduced using safeguarding, like, for example, a physical guard or an Active Opto-Electronic Protective Device (AOPD).
3. The risks that cannot be eliminated, or reduced with Safeguarding, **have to be managed** by informing the operator: Instruction manuals, Awareness Means on the machine, Procedures, and Training can all be used for this purpose. This third step is defined as **Information for use** in ISO 12100 language or **Administrative Controls** in North American B11.19 or CSA Z432 language.

Often, the second step, Safeguarding, requires the implementation of a safety function. The parts of the machinery control system that provide safety functions are called **Safety-related Parts of Control Systems** (SRP/CS) by ISO 13849-1 and **Safety-related Control Systems** (SCS) by IEC 62061. These can consist of hardware and/or software and can either be separated from the machine control system or an integral part of it. In other words, there can be an automation (or general purpose) PLC and a separate safety module, or both can be integrated in the same control system that has a shared safety CPU module and, usually, dedicated I/O cards for safety.

4.1.1 Cybersecurity

One recent subject, that regularly comes up during discussions inside Technical Committees, is about the impact of security on machinery safety. Hereafter one of the definitions:

> ***ISO TR 22100-4 [44] 3 Terms and definitions***
> ***3.10 IT-security****; information security; cyber security. Protection of an IT system from the attack or damage to its hardware, software or information, as well as from disruption or misdirection of the services it provides.*

Functional Safety of Machinery: How to Apply ISO 13849-1 and IEC 62061, First Edition. Marco Tacchini.

Cybersecurity includes all the activities necessary to protect the network and information systems of the machine control system, the users of such systems, and other persons affected by cyber threats, typically regarding the aspects of confidentiality, integrity, and availability.

Cyber threat means any potential circumstance, event or action that could damage, disrupt, or otherwise adversely impact network and information systems, the users of such systems, and other persons, typically exploiting vulnerabilities of a system.

The starting point is that **the manufacturer has to make sure the machine is safe when used according to what he has foreseen as correct behavior**. Moreover, incorrect behaviors, due to a reasonably predictable reason, have to be safe and therefore to be considered in a risk assessment.

In case the machine can be connected and somehow operated remotely, it is important to adopt precautions that prevent affecting the safety system and therefore creating a dangerous situation. Several standards deal with the subject: IEC 62443 series, ISO/TR 22100-4 [40], and IEC/TR 63074.

Moreover, the New Machinery Regulation, whose text will be finalised in 2023, contains a new E.H.S.R. that stresses the importance of avoiding dangerous situations due to cyberattacks:

[New Machinery Regulation] Annex III: 1.1.9. Protection against corruption

The machinery or related product shall be designed and constructed so that the connection to it of another device, via any feature of the connected device itself or via any remote device that communicates with the machinery or related product ***does not lead to a hazardous situation****.*

A hardware component transmitting signal or data, relevant for connection or access to software that is critical for the compliance of the machinery or related product with the relevant health and safety requirements shall be designed so that it is adequately protected against accidental or intentional corruption. The machinery or related product shall collect evidence of a legitimate or illegitimate intervention in the aforementioned hardware component, when relevant for connection or access to software that is critical for the compliance of the machinery or related product.

Software and data that are critical for the compliance of the machinery or related product with the relevant health and safety requirements shall be identified as such and shall be adequately protected against accidental or intentional corruption.

The machinery or related product shall identify the software installed on it that is necessary for it to operate safely, and shall be able to provide that information at all times in an easily accessible form.

The machinery or related product shall collect evidence of a legitimate or illegitimate intervention in the software or a modification of the software installed on the machinery or related product or its configuration.

4.1.2 Protective and Preventive Measures

In general, Risk Reduction happens through both **Preventive measures**, which reduce the frequency, and mitigating or **Protective measures**, which reduce the severity, as shown in Figure 4.2.

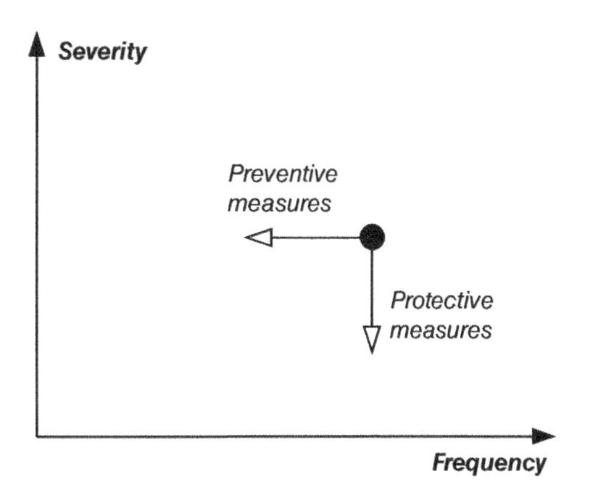

Figure 4.2 The risk analysis as required by ISO 12100.

The operation of **cutting trees in a forest** is an example of a Preventive measure since it prevents fires to spread. On the other hand, **a fire detector** is an example of a Protective measure since it cannot reduce the frequency at which a fire occurs, but it can reduce the severity of consequences by initiating a sprinkler system.

Preventive measures reduce the likelihood of a dangerous event, while **protective** measures reduce the severity of the damage.

The use of glycolic water in Forging Presses is another example of a preventive measure since it reduces the likelihood of fires.

The creation of a safeguarded space with an interlocked door can be seen as a protective measure; however, it only reduces the likelihood of the accident to happen and not the severity of the damage. For that reason, it is an example of a preventive and not of a protective measure. However, just as a reference, hereafter the definition of protective measures:

> ***ISO 12100 [41] 3 Terms and definitions***
>
> ***3.19 Protective Measures.*** *Measures intended to achieve risk reduction, implemented:*
>
> - *by the designer (inherently safe design, safeguarding and complementary protective measures, information for use) and/or*
> - *by the user (organization: safe working procedures, supervision, permit-to-work systems; provision and use of additional safeguards; use of personal protective equipment; training)*

The issue is that ISO 12100 has never made a distinction between Protective and Preventive Measures and that generated confusion during the risk reduction process, whereby a safeguarded space reduced the severity of the damage, which is not correct. Experts in ISO/TC 199 technical committees are now using the term Risk Reduction Measures instead of Protective Measures in order to eliminate possible ambiguities. That will be the term used in the new edition of ISO 12100 as well as in this book. **Risk Reduction Measures are means to eliminate hazards or reduce risks**. Therefore, glycolic water in Forging Presses is a type of **Risk Reduction by Inherently Safe Design**, while a safeguarded space with an interlocked door is a type of **Risk Reduction by Safeguarding**.

Let's consider a robot inside a safeguarded space; **the damage** in case it starts unexpectedly does not change, compared to a situation whereby the robot operates in an open area. However, since the SCS that keeps the robot in a "Safety-Related Stop" when a gate is open, has a low probability of failure, the likelihood that a person gets injured is reduced, even if it is not impossible.

Therefore, **the use of an SCS reduces the probability of the event, but not its Severity**. That is valid for most of the Risk Reduction Measures applied in Machinery Safety.

4.1.3 Functional Safety as Part of the Risk Reduction Measures

Machinery safety starts with **an open and sincere Risk Analysis**. That means:
- Determination of the limits of the machinery
- Hazard Identification
- Risk Estimation

At the end of the Analysis, a Risk Evaluation has to be done, to decide whether the risk is sufficiently low. In case it is not, a **Risk Reduction process** has to take place: that goes through three steps. Figure 4.3 shows all the steps of a Risk Assessment and Risk Reduction according to ISO 12100: **Functional Safety plays a role mainly in STEP 2** of the risk reduction activity.

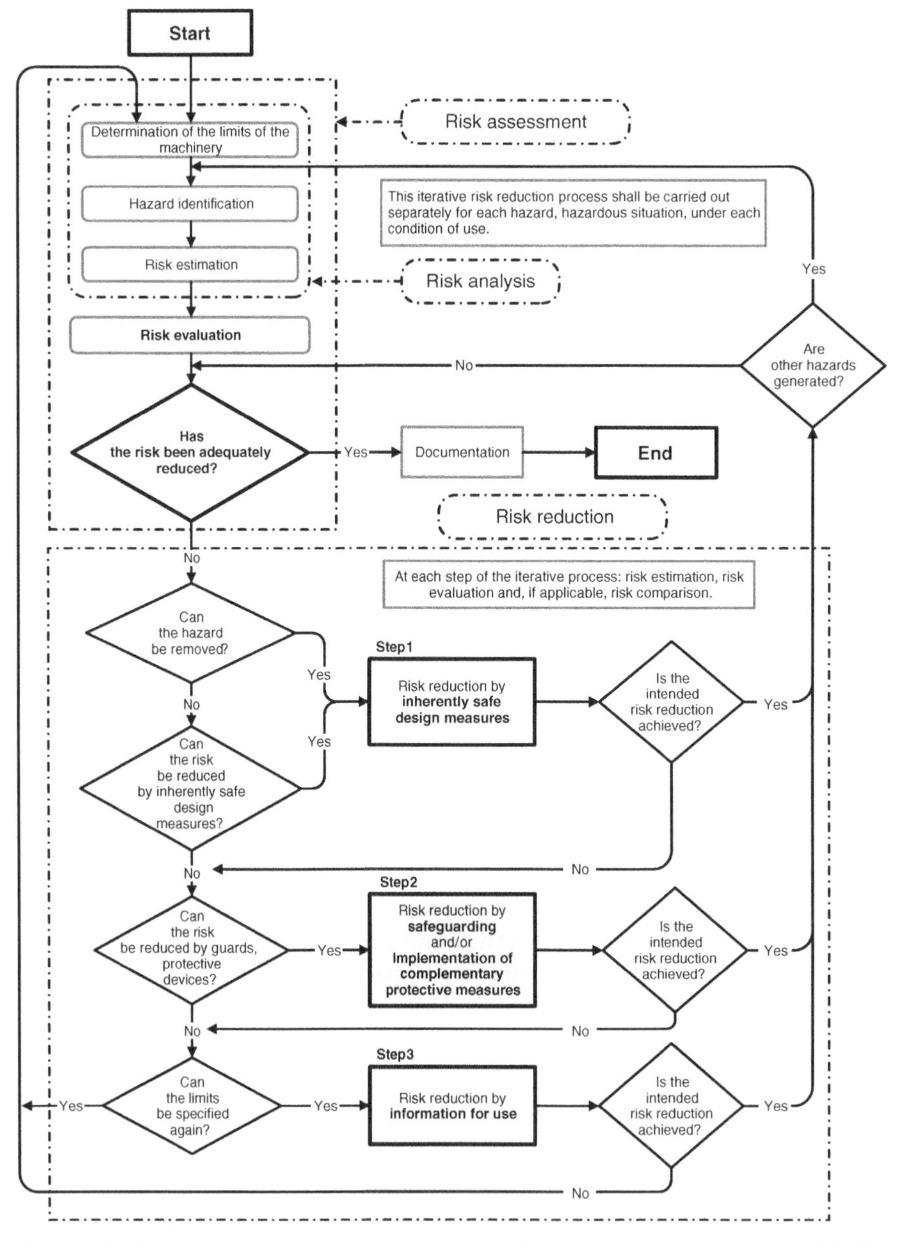

Figure 4.3 Representation of the risk assessment and reduction process, as described in ISO 12100.

As an example, **10 risks may have been identified** in a machinery application. We can start by eliminating three of them; for example, one of them may be the use of glycol water instead of oil in a Forging Press. Then, we may be able to reduce two more risks by installing fixed guards and four more by installing three movable guards and one AOPD (Active Opto-Electronic Protective Device). **For these last four Risk Reduction measures, we need to use the prescriptions stated in either ISO 13849-1 or IEC 62061**. In other terms, every time I need to reduce a risk by using an SCS, I need to follow the prescriptions of one of the two Functional Safety Standards previously mentioned, supposing the safety loops operate in high-demand mode.

A similar approach is valid in the United States. Table 4.1 shows the key steps in Risk Reduction according to B11.0 [45]. The most preferred are, of course, Elimination or Substitution.

Please notice that Administrative Controls are listed from the most to the least preferred.

Table 4.1 The hazard control hierarchy according to B11.0.

Classification	Risk reduction measures	Examples
Inherently safe by design	Design out (elimination or substitution)	• Eliminate pinch points (increase clearance) • Intrinsically safe (energy containment) • Automated material handling (robots, conveyors, etc.) • Redesign the process to eliminate or reduce human interaction • Reduce force, speed, etc. through the selection of inherently safe components • Substitute less hazardous chemicals
Engineering controls	Guards, control functions, and devices	• Guards • interlocking devices • Presence sensing devices (light curtains, safety mats, area scanners, etc.) • Two-hand control and two-hand trip devices
Administrative controls	Awareness means	• Lights, beacons, and strobes • Computer warnings • Signs and labels • Beepers, horns, and sirens
	Information for use	• Safe work procedures • Training
	Administrative safeguarding methods	• Safe-holding safeguarding method
	Supervision	• Supervisory control of configurable elements
	Control of hazardous energy	• Lockout, tagout, and alternative methods
	Tools	• Workholding equipment • Hand tools
	Personal protective equipment (PPE)	• Safety glasses and face shields • Ear plugs • Gloves • Protective footwear • Respirators

4.1.4 The Naked Machinery

When doing a risk assessment of Machinery, regardless of whether ISO 12100 or B11.0 is used, it is good to keep in mind the concept of the **Naked Machinery**.

If you are assessing a machine that is already in operation, you need to imagine it without any safeguard. The machine may have a fixed guard to protect from a dangerous mechanical movement: you need to imagine the machine without that guard and, for example, ask yourself the question: "How often do I need to open that guard?". If it is once a day, then that guard is not correct since an interlocked guard needs to be installed.

Without this approach, you would not have assessed the risk of that specific mechanical movement. That is clearly stated in an informative note of B11.0:

> ***[B11.0:2020] 6.3 Identify tasks and hazards***
> *The reasonably foreseeable tasks and associated hazards shall be identified for the applicable phases of the lifecycle of the machine.*
>
> ***Informative Note 3:*** *The risk assessment process includes identifying hazards regardless of the existence of risk reduction measures. The machine should not be considered harmless as shipped and guarded. To verify that all hazards are included, hazard identification should be conducted with all risk reduction measures (including engineering controls and specialized training) conceptually removed. This is to confirm that hazards are not ignored due to an assumption that the risk reduction measures supplied are adequate for all tasks, including reasonably foreseeable misuse. Existing risk reduction measures that help achieve acceptable risk can be retained after evaluating their performance. This decision will be confirmed during the validation/verification portion of the risk assessment (see § 6.8). If a thorough risk assessment is delivered with the machine, it may be used as a starting point for the user's risk assessment.*

4.2 SRP/CS, SCS, and the Safety Functions

4.2.1 SRP/CS and SCS

The parts of the machinery control system that provide safety functions are defined by ISO 13849-1 as SRP/CS, while IEC 62061 defines them as SCS. These can consist of hardware and/or software and can either be separated from the machine control system or be an integral part of it.

The previous edition of IEC 62061 called them Safety-related Electrical Control System (SRECS). Since the new edition is applicable to non-Electrical safety systems as well, the definition was changed to SCS. Hereafter the definitions from the two standards:

> ***[ISO 13849-1] 3.1 Terms and definitions***
> ***3.1.1 Safety-related Part of a Control System SRP/CS.*** *Part of a control system that performs a safety function, starting from safety-related input(s) to generating safety-related output(s).*
>
> ***Note 1 to entry:*** *The safety-related parts of a control system start at the point where the safety-related inputs are initiated (including, for example, the actuating cam and the roller of the position switch) and end at the output of the power control elements (including, for example, the main contacts of a contactor).*

> ***[IEC 62061] 3.2 Terms and definitions***
> ***3.2.3 Safety-related Control System SCS.*** *Part of the control system of a machine which implements a safety function.*

Note 1 to entry: *An SCS is the combination of one or more subsystems necessary to implement the respective safety sub-function(s).*

Note 2 to entry: *SCS is similar to SRECS of the previous edition of this document.*

4.2.2 The Safety Function and Its Subsystems

Both standards decompose a Safety-related part of the control System or SRP/CS into **Subsystems**, usually made of (Figure 4.4):

- Input (Sensor)
- Logic Solver
- Output (Final Element)

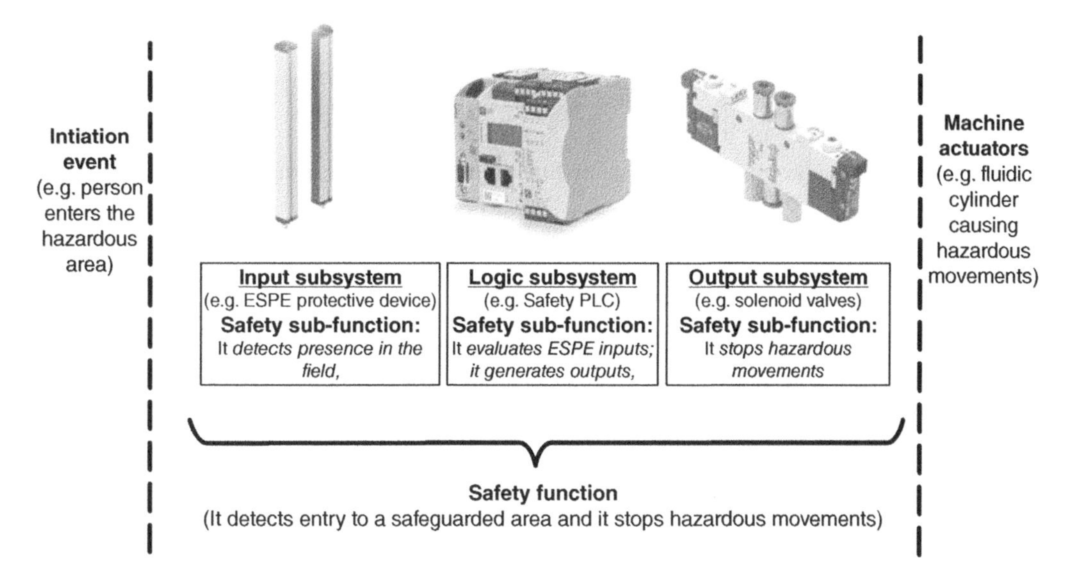

Figure 4.4 Example of an SRP/CS or SCS, divided into subsystems.

[IEC 62061] 3.2 Terms and definitions

3.2.4 Subsystem. *Entity of the top-level architectural design of a safety-related system where a dangerous failure of the subsystem results in dangerous failure of a safety function.*

4.2.3 The Physical and the Functional Level

Both an SRP/CS and an SCS are the "**physical**" aspect of a Safety Function: a clear example is an AOPD that detects the entering of a person in a safeguarded area and stops a dangerous movement by de-energizing a motor contactor. From a "**functional**" point of view, that is called, in both standards, a **Safety Function**. Here is the definition according to ISO 12100, ISO 13849-1, and IEC 62061:

[ISO 12100] 3.2 Terms and definitions

3.30 Safety Function: *function of a machine whose failure can result in an immediate increase of the risk(s).*

[ISO 13849-1] 3.1 Terms and definitions

3.1.27 Safety Function: *function of the machine whose failure can result in an immediate increase of the risk(s).*

***Note 1 to entry**: A safety function is a function to be implemented by a safety-related part of a control system, which is needed to achieve or maintain a safe state for the machine, in respect of a specific hazardous event.*

[IEC 62061] 3.2 Terms and definitions
***3.2.18 Safety Function**: function implemented by an SCS with a specified integrity level that is intended to maintain the safe condition of the machine or prevent an immediate increase of the risk(s) in respect of a specific hazardous event.*

***Note 1 to entry:** This term is used instead of "safety-related control function (SRCF)" of IEC 62061:2015. This definition differs from ISO 12100 because this document addresses risk reduction performed by SCS.*

A Safety Function typically starts with a detection and evaluation of an **initiation event** and ends with an output causing the reaction of a **machine actuator**. Therefore a Safety Function is not only the logic implemented in a microprocessor-based component, **but it includes sensors and actuators**.

Therefore, there are two levels:

- **A Logical or Functional one**: the Safety Function
- **A Physical one**: the SCS

The equivalent of a subsystem for the safety function is a **Sub-function**.

[ISO 13849-1] 3.1 Terms and definitions
***3.1.28 Sub-function.** Part of a safety function whose failure results in a failure of the safety function.*

The approach of IEC 62061 in a graphical representation is shown in Figure 4.5.

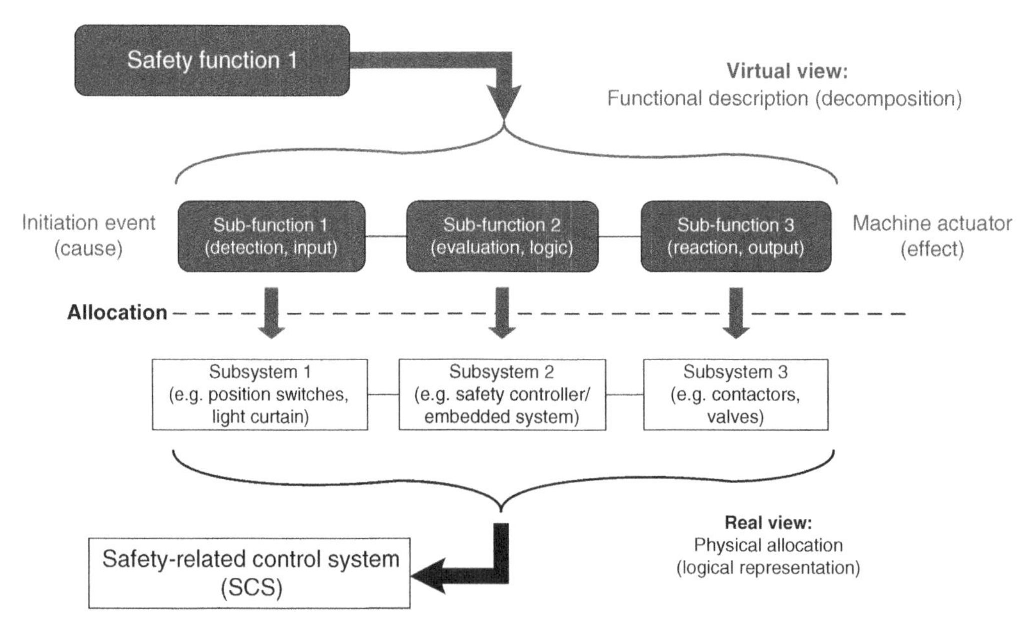

Figure 4.5 Decomposition of a safety function into sub-functions.

4.3 Examples of Safety Functions

Hereafter are examples of Safety functions and Sub-functions.

4.3.1 Safety-Related Stop

This is probably the most common Safety Sub-Function. A Rolling Mill Stand is a safeguarded space. When access is permitted (with the use of a trapped key, for example) all movements are stopped, thanks to the activation of an SCS: **the machine is then placed in a safe state.** That can be stated in a different way: the machine is placed in a **safety-related stop**.

A Safety-related stop can be activated by a Safeguard (an interlocking device or an Active Opto-electronic Protective Device) and can be "translated" in the de-energization of a contactor or in the activation of the Safe Torque off (STO) of a Variable Speed Drive.

As a result of the risk assessment, safe stopping sub-functions can be realized according to the stop categories in IEC 60204-1, § 9.2.2, and/or according to other similar Safety Functions as described in IEC 61800-5-2, § 4.2.

After a stop command is initiated, the stop condition shall be maintained until safe conditions for restarting are established.

4.3.2 Safety Sub-Functions Related to Power Drive Systems

IEC 61800-5-2 defines the characteristics of adjustable speed electrical **Power Drive System**, suitable for use **in Safety-Related applications**, using the acronym **PDS(SR)**.

Safety sub-functions associated with PDS(SR) are divided in two groups:

- Stopping Functions.
- Monitoring Functions.

4.3.2.1 Stopping Functions

Safe torque off (STO): *Power, that can cause rotation (or motion in the case of a linear motor), is not applied to the motor.* This safety sub-function, illustrated in Figure 4.6, corresponds to an uncontrolled stop in accordance with stop Category 0 of IEC 60204-1. When external influences (for example, falling of suspended loads) are present, additional measures (for example, mechanical brakes) may be necessary to prevent any hazard. Please bear in mind that an STO function provides de-energization but not electrical isolation.

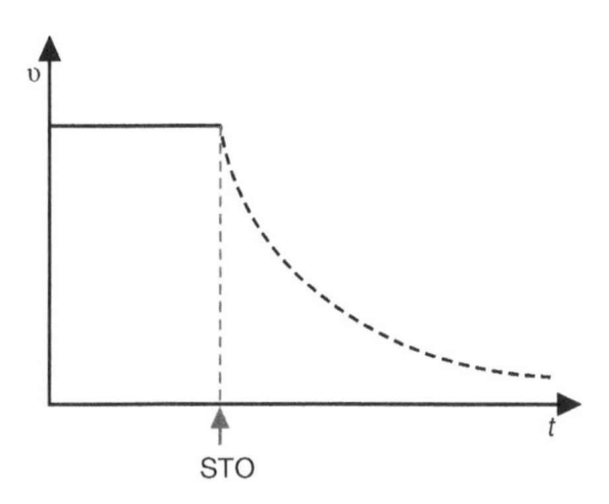

Figure 4.6 Speed-time graphical representation of the STO function.

BASIC CONCEPT: there is a fundamental difference between de-energisation and Isolation.

> ***[IEC 61140] 3 Terms and definitions***
>
> ***3.41 Isolation:*** *function intended to disconnect and maintain for reasons of safety adequate clearance from every source of electric energy.*

Disconnect switches, Circuit Breakers and Fuses, de-energise and provide isolation functions; PDS and Contactors only de-energise the load but do not provide a safe situation in case an electrician needs to work on live parts.

Safe stop 1 (SS1). This sub-function is specified as:

a) **Safe Stop 1 deceleration controlled (SS1-d):** *it initiates and* ***controls*** *the motor deceleration rate within selected limits to stop the motor and performs the STO function when the motor speed is below a specified limit*; or
b) **Safe Stop 1 ramp monitored (SS1-r):** *it initiates and* ***monitors*** *the motor deceleration rate within selected limits to stop the motor and performs the STO function when the motor speed is below a specified limit*; or
c) **Safe Stop 1 time controlled (SS1-t):** *it initiates the motor deceleration and performs the STO function after an application-specific time delay.* No speed monitoring is present; the manufacturer should then consider the risk that, after the elapsed time, the motor is still turning.

This safety sub-function, illustrated in Figure 4.7, corresponds to a controlled stop in accordance with **stop Category 1** of IEC 60204-1. SS1 should be used when there is the need to stop the motor as quickly as possible. Following the stop, unexpected start-up is prevented, since the STO is activated.

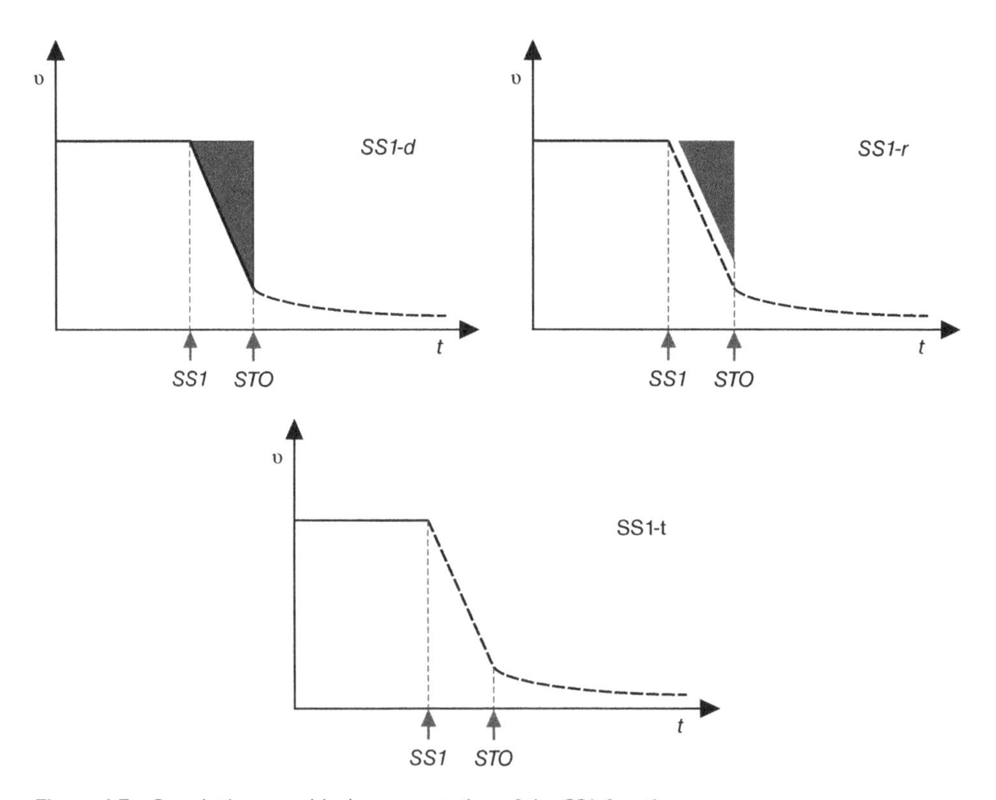

Figure 4.7 Speed–time graphical representation of the SS1 functions.

Safe stop 2 (SS2). This sub-function is specified as either

a) **Safe Stop 2 deceleration controlled** (SS2-d). *It initiates and controls the motor deceleration rate within selected limits to stop the motor and performs the* ***safe operating stop*** *function when the motor speed is below a specified limit; or*

b) **Safe Stop 2 ramp monitored (SS2-r).** *It initiates and monitors the motor deceleration rate within selected limits to stop the motor and performs the* ***safe operating stop*** *function when the motor speed is below a specified limit; or*

c) **Safe Stop 2 time controlled (SS2-t).** *It initiates the motor deceleration and performs the* ***safe operating stop*** *function after an application-specific time delay.*

This safety sub-function, illustrated in Figure 4.8, corresponds to a controlled stop in accordance with stop Category 2 of IEC 60204-1.

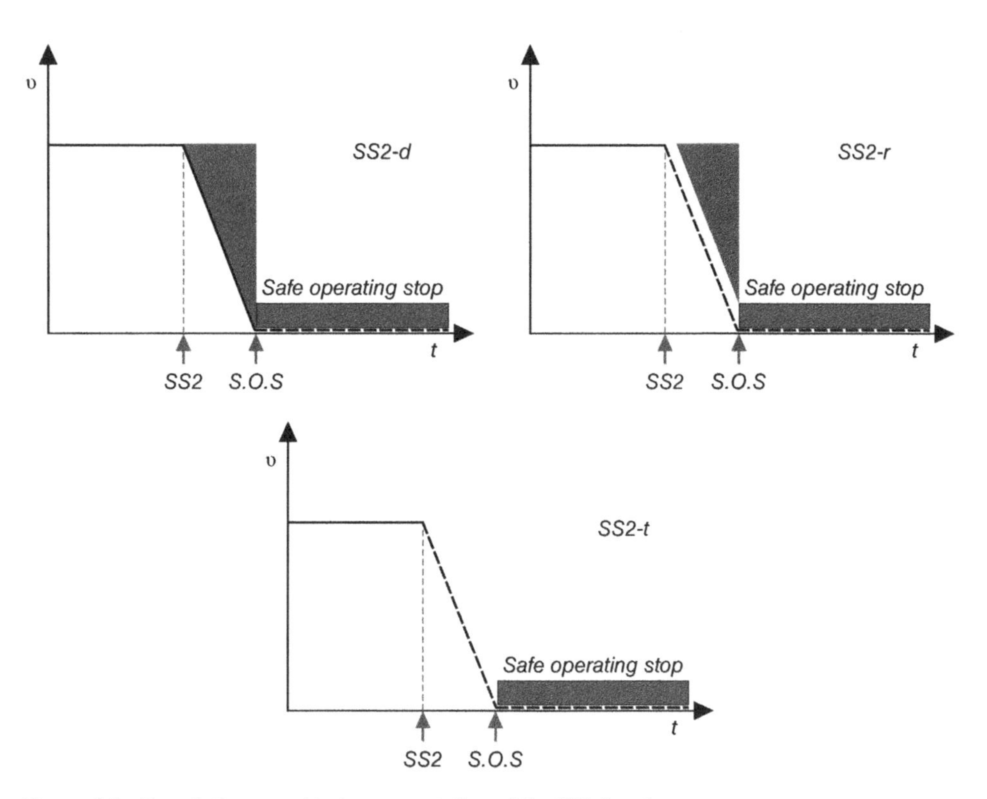

Figure 4.8 Speed–time graphical representation of the SS2 functions.

4.3.2.2 Monitoring Functions

Safe operating stop (SOS). *This function prevents the motor from deviating more than a defined amount from the stopped position. The PDS(SR) provides energy to the motor to enable it to resist external forces.*

In most applications, the motor does not turn, but it is stationary "in torque." A graphical representation is shown in Figure 4.9. A safe operating stop sub-function needs to be monitored and, in case of its failure, a Safe torque off has to be initiated in SIL 2 or PL d. Application examples are the set-up activities on Turning machines (lathes) or the measurements during machining, whereby the spindle cannot be de-energized, since the machine would lose its references.

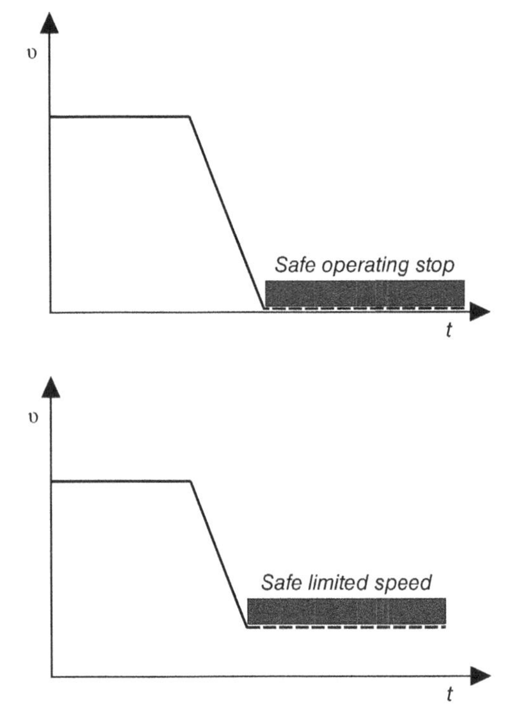

Figure 4.9 Speed–time graphical representation of the SOS and SLS functions.

Safely-limited speed (SLS). *This function prevents the motor from exceeding the specified speed limit.*

Safely-limited torque (SLT). *This function prevents the motor from exceeding the specified torque (or force, when a linear motor is used) limit.*

Safely-limited position (SLP). *This function prevents the motor shaft (or mover, when a linear motor is used) from exceeding the specified position limit(s).*

Safe speed monitor (SSM). *This function provides a safe output signal to indicate whether the motor speed is below a specified limit.*

Safe brake control (SBC) (Output function). *This function provides a safe output signal(s) to control an external brake(s).*

4.3.2.3 Information to be Provided by the PDS Manufacturer

According to IEC 61800-5-2 [22] § 7.2, the following information have to be indicated by the manufacturer in the ***Safety Manual*** (for the full list, please refer to the standard):

- *a detailed **description of the safety sub-function** (including the reaction(s) to a violation of limits);*
- ***the response time** of each safety-related function and of the associated fault reaction functions*
- ***the SIL** or SIL capability;*
- ***the PFH value for each safety sub-function;***
- ***PL and category according to ISO 13849-1**, when applicable.*

4.3.3 Manual Reset

The manual reset can be used to re-establish a safety function. In case of a safeguard, like an AOPD for example, the Manual Reset cancels the safety-related stop command.

BASIC CONCEPT – SAFEGUARDED SPACE: it is a relatively new terminology in Machinery Safety and it is a replacement for the term "hazard zone" used in ISO 12100 "*any space within or around the machinery, in which a person can be exposed to a hazard.*"

Hereafter the definition from ISO 11161 [23]:

> ***[IEC 11161] 3 Terms and definitions***
>
> ***3.1.7 Safeguarded Space:*** *area or volume enclosing (a) hazard zone(s) where guards and/or protective devices are intended to protect persons.*

BASIC CONCEPT – WHOLE BODY ACCESS. Whole Body access is also a relatively new terminology in Machinery Safety and it indicates a situation where a person can be completely inside a safeguarded space. If whole body access exists such that an operator can be completely inside the safeguarded space and the SCS can be restored or re-enabled, additional protective/risk reduction measures shall be used.

In case there is an issue of whole body access, manual reset can reduce the associated risk of unexpected start-up. **In this case, the Manual Reset is a Safety Function,** and it should have a PL_r or SIL_r; however, it is not always necessary to be the same PL_r or SIL_r of the associated safety function. Normally the level is lower. Hereafter the definition from ISO 13849-1:

> ***[ISO 13849-1] 3 Terms and definitions***
> ***3.1.16 Manual Reset.*** *Safety Function within the SRP/CS used to restore manually one or more safety functions before re-starting a machine.*

and a key language in the standard:

> ***[ISO 13849-1] 5.2.2.3 Manual reset function***
> *[...] When the function "manual reset" is required to be a safety function (e.g., prevention of unexpected start), the required performance level shall be determined. The PL of the manual reset function can be different from the PL_r of the associated safety function.*

In case, for example, there is no issue of whole body access, an automatic reset can be implemented since there is no need of a Manual Reset. That means, for example, that an emergency stop normally does not require any manual reset; it requires a disengagement, though, and that is the reason why **the actuator has to be of the self-latching type**.

> ***ISO 13850 [42] 4.1.4 Disengagement (e.g. unlatching) of the emergency stop device***
> *The effect of an activated emergency stop device shall be sustained until the actuator of the emergency stop device has been disengaged. This disengagement shall only be possible by an intentional human action on the device where the command has been initiated. The disengagement of the device shall not restart the machinery but only permit restarting.*

Here the reasoning that could be done in order to justify, normally, a lower Reliability level for the manual reset. **Let's consider the risk of unexpected start-up inside a safeguarded space having an access gate.** The maintenance person has its hands inside a dangerous area (a gear for example). If the safety function fails (Safety-related stop), a contactor closes and the movement starts: the person would get injured. The risk assessment may conclude the safety system should have PL_r d.

Let's now analyze the reset function. The maintenance person is working with his hands inside the gear and the door suddenly closes (that already is a condition with a certain probability, but let's suppose the door has been badly engineered and it closes automatically or an AOPD is used to define the safeguarded area). Now, the person is inside with its hands in the gear and the reset function has just failed. **The failure does not cause any movement**, since a start button has to be activated (a clear condition for the reset is that it does not initiate any movement). That means the Reliability level may be lower than its related safety function.

The manual reset function shall:

- Be provided through a **manually operated device that is separate from the start command**,
- Be achieved only if all affected safety functions and safeguards are operational,
- Not initiate any hazardous situation by itself,
- Enable the control system to accept a separate start command,
- Be accepted by monitored signal change, in order to avoid foreseeable misuse; please refer to § 4.3.3.2 on this aspect.

The reset actuator should be located **outside the hazardous area** and in a position from which there is a good visibility to ensure that no person is inside the safeguarded space. It should not be possible to activate the reset function from inside that area.

4.3.3.1 Multiple Sequential Reset

Where the visibility of the hazardous area is not complete, a special reset procedure is required. One solution is the use of a **Multiple sequential time-limited manual reset**. The reset function is initiated, for example inside the safeguarded space by the first actuator, in combination with a second reset actuator located outside the hazardous area (near the safeguard). This reset procedure needs to be activated within a limited time.

4.3.3.2 How to Implement the Reset Electrical Architecture

A few international standards highlight the issue that the reset function can be manipulated, for example by blocking the reset button in a pressed position. That risk has to be addressed by the machinery manufacturer. Several solutions are possible; one of them is the activation of the reset function on the electrical signal edge, normally the falling edge. Hereafter how the IEC 62046 [31] addresses the issue:

> ***[IEC 62046] 5.6 Restart interlock***
> *[...] Resetting a restart interlock of an ESPE application is always a safety-related function. Measures shall be provided to reduce the probability of the restart interlock being reset by a transient or steady-state fault condition. Such measures can include, for example, requiring both a rising and falling edge signal within a defined time (e.g. between 150 ms and 4 seconds) from a manually actuated reset device.*

4.3.4 Restart Function

In case the access to a safeguarded area is frequent and there is no risk of whole body access, the machine may restart as soon as the interlocked movable guard is closed again, or the AOPD is not engaged anymore: that is an example of a Restart Function. However, that can happen only if a hazardous situation cannot exist. The example is for an **Interlocked movable Guard with a start Function**; that may only be used, provided that all conditions stated in ISO 12100 § 6.3.3.2.5 are satisfied, among which, the cycle time of the machine is short.

4.3.5 Local Control Function

When a machine is controlled locally, e.g. by a portable control device or pendant, the following requirements apply:

- The mean for selecting the local controls shall be located outside the danger zone;
- It shall be possible to initiate a command by a local control, only in a zone defined by the risk assessment in order to avoid hazardous situations;
- Switching between local and another control station shall not create a hazardous situation;
- The control system shall be designed in such a way that the initiation of commands from different control stations do not lead to a hazardous situation. It is necessary to hinder the use of other controls when the local control station is active.

4.3.6 Muting Function

Muting is a temporary suspension of a safety function by the machine safety-related control system.

It can be used, for example, to allow the exit of material from a safeguarded area while keeping a safe access to a person. The muting function shall be initiated and terminated automatically. This shall be achieved by the use of appropriately selected and placed sensors or by signals from the machine control system.

The part of the control system that performs the muting function shall have an appropriate safety-related performance (SIL or PL). At the end of the muting, all affected safety functions of the SCS shall be reinstated.

The implementation of the muting functions for AOPD shall be in accordance with IEC 62046.

> ***[IEC 62046] 5.7 Muting***
> *[...] The part or parts of the control system that performs the muting function shall have an appropriate safety-related performance (SIL or PL, see IEC 62061 or ISO 13849-1) and shall not reduce the safety-related performance of the protective function below that required for the application.*

4.3.7 Operating Mode Selection

A machine can operate in different modes, among which:

- **Automatic:** it is the normal production mode, where all safety systems are active.
- **Manual:** in this mode, the operator is able to set up the machine or use it in a step-by-step mode.

To change from one mode to another, the operator acts on a mode selector like the one shown in Figure 4.10.

If that mode selector does not affect any safety function, it is not a safety function. For example, the selection of the manual mode allows the use of the robot teach pendant that has an enable switch. The switch allows robot movements with the robot cell access gate open: the enable switch shall be implemented in a safety system, while the mode selection can be handled in a normal General Purpose PLC. Hereafter the key language in the standard that clarifies what just stated:

> ***[ISO 13849-1] 5.2.2.9 Requirements for operating mode selection***
> *The following is required:*
> *[...] **c)** when changing from one operating mode to another, safety functions and/or risk reduction measures necessary for the selected operating mode shall be activated; without any loss of protection coverage during the transition.*

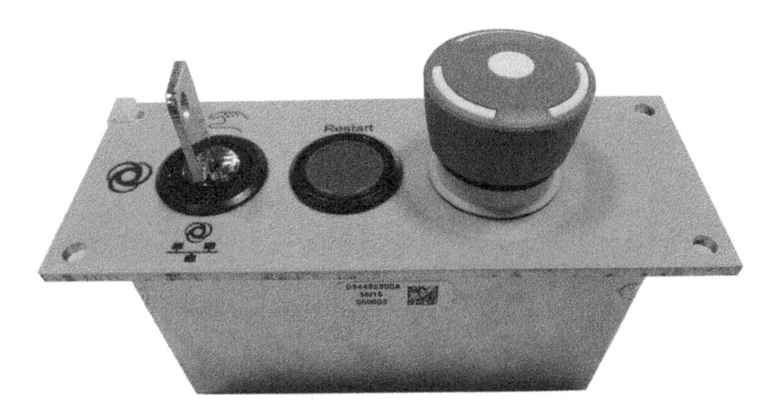

Figure 4.10 Key selector for automatic or manual mode of a robot.

4.4 The Emergency Stop Function

All Safety Functions described above contribute to a reduction of Risk. The Emergency Stop is a special one. It is defined as a **complementary protective measure** by ISO 12100. The language in the Machinery Directive Annex I, EHSR 1.2.4.3, is key to understand the real use of the function.

> ***[2006/42/EC] EHSR 1.2.4.3*** *[...] Emergency stop devices must be a back-up to other safeguarding measures and not a substitute for them.*

Emergency stop devices are intended to enable operators to stop hazardous movements of machinery as quickly as possible if, despite other protective measures in place, a hazardous situation, or event, arises. **However, the emergency stop does not, in itself, provide protection in a risk reduction analysis.** It is simply a backup to other risk reduction measures, such as guards and protective devices, not a substitute for them.

Of course, an emergency stop can enable operators to prevent an unexpected dangerous situation from resulting in an accident, or at least reduce the severity of the consequences of an accident. An emergency stop may also enable operators to prevent malfunctioning from damaging the machinery itself.

The same concept can be expressed in a different way: going back to the risk reduction described in § 4.1.3, of the 10 risks that were highlighted, 3 were first eliminated. We are now left with 7 risks: the emergency stop cannot be used to eliminate any of those risks. Let's suppose a conveyor is constructed in such a way that there are both drawing-in and impact risks. **It is incorrect to install a safety rope** (the rope is the actuator of an emergency stop device) **and state that the risks were reduced;** in this case, the risks have to be reduced by guarding the dangerous movements of the conveyor. Once the conveyor is safe, we can decide to install a safety rope, in case something happens to the goods that are transported and the operator needs a way to stop it safely. In summary: **a dangerous conveyor with a safety rope is not a safe conveyor**!

Since the Emergency Stop cannot be used to reduce any risk, its failure does not result in an immediate increase in the risk of the machinery. In this respect, according to ISO 12100 definition, **it is not a safety function.**

However, the function needs to have a Reliability level of at least $PL_r = c$ or SIL 1:

> ***[ISO 13850:2015] 4.1.5 Emergency stop equipment***
>
> *4.1.5.1 The safety related parts of the control system or subsystems which perform the emergency stop function shall comply with the relevant requirements of ISO 13849-1 and/or IEC 62061. Determination of the Performance Level (PL) or SIL required should take into account the purpose of the emergency stop function, but the minimum required is PL_r c or SIL 1.*
>
> ***Note:*** *The emergency stop function can share safety related parts with other safety functions taking into account the requirements of ISO 13849-1 and/or IEC 62061.*

Therefore, the requirements of either ISO 13849-1 or IEC 62061 also apply to this complementary protective measure; **it is correct to define a PL_r or SIL** for the emergency stop function, based upon the risk assessment, and to evaluate a PL or SIL as for "normal" safety functions; however, **it is not a "safety function" because it is not a Risk Reduction Measure** [74].

Some colleagues disagree with this last statement and may state that "*Emergency Stop is a Safety Function but not a Protective Device.*" In any case, most of us agree with the fact that the Emergency Stop function shall have a SIL or PL level and that **it cannot be used during a risk reduction process**.

4.5 The Reliability of a Safety Function in High Demand Mode

In high demand mode, the unreliability of a safety function is defined by the parameter PFH_D.

Historically both IEC 62061 and ISO 13849-1 defined it as the **Average Probability of dangerous failure per hour**. In reality PFH_D is a frequency, since it is measured in (s^{-1}).

The correct definition is given in the new edition of IEC 62061: **Average Frequency of Dangerous Failure per hour**.

> ***[IEC 62061] 3.2 Terms and definitions***
> ***3.2.29 average frequency of a dangerous failure per hour (PFH or PFH_D).*** *Average frequency of dangerous failure of an SCS to perform a specified safety function over a given period of time.*

PFH_D is therefore equivalent to the $F(t)$ function, or the probability of Failure function, or the unreliability function described in Chapter 1.

4.5.1 PFH_D and PFH

ISO 13849-1 has always used the acronym PFH_D to indicate the probability of dangerous failure per Hour. IEC 62061 2005 edition used the same acronym. The new edition of IEC 62061 uses the IEC 61508-4 [8] acronym, **PFH**, to indicate the same variable. For completeness, we report hereafter the definition as in IEC 61508-4.

> ***[IEC 61508-4] 3.6 Fault, failure and error***
> ***3.6.19 Average Frequency of a Dangerous Failure Per Hour*** *(PFH). Average frequency of a dangerous failure of an E/E/PE safety related system to perform the specified safety function over a given period of time.*

Therefore, don't get confused, because $\boldsymbol{PFH_D \equiv PFH}$.

4.5.2 The Performance Level

The reliability parameter is called **PL or Performance level** and it ranges from **PL a,** being the lowest level for a safety function, to **PL e,** the Highest level of Reliability.

Table 4.2 details the average values of PFH_D for each level.

Table 4.2 Performance levels (PL).

PL	Average probability of dangerous failure per hour (PFH_D) (h^{-1})
a	$10^{-5} \leq PFH_D < 10^{-4}$
b	$3 \times 10^{-6} \leq PFH_D < 10^{-5}$
c	$10^{-6} \leq PFH_D < 3 \cdot 10^{-6}$
d	$10^{-7} \leq PFH_D < 10^{-6}$
e	$PFH_D < 10^{-7}$

4.5.3 The Safety Integrity Level

IEC 62061 uses Safety Integrity levels, or SILs, to indicate the level of reliability of a safety function.

Table 4.3 is taken from IEC 61508-1 [5] table 3. SIL 4 is not applicable to the Machinery Sector.

Table 4.3 Safety integrity levels and the corresponding PFH_D.

Safety integrity level (SIL)	Average frequency of a dangerous failure of the safety function (h^{-1}) (PFH_D)
4	$\geq 10^{-9}$ to $<10^{-8}$
3	$\geq 10^{-8}$ to $<10^{-7}$
2	$\geq 10^{-7}$ to $<10^{-6}$
1	$\geq 10^{-6}$ to $<10^{-5}$

4.5.4 Relationship Between SIL and PL

Both IEC 62061 and ISO 13849-1 give requirements for the design and implementation of SCS of machinery. The methods developed in both standards are different, **but they achieve the same level of risk reduction**.

In principle, SRP/CS that are designed to the relevant PL in accordance with ISO 13849-1 can be integrated, as subsystems, into an SCS designed in accordance with IEC 62061.

Table 4.4 Indicates the Relationship between PL and SIL, based upon the average probability of dangerous failure per hour.

Table 4.4 Relationship between performance levels (PL) and safety integrity levels (SIL).

Performance level (*PL*)	Average probability of a dangerous failure per hour (h^{-1})	Safety integrity level (SIL)
a	$\geq 10^{-5}$ to $<10^{-4}$	No correlation
b (with Category 1)	$\geq 3 \times 10^{-6}$ to $<10^{-5}$	1
c	$\geq 10^{-6}$ to $<3 \cdot 10^{-6}$	1
d	$\geq 10^{-7}$ to $<10^{-6}$	2
e	$\geq 10^{-8}$ to $<10^{-7}$	3

PL a has no correlation on the SIL scale and is mainly used to reduce the risk of slight, normally reversible, injury.

In order to achieve a certain reliability level, both standards provide a **limited number of structures**, represented by Safety-related Block Diagrams. ISO 13849-1 defines them Categories, while IEC 62061 defines them as Architectures. Those are just ways to achieve the PL or SIL level. A PL d, reached with a Category 2, has the same "value" of a PL d reached with a Category 3; that means, from a reliability standpoint, the Architecture or the Category used is not important. **The reliability of a safety system is identified only by the PL or SIL level** obtained: the way used to reach it, meaning the category or the architecture used, is not relevant.

4.5.5 Definition of Harm

Please consider that not all standards in Functional Safety define harm in the same way. Hereafter the two key definitions:

> ***[IEC 61508-4] 3.1 Safety terms***
> ***3.1.1 Harm**: physical injury or damage to the health of people or damage to property or the environment.*
>
> ***[ISO 12100] 3 Terms and definitions***
> ***3.5 Harm**: physical injury or damage to health.*

As you can read, the process sector defines harm as damage to people, property, and the environment, while in Machinery, only the damage to people is taken into consideration. Both ISO 13849-1 and IEC 62061 follow this latter approach. The definition is important when doing the risk assessment, and therefore it has an impact on the number of risk reductions using a Safety Instrumented System or an SCS.

4.6 Determination of the Required PL (PL_r) According to ISO 13849-1

Annex A of ISO 13849-1 can be used to determine the required Performance level of a Safety Function: the so-called **PL_r**. The method is applicable to Safety Functions in high-demand mode only.

Annex A is an informative annex: that means other methods can be used; for example, the one of the IEC 62061.

First of all, it is important to state that **the method described in annex A has subjectivity built in it**; it is not an engineering tool, but rather it is a qualitative way to state the level of Reliability of a Safety System. That is the reason why annex A in both ISO 13849-1 and IEC 62061 is informative.

Harmonized standards to the Machinery Directive provide a **Presumption of Conformity to the EHSR**; informative annexes are not considered in this respect.

Regardless of the method used, it is important that the assessment is made by a team composed by different disciplines: like people from the mechanical and the electrical department, the commissioning, service, etc. Please also refer to annex C of B11.0 [45]. Being a competent team, there is a high probability that the level of reliability of the specific Safety Function is the correct one. You may have played once in the teambuilding activity of **Moon Landing**. In essence, you have to rank the importance of 15 items you can bring with you, to survive on the Moon. Items rank from a box of matches to two 100 lb. tanks of oxygen. Each team member does its own ranking and, afterward, they have to prepare one common agreed ranking. What comes out, normally, is that the team judgment is better than the one of each member. The same usually happens with the determination of the required performance level (PL_r) of a safety function.

4.6.1 Risk Parameters

The method is based on the estimation of **three parameters**:

- **Severity of injury (S):**
  S1: slight (normally reversible injury);
  S2: serious (normally irreversible injury or death).
- **Frequency and/or exposure to hazard (F):**
  F1: seldom-to-less-often and/or exposure time is short;
  F2: frequent-to-continuous and/or exposure time is long.
- **Possibility of avoiding hazard or limiting harm (P):**
  P1: possible under specific conditions;
  P2: scarcely possible.

A combination of those parameters allows the determination of the Required performance level (PL_r).

The principle is that, the higher the risk to be reduced by the Safety Function, the higher its required Performance Level.

4.6.1.1 S: Severity of Injury

To make a decision between S1 and S2, you can also refer to the indications given by the Rapex Directive (§ 4.7). For example, bruising and/or lacerations without complications would be classified as S1, whereas amputation or death would be S2.

4.6.1.2 F: Frequency and/or Exposure Time to Hazard

The frequency parameter should be chosen according to the frequency and duration of access to the hazard. In case of no other justification, F2 should be chosen, if the frequency is **higher than once per 15 minutes**. F1 may be chosen if the accumulated exposure time does not exceed 1/20 of the overall operating time and the frequency is not higher than once per 15 minutes.

Just to give an example, if we consider a machine with manual winding whose operator must cyclically reach the loading area, F2 is clearly the appropriate choice. For a machining center that operates automatically, F1 could be selected.

4.6.1.3 P: Possibility of Avoiding Hazard or Limiting Harm

It is important to know whether a hazardous event can be recognized before it can cause harm and be avoided. Important aspects which influence the selection of parameter P include:

- Speed with which the hazard arises (e.g. quickly or slowly);
- Possibilities to avoid the hazard (e.g. by escaping);
- Past experience related to the machine;
- Whether operated by trained and suitable operators;
- Operated with or without supervision.

When a hazardous event occurs, P1 should only be selected if there is a realistic possibility of avoiding a hazard or of significantly reducing its effect; otherwise, P2 should be selected. In the new edition of ISO 13849-1, there is a methodology that may be followed in order to decide whether P1 or P2 is the correct parameter.

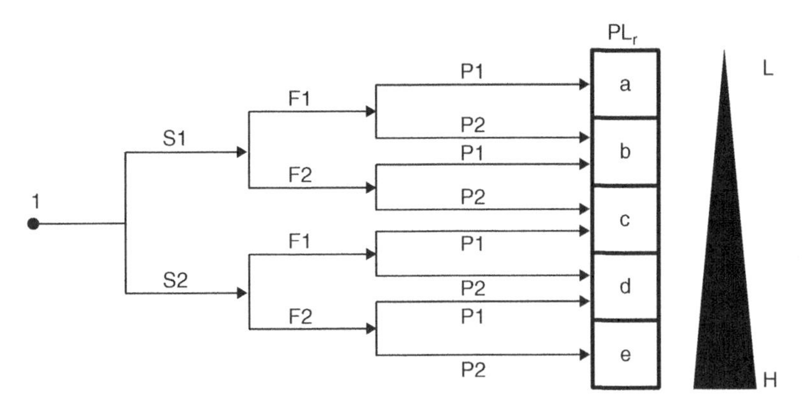

Figure 4.11 Graph for determining required PL_r for safety function.

The graph in Figure 4.11 shows the path to be followed, once the various parameters have been decided. Please note that **the analysis is based on the situation prior to the provision of the intended safety function**: please refer to the concept of the Naked Machinery in § 4.1.4.

In the previous edition of ISO 13849-1, the parameter *P* was divided into two parameters: one for the probability of occurrence of a hazardous event, called "O," and one for the possibility of avoiding it, called "P."

4.6.1.4 An Example on How to Use the Graph

With reference to a manually loaded Press

- The consequence of the dangerous event is a serious irreversible injury ➔ S2
- An operator is exposed to the hazard several times a day ➔ F2.
- It is not possible to avoid *hazard or limiting harm* caused by the dangerous event ➔ P2

The analysis of table A.1, illustrated in Figure 4.12 **shows that the PL_r value is e.**

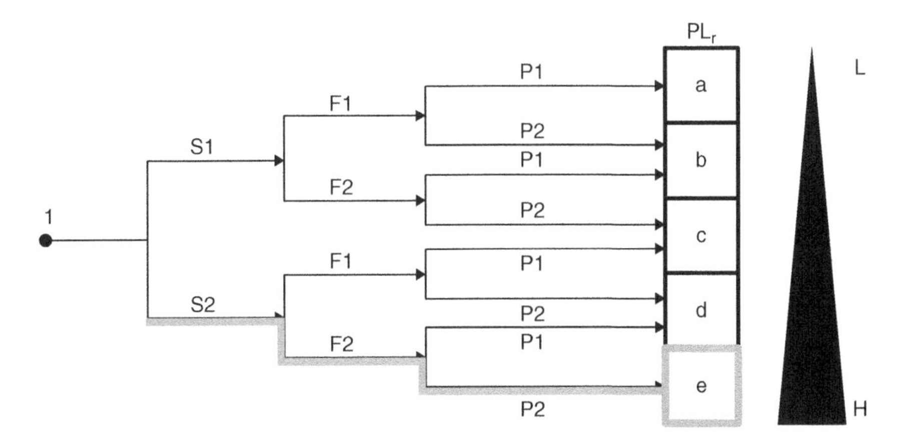

Figure 4.12 Performance Levels required (PL_r) in case of a manually loaded press.

4.7 Rapex Directive

The Rapid Information System "RAPEX" supports Directive 2001/95/EC: **the General Product Safety Directive**. It addresses consumer products, but its risk assessment methodology could be used for Machinery as well.

As seen in the previous paragraph, the decision whether an injury is severe or not can significantly change the required PL or SIL level. That is the reason why we refer to the Rapex guide: it contains good insights on how to estimate the Severity parameter "S." That will be useful also when we will be discussing IEC 62061. Four levels are defined, similarly to IEC 62061:

1. Injury or consequence that after basic treatment (first aid, normally not by a doctor) does not substantially hamper functioning or cause excessive pain; usually **the consequences are completely reversible**.
2. Injury or consequence for which a visit to A&E (Accident and Emergency Department) may be necessary but, in general, hospitalization is not required. **Functioning may be affected for a limited period, no more than about 6 months**, and recovery is more or less complete.
3. Injury or consequence that normally requires hospitalization and will affect functioning for more than six months or lead to a **permanent loss of function**.
4. Injury or consequence that is or could be fatal, including brain death; consequences that affect reproduction of offspring; **severe loss of limbs and/or functions**, leading to more than approximately 10% of disability.

Table 4.5 Hereafter is an abstract from the RAPEX guide table. We only list the typical injures caused by machines.

Table 4.5 Abstract from the RAPEX guide table.

Type of injury	Severity of injury			
	1	2	3	4
Laceration; cut	Superficial	External (deep) (>10 cm long on body) (>5 cm long on face) requiring stitches Tendon or into joint; White of eye or cornea	Optic nerve Neck artery Trachea Internal organs	Bronchial tube; esophagus; aorta; spinal cord (low) Deep laceration of internal organs; severed high spinal cord; brain (severe lesion/dysfunction)
Fracture	—	Extremities (finger, toe, hand, foot) Wrist; arm; rib Sternum; nose; tooth Jaw; bones around eye	Ankle; leg (femur and lower leg); hip; thigh; skull Spine (minor compression fracture); jaw (severe); larynx Multiple rib fractures Blood or air in chest	Neck Spinal column

Table 4.5 (Continued)

Type of injury	Severity of injury			
	1	**2**	**3**	**4**
Crushing	—	—	Extremities (fingers, toe, hand, foot) Elbow; ankle; wrist; forearm Leg; shoulder; trachea Larynx; pelvis	Spinal cord Mid-low neck Chest (massive crushing) Brain stem
Amputation	—	—	Finger(s); toe(s) Hand; foot (Part of) arm; leg; eye	Both extremities
Eye injury, foreign body in eye	Temporary pain in eye without need for treatment	Temporary loss of sight	Partial loss of sight Permanent loss of sight (one eye)	Permanent loss of sight (both eyes)
Hearing injury, foreign body in ear	Temporary pain in ear without need for treatment	Temporary impairment of hearing	Partial loss of hearing Complete loss of hearing (one ear)	Complete loss of hearing (both ears)

For further details, please refer to the Rapex Guide.

Another classification, this time also linked to the number of **lost working days**, is the following:

1. **Minor injury:** Slightly healing health damage. Max. two day work loss, first aid is sufficient.
2. **Reversible injury**: No permanent health damage, reversible. Max. six months Functional impairment.
3. **Irreversible injury:** Severe permanent health damage, irreversible. More than six months Functional impairment.
4. **Death or disabling injury**: Loss of limb or fatal injury or disability >10%.

4.8 Determination of the Required SIL (SIL_r) According to IEC 62061

Annex A of IEC 62061 gives a method to estimate the required SIL or Safety Integrity Level of an SCS. The method fits well to the high-demand mode of operations, and it is not recommended for Safety Functions in low-demand mode, where other methods like the LOPA (Layer of Protection Analysis) [71] can be used.

Also, in IEC 62061, annex A is informative and exactly the same considerations about subjectivity, as described in the previous paragraph, apply.

4.8.1 Risk Elements and SIL Assignment

The following parameters apply:

- Severity of harm (**Se**), equivalent to the Severity of Injury of ISO 13849-1
- Probability of occurrence of that harm, which is a function of:
  - Frequency and duration of the exposure of persons to the hazard (**Fr**)
  - probability of occurrence of a hazardous event (**Pr**)
  - possibilities to avoid or limit the harm (**Av**)

The severity of harm (Se) is given a score from 1 to 4, with 4 being the most severe.

The probability of harm occurring is broken down into three parameters; each of these parameters is scored from 1 to 5, with 5 being the "worst" situation, and their scores are added to determine a class (Cl).

The SIL rating is then chosen from a matrix that plots the severity score (Se) and the class (Cl). The SIL is determined using Table 4.6. The class (Cl) is calculated as follows $Cl = Fr + Pr + Av$.

The dark area indicates the SIL assigned as the target for the SCS. The lighter shaded areas indicate that a safety function with value less than SIL 1 can be used. That is the equivalent of a PL_r a according to ISO 13849-1. In case a control system is used to reduce the risk, only Basic Safety Principle should be used. The concept is also called "Other Measures" (OM).

Compared to the first edition of IEC 62061, SIL 2 at Class 3 and 4 is now reduced to SIL 1, because of the low score for the classes of Frequency, Probability, and Avoiding Harm.

As indicated in Table 4.4, there is a relationship between SIL and PL. Moreover, a machinery engineer can use annex A from IEC 62061 for the determination of the required PL_r and use ISO 13849-1 for the calculations of the reached reliability. That is the reason why, in Table 4.6 (same as table A.6 of IEC 62061), both SIL and PL levels are indicated.

Table 4.6 Matrix for the determination of the required SIL or PL.

		Class $Cl = Fr + Pr + Av$				
Consequences	**Severity *Se***	**3 4**	**5 6 7**	**8 9 10**	**11 12 13**	**14 15**
Death, losing an eye or arm	4	SIL 1 PL_r c	SIL 2 PL_r d	SIL 2 PL_r d	SIL 3 PL_r e	SIL 3 PL_r e
Permanent injury, losing fingers	3		OM PL_r a or PL b (Category B)	SIL 1 PL_r b (Category 1) or PL_r c	SIL 2 PL_r d	SIL 3 PL_r e
Reversible injury, medical attention	2	No SIL, nor PL, required		OM PL_r a or PL b (Category B)	SIL 1 PL_r b (Category 1) or PL_r c	SIL 2 PL_r d
Reversible injury, first aid	1	**OM**: Other Measures (e.g. basic safety principles)			OM PL_r a or PL b (Category B)	SIL 1 PL_r b (cat 1) or PL_r c

4.8.2 Severity (Se)

Similarly to ISO 13849-1, the severity of harm can be estimated by choosing the appropriate value based upon the consequences of the accident. This time, as shown in Table 4.7, four levels are given, similar to the Rapex guidelines:

- **4** is a fatal or significant irreversible injury: limb loss, permanent lung damage, loss of an eye or partial or total loss of vision;
- **3** is a serious or irreversible injury, but it is possible to continue work after healing; examples are loss of fingers or toes, but also broken limbs;
- **2** is a more serious reversible injury that requires medical attention. It is possible to resume work after a short period of time, e.g. serious lacerations, excruciating and severe bruising;
- **1** is a slight injury in which first aid care without medical intervention is sufficient.

Compared to ISO 13849-1, a score of 3 or 4 corresponds to S2 and a score of 1 or 2 corresponds to S1.

Table 4.7 Severity (Se) classification.

Consequences	Severity (*Se*)
Irreversible: death, losing an eye or arm	4
Irreversible: broken limb(s), losing a finger(s)	3
Reversible: requiring attention from a medical practitioner	2
Reversible: requiring first aid	1

4.8.3 Probability of Occurrence of Harm

Each of the three parameters *Fr*, *Pr*, and *Av* must be estimated separately, using the most unfavorable situation.

4.8.3.1 Frequency and Duration of Exposure (Fr)

The parameter Fr is linked to:

- the frequency of presence of the person in the hazardous area and
- the average duration of presence.

The standard, as shown in Table 4.8, gives the frequency and duration of exposure classification in five levels.

As shown in the table, if the duration is less than 10 minutes, the value can be rounded down to the next level. This does not apply to the frequency of exposure of one hour, which should not be decreased in any case.

Table 4.8 Frequency and duration of exposure (Fr) classification.

	Value of parameter *Fr*	
Frequency of exposure	**Duration ≥10 minutes**	**Duration <10 minutes**
Greater or equal (≥) to once every hour	5	5
Less than h^{-1} but ≥1/day	5	4
Less than once every day but ≥2/weeks	4	3
Less than once twice per week but ≥1/year	3	2
Less than once in a year	2	1

4.8.3.2 Probability of Occurrence of a Hazardous Event (Pr)

This parameter is not explicit in ISO 13849-1.

The probability of occurrence of a hazardous event, as shown in Table 4.9, is in a scale between 1, for negligible probability, and 5, in case of very high probability.

This is probably the most difficult parameter to estimate because of the influence of Automation, or Control system, that has no Reliability data.

Table 4.9 Probability (Pr) classification.

Probability of occurrence	**Probability (*Pr*)**
Very high	5
Likely	4
Possible	3
Rarely	2
Negligible	1

Let's consider the presence of both a person and a robot inside a safeguarded space. In order to determine the required SIL, one may ask what is the probability that the person could be hit by the robot, in case of an unexpected start up, obviously without SCS.

Someone may observe that, even without a safety system, the Automation keeps the robot still: the probability would then be Rarely or Negligible. The issue is that **the Automation system** has no Reliability data and, for a conservative approach, it **cannot be relied upon in this analysis**. If the person remains close to the Robot all the time, while he is inside the area, the probability of occurrence would then be **Very high**. That being said, one should then ask where the person is normally working and consider that, in case of an unexpected start-up, the person may be in the robot Operating Space. The narrower is the operating space, the higher is the probability.

4.8.3.3 Probability of Avoiding or Limiting the Harm (Av)

This parameter describes whether or not harm could be avoided or limited in case of a hazardous event.

The possibility can be estimated by considering the following aspects:

- Skills of the machine user.
- Speed of the hazard.
- Risk awareness.
- Ability to react.

Table 4.10 Probability of avoiding or limiting harm (Av) classification.

Probability of avoiding or limiting harm	Avoiding and limiting (*Av*)
Impossible	5
Rarely	3
Probable	1

Regarding skills and abilities, the standard clarifies that human abilities cannot be accounted more than once for each safety function. As shown in Table 4.10, there are three levels for Av.

4.8.3.4 Example of the Table Use

Still considering a manually loaded press, the consequence of the dangerous event is an irreversible injury, with possible loss of a hand: $Se = 4$.

All other parameters must be added together in order to select the class.

- An operator is exposed to hazard several times a day ➔ $Fr = 5$
- the hazardous event may occur ➔ $Pr = 3$
- The danger can be avoided ➔ $Av = 3$

The sum of *Fr*, *Pr,* and *Av* $(5 + 3 + 3) = 11$

A level of SIL 3 must be achieved by the SCS.

Document [55], is a good analysis of the differences between ISO 13849-1 and IEC 62061 for the determination of the required Reliability level.

4.9 The Requirements Specification

At the end of the Risk Assessment, the manufacturer may have decided that certain risks have to be reduced by using SCS. Now, he needs to define the behavior of each SCS in a document called **Safety Requirements Specification,** SRS (ISO 13849-1) or **Functional Requirements Specification**, FRS (IEC 62061).

The document describes what each safety function has to do and the reliability level required. The SRS or FRS **is important to avoid mistakes** at the transition from the risk reduction process to the SRP/CS or SCS design and evaluation process, especially if these two activities are performed by different people or departments.

4.9.1 Information Needed to Prepare the SRS or the FRS

The following information should be available, to define the safety requirements specification:

a) **Results of the risk assessment for the machine**, including all safety functions needed by the risk reduction process. For example, the opening of an interlocked movable guard, of a safeguarded space, has to stop the robot operating inside.
b) **Machine operating characteristics** that include:

- The intended use of the machine
- The reasonably foreseeable misuse
- The effect of overlapping hazards
- The operating modes, for example, local or automatic
- The modes of operation during which the safety function has to be active
- The cycle time
- The response time before a safe state is achieved.

c) **The emergency operations** that are required like, for example, an Emergency Start;
d) **Description of the interaction** of different working processes and manual activities like, for example, modes of operation with the safeguards suspended;
e) **Ergonomic aspects** to minimize incorrect operation or defeating;
f) **Limits of use** in relation to environmental conditions;
g) **Effect of overlapping hazards**.

4.9.2 The Specifications of All Safety Functions

The SRS or FRS shall have the following information, for each safety function, in relation to the specific application:

a) **Title and brief description** of the Safety Function: to have a clear reference.
b) **Initiation event** that triggers the safety function: for example, the gate opening.
c) **The reaction of the machine** caused by the outputs of the safety function: for example, the robot stops and pneumatic air is cut.
d) **The required** Performance Level (PL_r) in case of ISO 13849-1 or Safety Integrity Level (SIL_r) in case of IEC 62061. It could be $PL_r = d$ or SIL required $(SIL_r) = 2$. Please remember that the category or the Architecture should not be defined, since it is a characteristics of the Subsystem only.
e) **The Response time** before a safe state is achieved. In case of a saw, it may take 30 seconds before the hazard is eliminated; a brake may be needed.
f) **Modes of operation of the machine** during which the safety function is to be active. A machine can have an automatic mode and a manual mode; the automatic mode should, for example, be disabled when the gate opens, while in the manual mode it should be possible to move an element of the machine.
g) **All interfaces** between the safety functions and between safety functions and any other non-safety function. For example, a safety function may be muted to allow the exit of material from inside the safeguarded space.
h) **The behavior of the machine on loss of power**. For example, when it is necessary to hold a vertical axis, to prevent a fall under gravity, this can require two separate safety functions: with power available and without power available.
i) **The demand rate** upon the safety function. For example, the gate is opened 4 times every hour.
j) **The priority of the safety functions** that can be simultaneously active and that can cause conflicting actions. For example, an emergency stop function has priority over all other functions.
k) The way **the manual reset function** has to be performed; for example, in a Multiple Sequential time-limited way.

4.10 Iterative Process to Reach the Required Reliability Level

Both standards detail the steps to be followed to reach the required risk reduction, using either an SCS (IEC 62061) or an SRP/CS (ISO 13849-1). The starting point is Step 2 of the Risk Reduction Process of ISO 12100 (Figure 4.3 of this book). Figure 4.13 shows the design process of a safety function and the verification of whether the SRP/CS or SCS achieves the intended risk reduction.

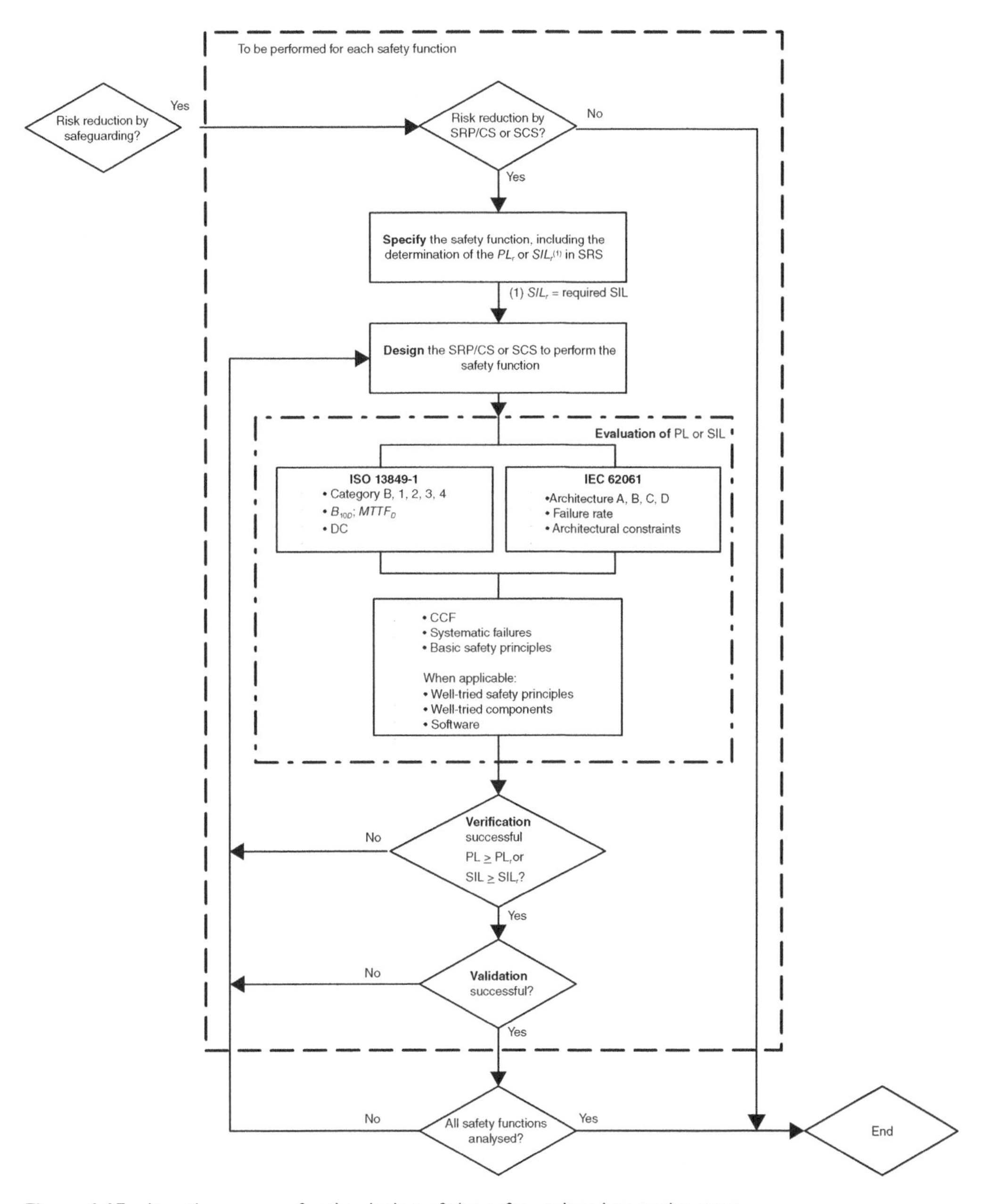

Figure 4.13 Iterative process for the design of the safety-related control system.

4.11 Fault Considerations and Fault Exclusion

4.11.1 How Many Faults Should be Considered?

A Safety system has to be reliable; but how many faults should be taken into consideration? A redundant system can be designed with two sensors and two redundant final elements but, in case of two faults, we are in trouble! In general, the following fault criteria shall be taken into account:

- If, because of a fault, further components fail, the first fault together with all following faults **shall be considered as a single fault** (known as a dependent fault).
- Two or more separate faults, **having a common cause,** shall be considered as a single fault. This situation is analyzed in detail in § 3.6.
- The simultaneous occurrence of two or more faults having separate causes **is considered highly unlikely and therefore needs not be considered** (*[ISO 13849-1] 6.1.10.2 Fault consideration*).

That means, in machinery, due to the advantage of having, most of the time, a high-demand mode of operations, only one independent fault has to be considered. **Two faults are considered very unlikely during the time between two demands upon the safety function**. The assumption becomes less sustainable in case of demands of few times a year or in case of a safety system in low demand mode.

4.11.2 Fault Exclusion and Interlocking Devices

4.11.2.1 Fault Exclusion Applied to Interlocking Devices

A loss of a safety function in the absence of a hardware fault **is due to a systematic failure** or to a Common Cause Failure. The latter is discussed in § 3.6, while the former can be caused by errors made during the design or the integration stages. Some of these systematic failures will be revealed during the design process, while others will be revealed during the validation of the safety system.

Regarding **hardware faults**, considerations on fault exclusion arise when discussing the need for redundancy. In machinery, a very common sensor is an **Interlocking Device with guard locking** mechanism.

In Figures 4.14 and 4.15, an example and a graphical representation of an interlocking device are shown with its key elements:

1. **The Actuator:** in the picture, it is of "tongue" type. According to ISO 14119 [30], that is a Type 2, low coded, interlocking device.
2. The Actuating Head
3. Interlocking Plunger
4. **Guard Locking Solenoid**. It allows the blocking of the actuator so that the door cannot be opened, unless all dangerous movements inside the safeguarded area have been stopped, for example.
5. **Interlocking Monitoring Contact**. It is a normally closed contact in the sense that when the door is closed, the contact is closed, and therefore the input circuit is energized.
6. **Guard Locking Monitoring Contact**. It gives the status of the locking mechanism: usually, if the door is locked, the contact is closed.
7. Housing.

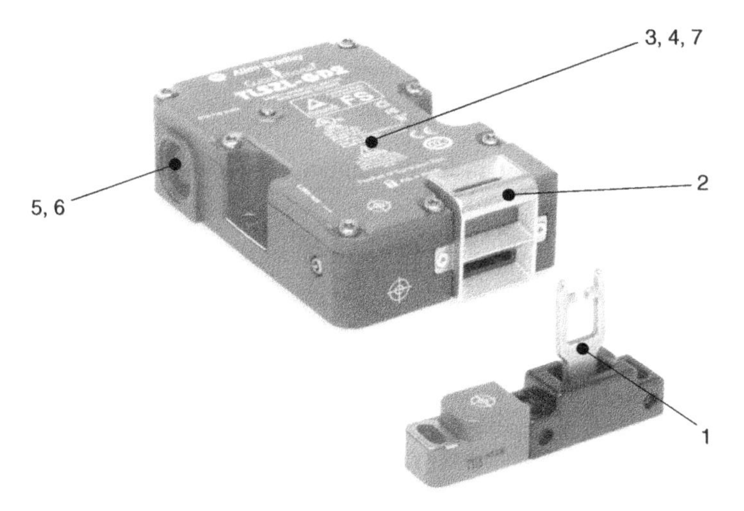

Figure 4.14 Example of Type 2 Interlocking device with guard locking. *Source:* Rockwell Automation, Inc.

If a complete redundancy has to be achieved on the input subsystem (the access door), two interlocking devices have to be installed, on that specific door, by the machinery manufacturer. That is rarely done. What normally happens is that the component manufacturer works on its interlocking device, **trying to increase the redundancy inside the component itself.**

The first thing that manufacturers do is to provide two voltage-free contacts for the Interlocking monitoring. They may even provide two contacts for the Guard Locking Monitoring. What they would never do is to have two guard locking solenoids: in this case, the manufacturer would make some fault considerations and come to the conclusion that he can do a fault exclusion on the solenoid.

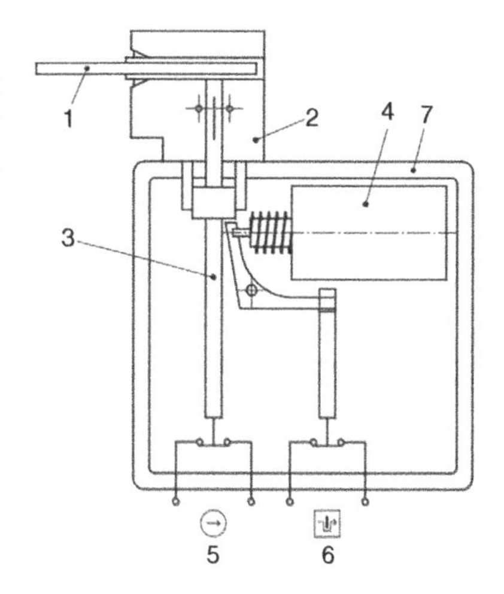

Figure 4.15 Schematics of Type 2 Interlocking device with guard locking.

In general, certain hardware faults may be excluded because **if an element clearly has a very low probability of failure** by virtue of properties inherent to its design and construction then, normally, it would not be considered necessary to constrain (on the basis of the hardware fault tolerance) the safety integrity of any safety function that uses that element.

In other words, it is not always possible to evaluate subsystems without assuming that certain faults are excluded. Fault exclusion is a **compromise between technical safety requirements and the possibility of the occurrence of a fault.**

Fault exclusion can be based upon:

- the **technical improbability** of occurrence of some faults,
- generally accepted technical experience, independent of the considered application, and
- technical requirements related to the application and the specific hazard.

4.11.2.2 Fault Exclusion on Pre-defined Subsystems

In general, Fault Exclusions made by the component manufacturer are defined as made on **"pre-defined"** or **"pre-designed" subsystems**. The user may buy a Type 4 interlocking device, declared PL e, whereby the manufacturer has applied fault exclusions for some part of its component. In a certain sense, that is "transparent" for the user. That means the limitations applicable to a Safety Function when Fault Exclusions are made, are not valid in case the fault exclusion is made by the manufacturer of one of the components of the safety system.

> ***[ISO 14119] 9.2.2 Fault exclusion***
> ***9.2.2.1 General.*** *[...] In case of a fault exclusion for interlocking functions intended to reach PL e or SIL 3, the interlocking device shall, exhibit a dual channel structure or a Category 4 behaviour to the majority of its architecture. Individual parts in the architecture of an interlocking device may be of single channel structure. If it can be proven that the single channel part cannot fail before other dual channel parts, e.g. through over dimensioning, a fault exclusion is permissible and will not limit the PL or SIL.*

4.11.2.3 Fault Exclusion Made by the Machinery Manufacturer

Different is the situation whereby the Machinery manufacturer uses a Type 2 interlocking device, shown in Figure 4.14 and applies a fault exclusion on the Actuator. When he does it, he cannot claim PL e or SIL 3 for that safety Function.

> ***[IEC 62061] 7.3.3 Fault consideration and fault exclusion***
> ***7.3.3.3 Fault Exclusion*** *[...]* ***LIMITATION:*** *For some applications, it is not expected that all failures can be excluded with sufficient confidence for SIL 3. The following non exhaustive list provides an indication of (non-predesigned) subsystems with a hardware fault tolerance of zero and where fault exclusions have been applied to faults that could lead to a dangerous failure where a maximum of SIL 2 can be appropriate, provided that sufficient justification is given:*
>
> - *position switch with mechanical aspects with HFT of 0;*
> - *leakage of a fluid power valve (where leakage is dangerous failure).*
>
> ***Note: This limitation does not apply to pre-designed subsystems used within their specification.***

Similarly, the following is stated in ISO 13849-2 [14] table D.8:

> ***[ISO 13849-2] D.2.4 Fault exclusions and integrated circuits***
> ***Table D.8*** *[...] For PL e, a fault exclusion for mechanical (e.g. the mechanical link between an actuator and a contact element) and electrical aspects is not allowed. In this case redundancy is necessary. For emergency stop devices in accordance with IEC 60947-5-5, a fault exclusion for mechanical aspects is allowed if a maximum number of operations is considered.*

Please also refer to the following considerations made in ISO 14119 [30] on the possibility, by a machinery manufacturer, to apply a fault exclusion on interlocking devices.

> ***[ISO 14119] 9.2.2 Fault exclusion***
> ***9.2.2.3 Mechanical Fault Exclusions for Type 2 Interlocking Devices Without Guard Locking.***
> *For Type 2 interlocking devices, the following faults of their mechanical parts can be excluded.*

Damage (breaking) and wearing of the actuator and the actuating system due to misalignment, only if additional mechanical alignment elements prevent the actuation of the position switch outside the limits of misalignment specified by the manufacturer. The additional mechanical alignment elements shall be designed and constructed as to be effective when subjected to a load equal to 2 times the maximum force expected during the operation of the guard for the intended lifetime (mission time) of the interlocking device.

In essence, if you are a machinery Manufacturer and decide, for good reasons, that you can apply the fault exclusion to the Actuator of Type 2 interlocking device, you need to implement certain solutions in order to prevent its damage, as described in ISO 14119, and the maximum Reliability level you can reach is PL d or SIL 2. However, if your considerations are not sustainable (for example, the actuator does not have a correct "invitation" towards the Actuating Head), that safety function can only reach PL c or SIL 1.

[ISO 14119] 9.2.2 Fault exclusion
9.2.2.3 Mechanical Fault Exclusions for Type 2 Interlocking Devices Without Guard Locking *[...]*
Where not all mechanical faults can be excluded, an interlocking system applying Type 2 interlocking devices and requiring at least PL d in accordance with ISO 13849-1:2021 or SIL 2 in accordance with IEC 62061:2021 shall be implemented by the integration of an additional interlocking device of any of the Types 1–4. Application of diversity is recommended.

Bottom line, fault exclusion is only applicable to certain faults of an element, and it is up to the designer (manufacturer or integrator) to prove the exclusion of the respective faults, based on the limits set forward by its design and use. Such fault exclusions are only possible provided that the technical improbability of them occurring can be justified based upon the known laws of physical science. Any such fault exclusions shall be documented and justifiable under all expected industrial environments, including temperature, pressure, vibration, pollution, corrosive atmosphere, etc.

A fault exclusion can only be applied to the entire subsystem when all dangerous failures of the subsystem can be excluded. Please consider that the component manufacturer can apply a fault exclusion during the component Reliability assessment. Useful information on fault exclusions are available in ISO 13849-2:2012, annex A–D.

4.11.2.4 Types of Guard Locking Mechanism

Before we leave the subject, it is important to clarify a few more aspects.

There are two reasons to choose a guard interlocking with guard locking:

- Either **to protect people**. For example, inside a safeguarded area there are dangerous movements having inertia. The door is unlocked only when all movements are stopped.
- **For manufacturing** or **Process reasons**.

There are four possible ways to lock a door (guard lock) [75]:

1. **Spring applied – Power-ON released**. It is also called "mechanical guard locking". It means that the guard locking device is moved to the "locked" position by a spring at the removal of power. It is a closed-circuit current principle, in relation to the locking function. When power is provided, the device is unlocked. **In case of a black out the door remains locked.**

2. **Power-ON applied – Spring released**. It operates in the opposite manner and is called "electrical guard locking". It is an open-circuit current principle. In order to keep the door locked, power must be present all the time. **In case of a blackout, the spring is released, the door unlocks and it can be opened.**
3. **Power-ON applied – Power-ON released.** It is a principle that does not change position on the removal of power. It is also called **the bistable principle**. Power must be applied to change it to the other state. As the removal of the power does not change the position of the guard locking device, this principle is considered a closed-circuit current principle. **In case of a blackout, the door lock stays in its last position.**
4. **Power-ON applied – Power-OFF released**. It corresponds to an open-circuit current principle, as the guard locking device opens on the removal of the power. It has the same behavior as the second case, but in this one there is no spring. The door is kept closed thanks to an electromagnet. **In case of a blackout, the magnet is de-energized, the door unlocks and it can be opened.**

Which guard locking principle shall be selected? If the lock is for production reasons, all four are suitable: the second and the fourth are probably more "flexible." **For machinery protection,** the design engineer is completely free to decide which type of guard locking is selected, since it does not represent a safety function.

If a **guard locking is for personnel protection, solution 1 and 3 are the one recommended**.

4.11.2.5 What Are the Safety Signals in an Interlocking Device with Guard Lock?

The component has the following inputs and outputs:

- **One input signal**: the one that locks the interlocking device by acting on the Guard Locking Solenoid. In case the locking principle is chosen for people protection, the signal shall come from a safety system. In case of process reasons, it can come from a non-safety system.
- **Two output signals**: Interlocking Monitoring Contacts. They should always be routed to a safety system.
- **One or two output signals** for the Guard Locking Monitoring Contact. For process reasons, the status can be managed by a General Purpose PLC, otherwise it must be managed by a safety system.

4.11.2.6 What Safety Functions are Associated to a Guard Interlock

A Guard Interlock can be used on a door that gives access to a safeguarded space (§ 4.3.3). When the interlocking device is activated, all dangerous movements inside the area must be stopped. There are actually two safety functions to be analyzed with a Risk Assessment:

- The Safety related Stop function, when the door is opened.
- The Prevention of unintended start-up while the door remains open.

The two functions may require, in principle, different Performance or SIL levels.

Between the two, the latter is probably the most important. If inside the area there is a dangerous movement, normally it is visible. Therefore, in case it is not stopped, when the door is opened, the operator has a good chance to see it and protect himself. A more dangerous situation is when the movement is stopped, the operator is working on the dangerous part that suddenly restarts: in this case, the person may not have enough time to place himself in a safe position.

Also, the Guard lock has two safety functions to be analyzed in terms of required Performance or SIL level:

- The release of the guard locking device: in other terms, when the door can be unlocked.
- The Safety related Stop function when releasing the guard locking device: in other terms, what has to be stopped, inside the safeguarded space, in case the door unlocks (but it still stays closed)

4.11.3 Other Examples of Fault Exclusions

4.11.3.1 Short Circuit Between any Two Conductors

Figure 4.16 shows an output subsystem in Architecture D or Category 4. Provided both contactors and the Safety Module are inside the same control panel, it is possible to use the scheme shown in Figure 4.17 and do a fault exclusion for the probability of a short circuit of the cable connecting the output of the Safety Module with the contactors K_{P1} and K_{P2}.

That possibility is stated in ISO 13849-2, table D-4 and shown in Table 4.11.

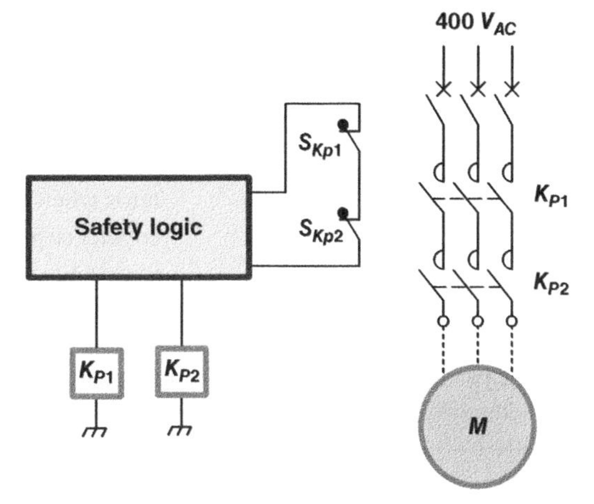

Figure 4.16 Category 4 output subsystem.

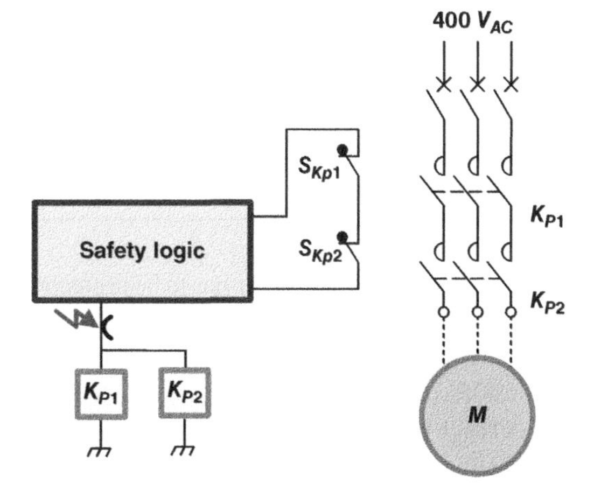

Figure 4.17 Category 4 output subsystem with Fault exclusion.

Table 4.11 Abstract from table D.4 of ISO 13849-2: Faults and Fault exclusions – conductors/cables.

Fault considered	Fault exclusion	Remarks
Short circuit between any two conductors	Short circuits between conductors which are • permanently connected (fixed) and protected against external damage, e.g. by cable ducting, armoring, • separate multicore cables, • within an electrical enclosure (see remark), or • individually shielded with earth connection.	Provided both the conductors and enclosure meet the appropriate requirements (see IEC 60204-1)

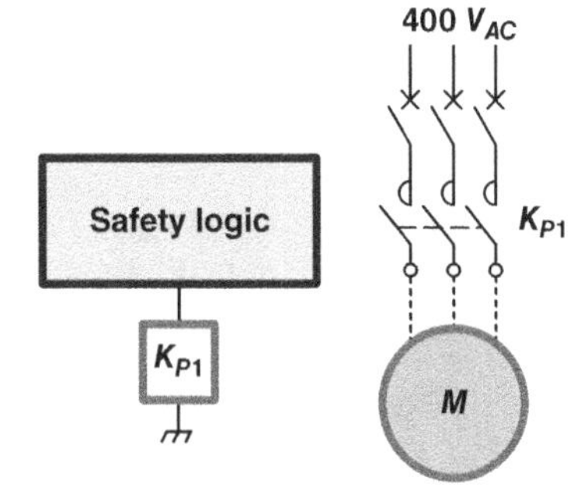

Figure 4.18 Single channel subsystem.

4.11.3.2 Welding of Contact Elements in Contactors

Manufacturers may consider to adopt the architecture shown in Figure 4.18 and claim that a PL = d or a SIL 2 level of reliability can be reached. In order to make that possible, a fault exclusion on the contactor K_{P1} is needed. However, that is not allowed by neither IEC 62061 nor ISO 13849-1. That is clearly stated in ISO 13849-2, table D.9, as shown in Table 4.12.

In other words, it is not possible to claim compliance with neither of the two standards for high-demand mode Safety systems when a fault exclusion is applied on the *Non-opening of contact elements due to permanent welding.* That is valid even in case the value of the nominal current of the contactor overcurrent protective device **has a safety factor of 0.6 or lower**.

Table 4.12 Abstract from table D.9 of ISO 13849-2: Faults and Fault exclusions – switches – electromechanical devices (e.g. relay, contactor relays).

Fault considered	Fault exclusion	Remarks
All contacts remain in the energized position when the coil is de-energized (e.g. due to mechanical fault)	None.	—
All contacts remain in the de-energized position when power is applied (e.g. due to mechanical fault, open circuit of coil)	None.	
Contact will not open	None.	
Contact will not close	None.	

The reason is that a contactor has to be protected from overloads and short circuits **with a proper overcurrent protective device** (Branch Circuit Protective Device, in North American language): the goal is to avoid **systematic failures**. Once that is done, the contactor random failure rates can be considered "real." The fact of "over protecting" the contactor may increase its B_{10D} value but **it is not correct to claim an infinite value of B_{10D}**. ISO 13849-1, table C.1, states a value of B_{10D} of 400 000 in case the contactor is subject to a current equal to its "nominal load". It also indicates a

B_{10D} of 20 000 000 in case the contactor is subject to a "small load" (meaning 20% of its nominal load). Please also refer to § 4.12.3.2.

Those considerations are also valid for auxiliary contactors used as inputs of a safety logic. Figure 4.19 shows two examples of input subsystems (Pressure Transmitter, a safety module with an internal threshold and auxiliary contactors) that can reach PL d or SIL 2.

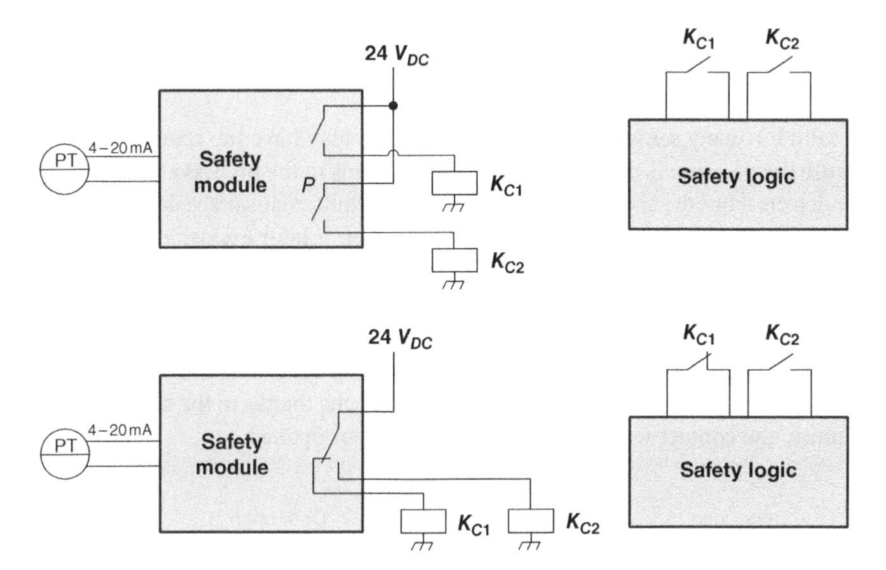

Figure 4.19 Examples of redundant input safety subsystems.

However, using the architecture shown in Figure 4.20, a maximum of PL c or SIL 1 can be reached.

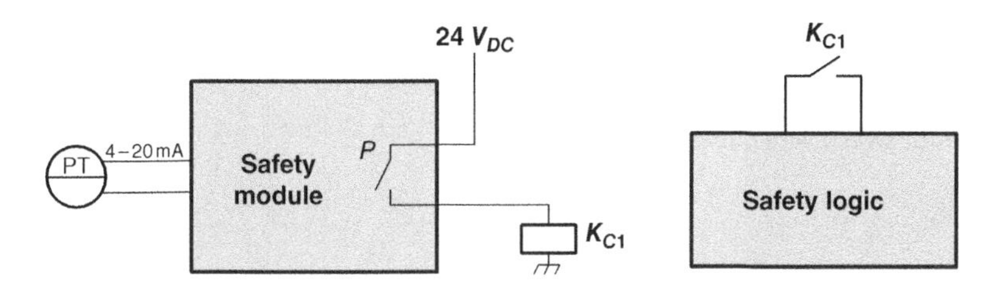

Figure 4.20 Example of a single channel input safety subsystem.

4.12 International Standards for Control Circuit Devices

4.12.1 Direct Opening Action

The electrical output of an interlocking device has two types of technologies

- It can be a **Voltage Free** Contact (VFC), or
- It can be an **OSSD** type (Output Signal Switching Device)

Type 2 interlocking devices often have VFC, but of a "special type": they are with a **Direct Opening Action**. The concept is defined in annex K of IEC 60947-5-1 [32]. Please refer to Figure 4.21; the opening of the door causes the following steps:

- The actuator moves out of the actuating head
- That causes the movement of the Interlocking plunger
- That opens the interlocking monitoring contact

Something may go wrong and, for example, the actuator movement does not open the electrical contact.

In general, that is valid for many sensors. A low-pressure switch may have the same issue: the contact is closed because the pressure is normal. If something happens in the process and the pressure drops, the sensor detects it but the contact does not open. Can a fault exclusion be done? Unless some considerations are made by the pressure switch manufacturer, a fault exclusion cannot be done by the user!

In case of the interlocking device, or other sensors using the same technology, **a fault exclusion can be done if the Electrical contact has a Direct Opening Action**. How does it work? As shown in Figure 4.21, left-hand side, the opening of the contact is through non-elastic elements. In the Non-direct opening example, the opening of the contact happens thanks to the spring force. If the spring has a failure, the contact will not open, despite the door opened.

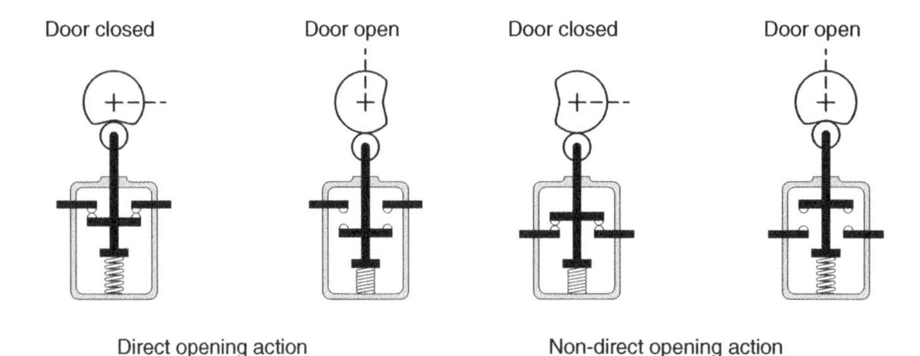

Figure 4.21 Example of control switch with direct opening action.

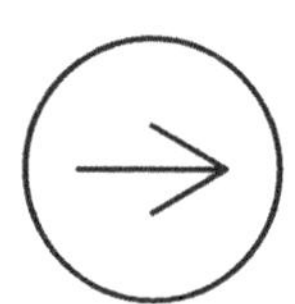

Figure 4.22 Direct opening action symbol.

[IEC 60947-5-1] Annex K

K.2.2 Direct Opening Action (of a Contact Element). *Achievement of contact separation as the direct result of a specified movement of the switch actuator through non-resilient members (for example not dependent upon springs).*

Direct Opening Action contacts are identified with symbol shown in Figure 4.22.

Recent interlocking devices have OSSD outputs: those are solid-state outputs. The signal goes to an OFF state when the door is open or when the component detects an internal failure. It has to comply with the standard IEC 60947-5-3 [33].

4.12.1.1 Direct and Non-Direct Opening Action

Figure 4.23 shows the combination of direct and non-direct mechanical action of the position switches of Type 2 interlocking devices. That is a way to avoid common cause failures of two mechanically actuated position switches by using associated direct and non-direct mechanical action.

In general, when using Voltage Free Contacts with Direct Opening action, both contacts are closed when the door is closed. In Figure 4.23 instead, D_1 is a position switch with normally closed contacts and direct opening action, while D_2 has normally open contacts, which means that direct opening action is not possible. When the safeguard is closed D_2 has its contacts closed. When the safeguard is opened, both D_1 and D_2 open their contacts because D_1 is actuated and D_2 is no longer actuated.

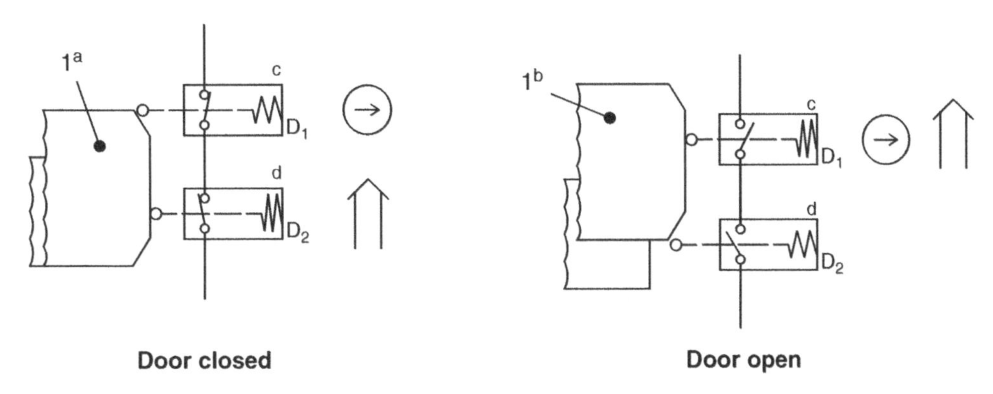

Figure 4.23 Example of direct and non-direct opening action.

4.12.2 Contactors Used in Safety Applications

There are three types of contactors used in SCS:

- **Power Contactors**: they have auxiliary **Mirror contacts.**
- **Auxiliary Contactors, also called Relays**: they have auxiliary **Mechanically Linked contacts.**
- **Electromechanical elementary relays**: among which there are relays with **Forcibly Guided** (mechanically linked) contacts. They are designed according to the IEC 61810-3 [35].

4.12.2.1 Power Contactors

They are used as motor controllers in Branch Circuits (Figure 4.24).

They have three parts (please refer to Figure 4.25):

- **The control part:** it is the coil that moves both the power and the auxiliary contacts. The terminals are indicated as A1 and A2. When the coil is energized, the power contacts close.
- **The Power contacts:** normally there are 3 input and 3 output terminals, numbered from 1 to 6. Those are used for the power lines L1, L2, and L3.
- **The Auxiliary contacts:** they indicate the status of the power contacts. Those contacts are called **Mirror contacts** and they are designed according to **IEC 60947-4-1** [34] **annex F.**

Figure 4.24 Example of a power contactor. *Source:* Siemens.

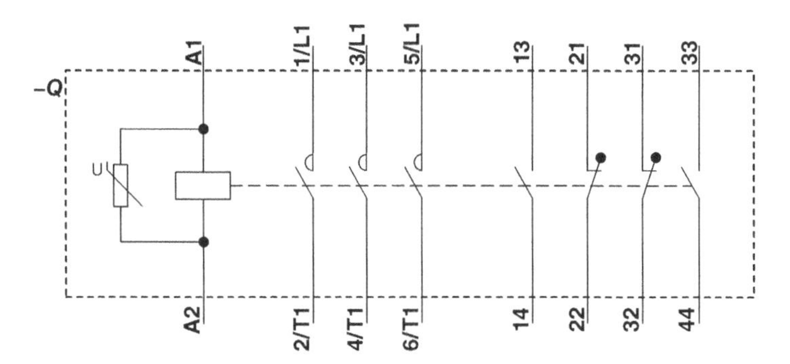

Figure 4.25 Schematic of a power contactor.

Figure 4.26 Electrical symbol of the mirror contact.

The Mirror contacts have the characteristics of being guaranteed that **if the power contacts are all in the open position, the Mirror contacts are closed**.

[IEC 60947-4-1] Annex F: Requirements for auxiliary contact linked with power contact (mirror contact)

F.2.1 Mirror Contact*. Normally closed auxiliary contact which cannot be in closed position simultaneously with the normally open main contact under conditions defined in Clause F.7*

Note*: One contactor may have more than one mirror contact.*

The symbol of a mirror contact is shown in Figure 4.26:

When the power contactor has these types of auxiliary contacts, Fault exclusion on those contacts can be applied.

Figure 4.27 Example of an auxiliary contactor. *Source:* Siemens.

4.12.2.2 Auxiliary Contactors

They are also called **Auxiliary Relays or Contactor-relays** and are used in the control part of the electrical equipment of Machinery (Figure 4.27).

They have two parts only:

- **The control part:** it is the coil that moves the auxiliary contacts. The terminals are indicated as A1 and A2.
- **The Auxiliary contacts:** some are normally open and some are normally closed.

Figure 4.28 shows a schematic of the contactor. The auxiliary contacts are shown when the coil is de-energized. That means:

- **Normally open** contacts means that when the coil is de-energized, the contacts are open. They are normally indicated as 13, 14, 43, 44.
- **Normally closed** contacts means that when the coil is de-energized, the contacts are closed. They are normally indicated as 21, 22, 31, 32.

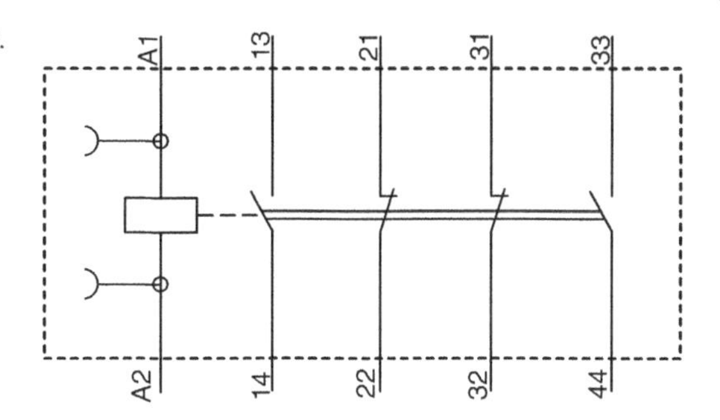

Figure 4.28 Schematic of an auxiliary contactor.

When the auxiliary contacts are indicated as Mechanically linked, it means that it is not possible that a NO contact has the same position of an NC one. They are designed according to **IEC 60947-5-1** [32] **annex L.**

> ***[IEC 60947-5-1] Annex L: Special requirements for mechanically linked contact elements***
> ***L.2.1 Mechanically Linked Contact Elements.*** *Combination of n Make contact element(s) and m Break contact element(s) designed in such a way that they cannot be in closed position simultaneously under conditions defined in L.8.4.*
>
> ***Note 1 to entry****: One control circuit device may have more than one group of mechanically linked contact elements*

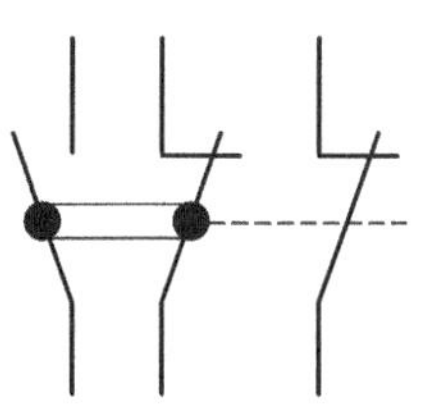

Figure 4.29 Example of representation of NO and NC contacts which are mechanically linked.

The mechanical linkage shall be identified in circuit diagrams by a double parallel line connecting a filled circle on each of the mechanically linked contact symbols: please refer to Figures 4.29 and 4.30.

4.12.2.3 Electromechanical Elementary Relays

They are of two types [35]:

- **Type A:** relay in which all contacts are mechanically linked (or Forcibly Guided).
- **Type B:** relay containing contacts that are mechanically linked to each other as well as contacts that are not mechanically linked.

They also have two parts only:

- **The control part:** it is the coil that moves the auxiliary contacts. The terminals are indicated as "A1" and "A2".
- **The Auxiliary contacts:** some are normally open and some are normally closed.

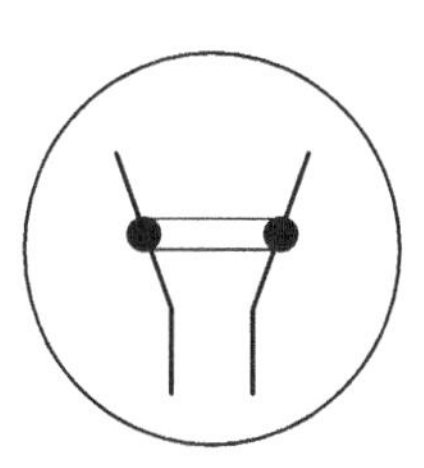

Figure 4.30 Symbol for device containing mechanically linked contacts.

Figure 4.31 shows an example of an Electromechanical elementary relay, while Figure 4.32 shows its electrical representation.

Figure 4.31 Example of an elementary relay. *Source:* OMRON Corporation.

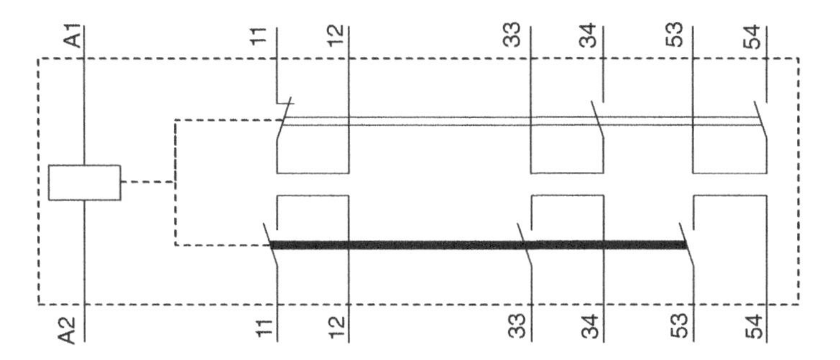

Figure 4.32 Schematic of an elementary relay with forcibly guided contacts.

Figure 4.33 Symbol for a forcibly guided (mechanically linked) contact set, Type A.

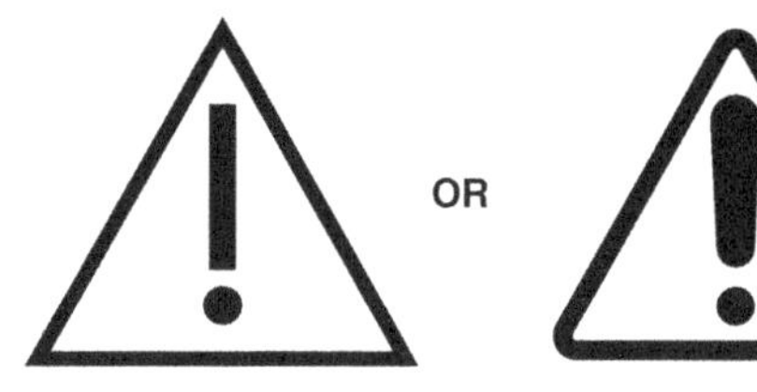

Figure 4.34 Symbols for use on Type B relays.

Type A relays with **Forcibly Guided** (mechanically linked) contacts are marked either with the words "Type A" or with the symbol shown in Figure 4.33.

Type B relays with Forcibly Guided (mechanically linked) contacts are marked either with the words "Type B" or with the symbols given in Figure 4.34.

4.12.3 How to Avoid Systematic Failures in Motor Branch Circuits

4.12.3.1 How to Protect Contactors from Overload and Short Circuit

A typical Motor Branch Circuit, shown in Figure 4.35, is made of three elements:

- The **Branch Circuit Protection Device (BCPD)**: normally a fuse or a circuit braker. It protects the contactor, the cable and the motor from both a short circuit and, in many cases, from a ground fault. For that reason, it is also called branch-circuit, short-circuit and ground-fault protective device.

- A **Power Contactor** (**Motor Controller** in North American language).
- An **Overload Protection,** also called Thermal protection.

Therefore, the contactor is protected by both the BCPD and the Thermal element. The thermal protection is normally based upon the AC-3 nominal current of the Power Contactor while, for the BCPD, IEC standards provide two types of "coordination": Type 1 and Type 2. Hereafter their definitions.

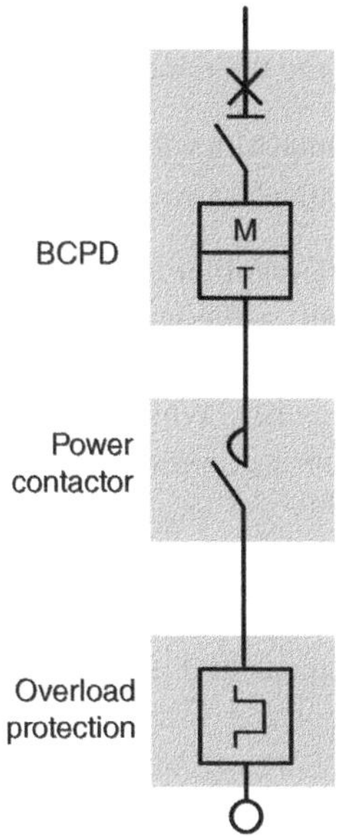

BCPD

Power contactor

Overload protection

Figure 4.35 Motor branch circuit.

> ***[IEC 60947-4-1] 8.2.5.1 Performance under short-circuit conditions (rated conditional short-circuit current)*** *[34]*
> *The rating of the SCPD shall be adequate for any given rated operational current, rated operational voltage and the corresponding utilization category.*
> *Two types of co-ordination are permissible, "1" or "2." The test conditions for both are given in § 9.3.4.2.1 and § 9.3.4.2.2.*
> ***Type "1"*** *co-ordination requires that, under short-circuit conditions, the contactor or starter shall cause no danger to persons or installation and may not be suitable for further service without repair and replacement of parts.*
> ***Type "2"*** *co-ordination requires that, under short-circuit conditions, the contactor or starter shall cause no danger to persons or installation* ***and shall be suitable for further use.*** *The risk of contact welding is recognized, in which case the manufacturer shall indicate the measures to be taken as regards the maintenance of the equipment.*
>
> ***Note:*** *Use of an SCPD not in compliance with the manufacturer's recommendations can invalidate the co-ordination. These tests are applicable to AC motor ratings only.*

In general, a short circuit will be reliably and safely cleared regardless of which type of coordination is applied. Assemblies of coordination Type "2" can therefore be considered as being more qualitative. In addition, they are often instantly available for further operation after a short circuit.

Bottom line is that the machinery manufacturer has to properly protect the contactor with either a Type 1 or a Type 2 coordination for short circuit and not only for the effects of overloads. If he fails to do that, the contactor may be subject to a systematic failure, and therefore all Reliability data are meaningless.

4.12.3.2 Contactor Reliability Data

In case of a correct design, the Reliability data indicated by the contactor manufacturer can be used. Normally they are given as a B_{10} or B_{10D} value; the procedure to determine data for electromechanical contactors used in functional safety applications is detailed in IEC 60947-4-1 [34] **annex K**.

Those values are different in case the contactor is used at a fraction of its nominal current. This idea of using the contactor at a fraction of the current is coming from the Functional Safety domain, not the electrical safety one. For Example, ISO 13849-1 table C.1, same as Table 3.16 in this book, indicates conservative values for B_{10D} for some key components used in SCS. For **power contactors** the table indicates:

- Contactors with small load: $B_{10D} = 20\,000\,000$
- Contactors with nominal load: $B_{10D} = 1\,300\,000$

The nominal load is the nominal current of the contactor in AC-3, if used in a Motor Branch Circuit, while in AC-1 if used ahead of a variable speed drive.

As a small load, the standard indicates **20% of the nominal load**. Some companies defines small load as 30%. When looking for the B_{10D} value of a contactor, please always look for the maximum current value they assume.

Example

Siemens 3RT1034 is a power contactor that can drive up to 32 A in AC-3 at 400 V_{ac}; let's suppose it is used in a Direct on Line application to control a motor. If it is protected by a 3RV1031-4EA10, overload relay with a range between 22 and 32 A, we are in a situation of a contactor with nominal load.

For **Auxiliary relays,** table C.1 of ISO 13849-1 indicates:

- Relays and contactor relays with small load: $B_{10D} = 20\,000\,000$;
- Relays and contactor relays with nominal load: $B_{10D} = 400\,000$.

Other standards, when dealing with the risk of component failures, limit the maximum current that can flow through a Contactor:

> ***[IEC 60730-2-5]*** [36] ***11.3.5.2.1 Measures to protect against common cause failures***
> *Designs where relays are used as switching elements, a non-replaceable fuse (see table H.21 Note 7) in series with two independent relay contacts with* $\mathbf{I_{N\text{-}fuse} < 0.6 * I_e}$ ***relay****, are considered to comply with the following requirements for prevention of common cause failure, without performing the following tests.*

4.12.4 Implications Coming from IEC 60204-1 and NFPA 79

Bonding is one of the most discussed topics in Machinery. There are several reasons to bond a conductive metal part; one of those reasons is to protect control circuits against malfunctioning.

The prescription is in both IEC 60204-1 [3] and NFPA 79 [44].

> ***[IEC 60204-1] 9.4.3 Protection against malfunction of control circuits. 9.4.3.1 Insulation faults 9.4.3.1.2 Method a) – Earthed control circuits fed by transformers****. The common conductor shall be connected to the protective bonding circuit at the point of supply. All contacts, solid state elements, etc., which are intended to operate an electromagnetic or other device (for example, a relay, indicator light) are to be inserted between the switched conductor of the control circuit supply and one terminal of the coil or device. The other terminal of the coil or device is connected directly to the common conductor of the control circuit supply without any switching elements.*

[NFPA 79] 8.3 Control Circuits.
Control circuits shall be permitted to be grounded or ungrounded. [...]
8.3.1 *If the control system is grounded, the output shall be grounded as near as practicable to the control power source and before the first control device. Switching devices shall not be permitted in a grounded conductor(s) unless the control circuit conductor(s) is opened simultaneously.*

[NFPA 79] 9.1.4 Connection of Control Circuit Devices.
9.1.4.1 *All operating coils of electromechanical magnetic devices and indicator lamps (or transformer primary windings for indicator lamps) shall be directly connected to the same side of the control circuit. All control circuit contacts shall be connected between the coil and the other side of the control circuit.*

Two possible faults are prevented with the above language, both linked to a ground fault in the control circuit.

4.12.4.1 Wrong Connection of the Emergency Stop Button

In Figure 4.36, the control circuit electrical contact, an emergency push button in this case, is connected between the contactor coil and ground: that is not allowed! All control circuit electrical contacts must be connected as indicated in Figure 4.38. The reason is that, in case the emergency pushbutton is connected like in Figure 4.36, in case of a ground fault, as indicated with the arrow, its function is bypassed, or in other terms, the emergency pushbutton becomes inactive!

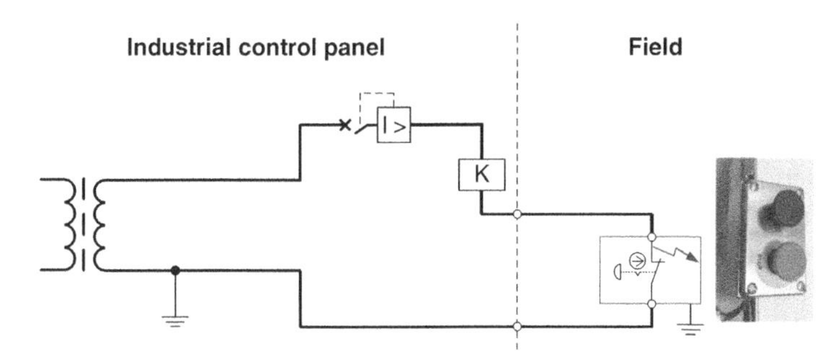

Figure 4.36 Malfunction of a safety part of the control circuit due to a ground fault.

4.12.4.2 Situation in Case of Two Faults: Again a Wrong Connection!

In case the same emergency push button is connected like indicated in Figure 4.37, in case of two faults in the field, as indicated by the arrows, the emergency pushbutton is bypassed again!

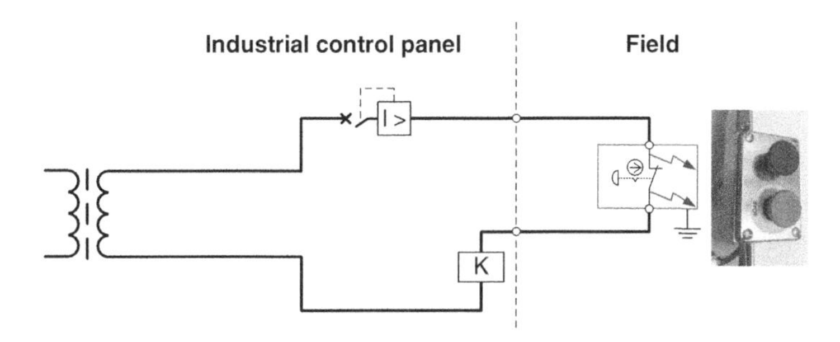

Figure 4.37 Malfunction of a safety part of the control circuit due to two ground faults.

4.12.4.3 Correct Wiring and Bonding in a Control Circuit

In order to avoid possible systematic failures in control circuits, it is important to ground (bond) one of the two poles of the transformer and all metal parts that contain control circuits, **even in case of Extra-low voltage sources**. By doing that, when the first ground fault happens in the field, like shown in Figure 4.38, the overcurrent protection, indicated as "I>", de-energizes the circuit and places the machinery into a safe state.

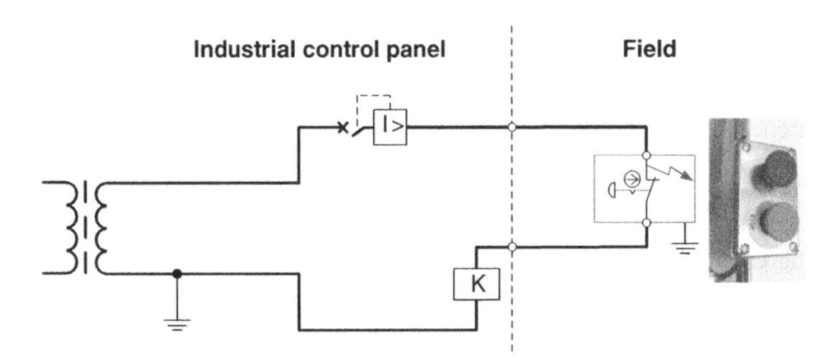

Figure 4.38 Correct bonding of the control circuit and detection of the ground fault.

4.12.5 Enabling and Hold to Run Devices

There are situations, for example during maintenance, whereby a person has to observe a movement, being close to it. Normally, that part of the machine is inside a safeguarded space and, for example, protected by an interlocked movable guard that stops dangerous movements when the guard is opened. The problem is how to handle that maintenance operation, safely.

In general, the design of the machine shall take into account maintenance tasks and, if needed, provide safety functions so that these tasks are done with acceptable low risk.

Two useful safety functions, related to maintenance, or machine setup, are:

- Enabling device function;
- Hold-to-run function.

4.12.5.1 Enabling Devices

An enabling device **can be a two or three-position switch** that allows the movement of a dangerous part of the machinery, normally at reduced speed.

Normally, the one used in safety applications is the three-position one (for example, the robot teach pendant), despite some Type C standards allow the use of a two-position switch (also defined as Pushbutton). The reference standard for the three-position switch is IEC 60947-5-8 [37]. Hereafter two important definitions:

> ***[IEC 60947-5-8:2020] 2 Terms and definitions***
> ***2.1 Enabling Device****. Manually operated control device used in conjunction with a start control and which, when continuously actuated, allows a machine operation.*
> ***2.2 Three-position Enabling Switch.*** *Switch having three sequential actuator positions, in which the contacts are closed when the actuator is in the mid position (partly depressed) and are open when the actuator is in the rest (not pressed) position and in the fully depressed position.*

In case the device has three positions, it allows the dangerous movement only when it is in the middle one. The three positions are designated as follows (see Figure 4.39):

- **Position 1:** OFF state of the contact (actuator is not pressed);
- **Position 2:** ON state of the contact (actuator is pressed to the normal enabling position);
- **Position 3:** OFF state of the contact (actuator is fully pressed).

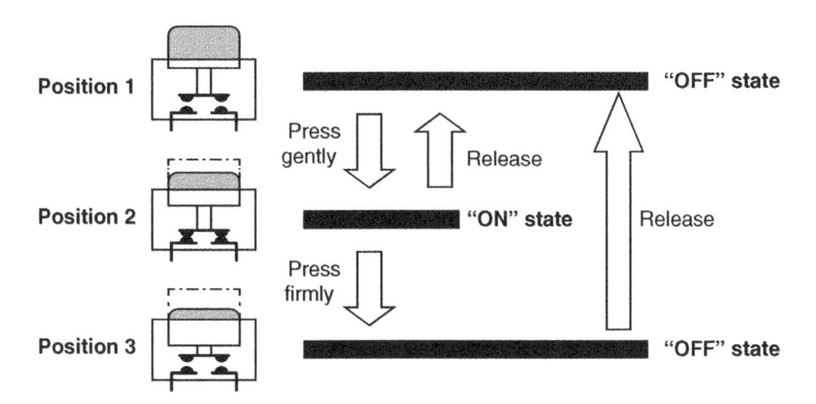

Figure 4.39 Operation of three-position enabling switches.

The three-position enabling switch pressed to Position 2 shall return to Position 1 when released.
The three-position enabling switch shall change from Positions 2 to 3 when pressed further.
When released from Positions 3–1, the switching element shall not close when the actuator passes through Position 2. Please look at the illustration in Figure 4.40.

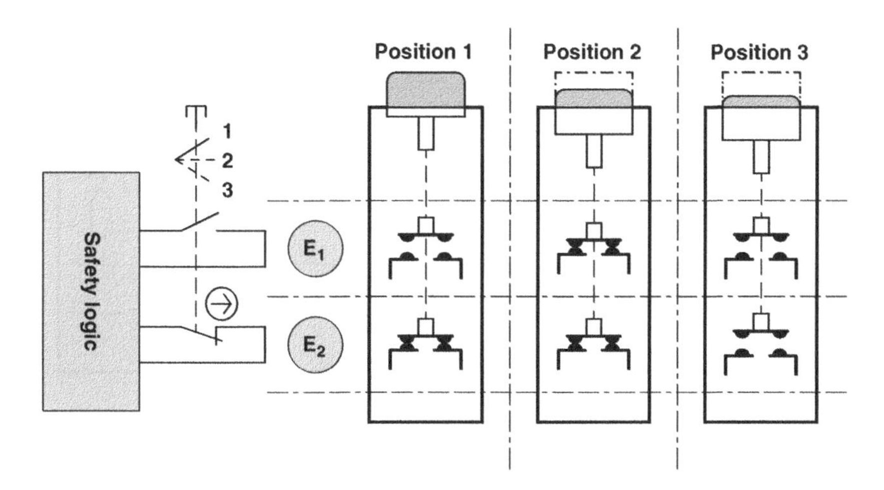

Figure 4.40 Status of the contacts in a three-position enabling device.

It is recommended that one of the two contacts of the three-position switch has a Direct Opening Action (please refer to § 4.12.1); that means it should be in compliance with IEC 60947-5-1 annex K. The contact status is illustrated in Figure 4.41.

According to IFA report [53], paragraph *D.2.5.5 Enabling switches*, the following PL can be achieved by the different types of **three-position enabling devices**:

- Enabling device compliant to both IEC 60947-5-8 and IEC 60947-5-1: PL = b.
- Enabling device compliant to both IEC 60947-5-8 and DGUV test report GS-ET-22E: PL = c.
- Enabling device compliant to both IEC 60947-5-8 and DGUV test report GS-ET-22E, with redundant contacts: PL = d.

Conceptual schematic circuit diagram	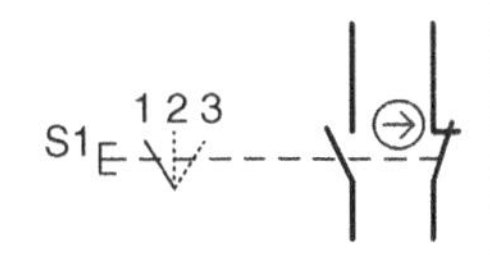 S1 1 2 3	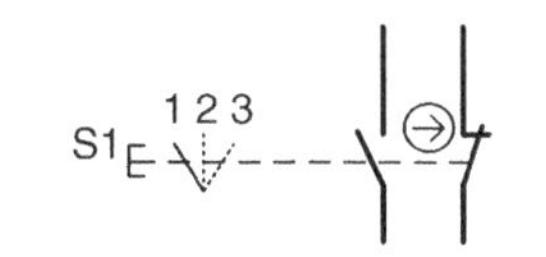 S1 1 2 3	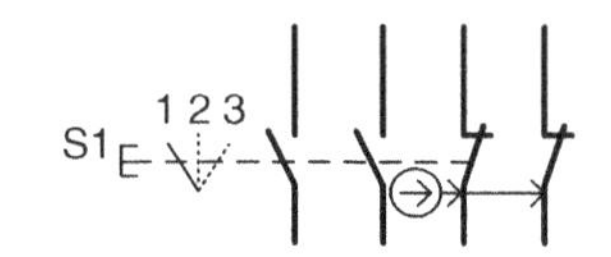 S1 1 2 3
Condition	Break contact to EN 60947-5-1 annex K	Enabling button to GS-ET-22E	Enabling button to GS-ET-22E
Category and PL	Category B Max. PL b	Category 1 Max. PL c	Category 3 Max. PL d

Figure 4.41 Modeling of a three-position enabling device.

A two-position switch can be used in a safety application, provided the risk assessment justifies it. Please consider that, in case of panic, people may keep the switch pushed instead of releasing it.

According to IFA report [53], paragraph *D.2.5.5 Enabling switches*, the PL indicated in Figure 4.42 can be achieved by the different types of two-position enabling devices.

Conceptual schematic circuit diagram	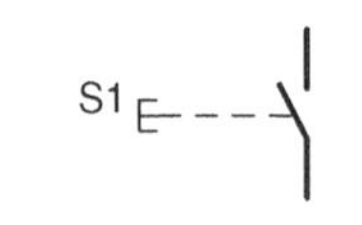 S1	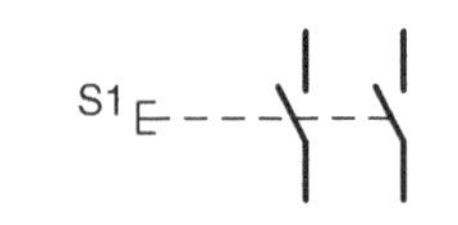 S1	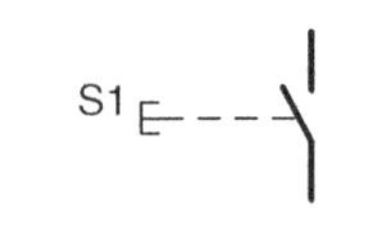 S1	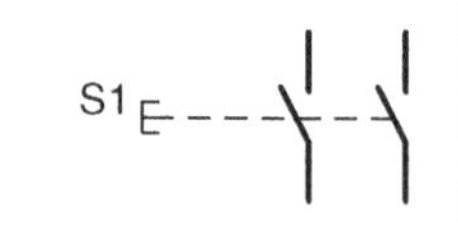 S1
Condition	According to EN 60947-5-1	According to EN 60947-5-1	According to GS-ET-22E	According to GS-ET-22E
Category and PL	Category B Max. PL b	Category B Max. *PL* b	Category 1 Max. PL c	Category 3 Max. *PL* d

Figure 4.42 Modeling of a two-position enabling device (pushbutton).

The three-position enabling device is usually available in three forms:

- push-button enabling devices (Figure 4.43);
- grip actuated enabling devices (Figure 4.44);
- foot actuated enabling devices (Figure 4.45).

Figure 4.43 Push-button enabling devices. *Source:* EUCHNER.

Figure 4.44 Push-button enabling devices. *Source:* IDEC Corporation.

Figure 4.45 Push-button enabling devices. *Source:* BERNSTEIN AG.

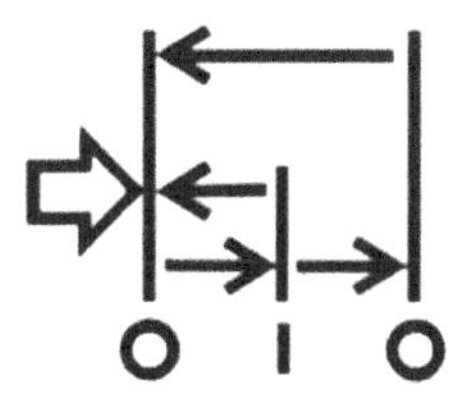

Figure 4.46 Electrical symbol of a three position enabling device.

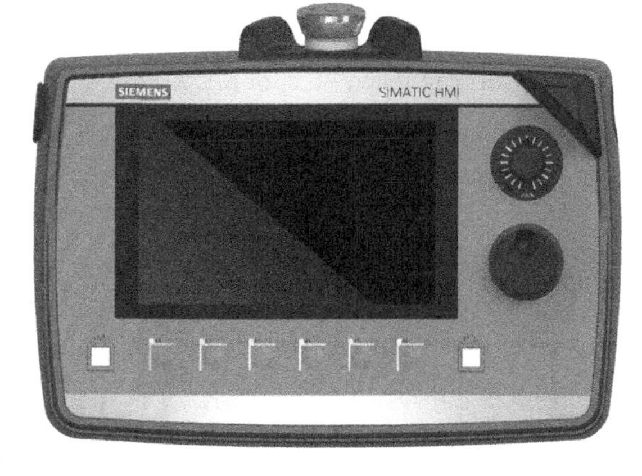

Figure 4.47 Push-button enabling devices. *Source:* Siemens.

The symbol of the three position enabling device is shown in Figure 4.46.

The enabling device is also available on mobile panels (Figure 4.47), to give the operator or the maintenance person the possibility to activate the dangerous movement at reduced speed. Please notice that, while the enabling device requires safety performances, the rest of the commands do not (except for the emergency push button).

4.12.5.2 Hold to Run Device

The simultaneous actuation of two buttons allows a hazardous movement: the safety is guaranteed by the fact that both operator's hands are on the device while the dangerous movement is happening. An example is in Figure 4.48.

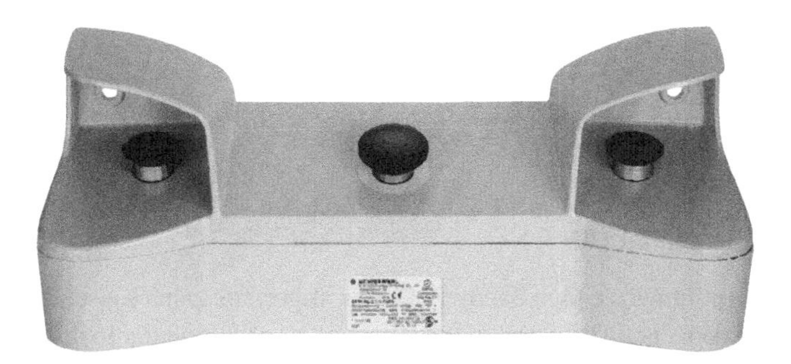

Figure 4.48 Example of a hold-to-run device. *Source:* Block Transformers Electronics GmbH.

The reference standard is ISO 13851 [38].

> ***[ISO 13851:2019] 2 Terms and definitions***
> ***2.1 Two-hand Control Device THCD.*** *Device which requires simultaneous actuation by the use of both hands in order to initiate and maintain hazardous machine functions, thus providing protective measure only for the person who actuates it.*

There are three types of devices. They differ for the level of synchronization while pushing the two actuators:

- **Type I.** The operator could keep one of the actuators always pushed and, by releasing and activating the other button, he can trigger the movement. This type can reach PL c, or SIL 1. I personally would not recommend its use.
- **Type II.** To trigger the movement, the operator has to push both actuators (Re-initiation of the output signal) but without any particular timeframe (no Synchronous actuation). This type can reach PL d, or SIL 2
- **Type III.** It requires a Synchronous actuation of the two buttons.
  - Type IIIA can reach PL c, or SIL 1
  - Type IIIB can reach PL d, or SIL 2
  - Type IIIC can reach PL e, or SIL 3

ISO 13851 considers a synchronous operation when the difference on the simultaneity of actuation is less than or equal to 0.5 seconds between the push of the first control actuating device and the second one. Figure 4.49 details the Synchronous Actuation.

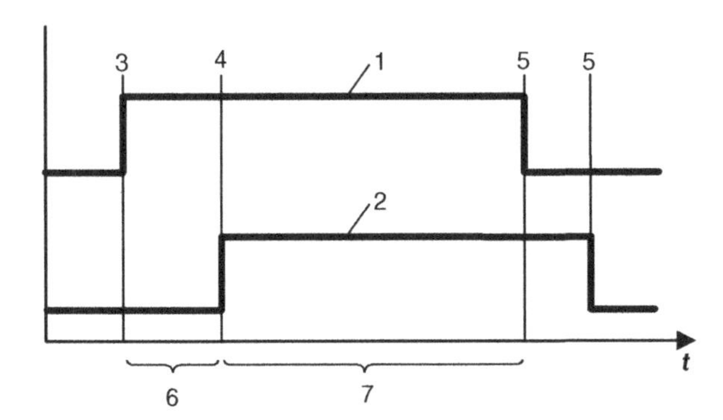

Figure 4.49 Example of hold-to-run device.

Key to the figure:
1. First hand
2. Second hand
3. Initiation of the first input signal
4. Initiation of the second input signal
5. Cessation of input signals
6. Time lag ≤ 0.5 seconds in case of synchronous actuation
7. Time period of simultaneous actuation

4.12.6 Current Sinking and Sourcing Digital I/O

When choosing the type of input or output module for a Safety PLC, it is important to have a solid understanding of sinking and sourcing concepts. The usage of these terms occurs frequently in discussions of input or output circuits.

Sinking and sourcing terminology applies only to DC input or output circuits. Input and output points that are sinking or sourcing can conduct current in one direction only. Here are the definitions and a representation is in Figure 4.50:

> ***[IEC 61131-2]** [39] **3 Terms and definitions***
> ***3.10 Current Sinking.** Property of receiving current.*
> ***3.11 Current Sourcing.** Property of supplying current.*

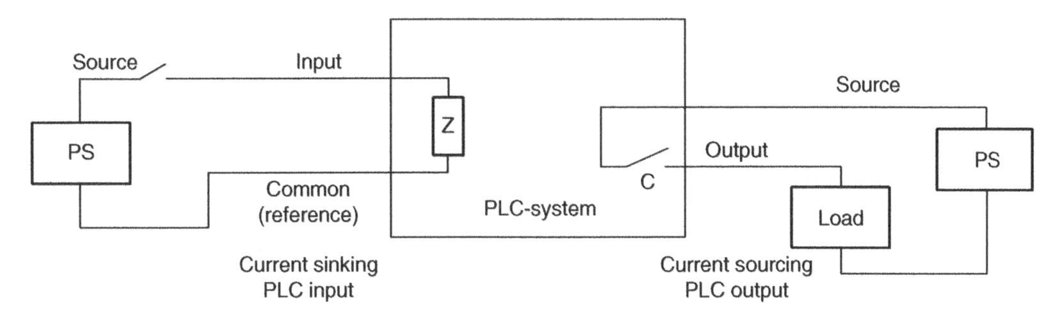

Figure 4.50 Illustration of sinking and sourcing.

Figures 4.51 and 4.52 show examples of interconnections of a Digital input module in a Sinking or a Sourcing configuration, respectively. Each dashed rectangle contains a field contact.

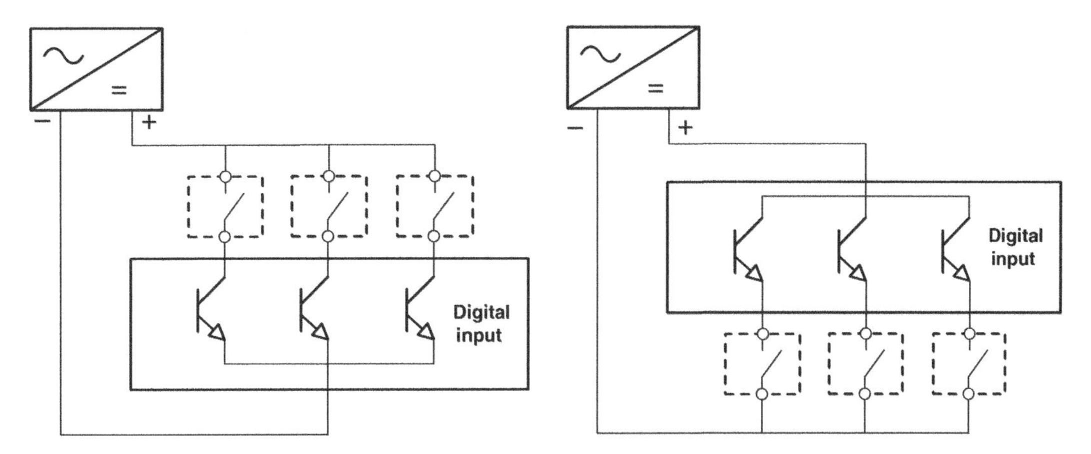

Figure 4.51 Illustration of sinking digital input card.

Figure 4.52 Illustration of sourcing digital input card.

Figure 4.53 shows all four possible cases; the dashed rectangle represents the PLC Input or Output module while the contact or the load are in the field:

- Sinking = provides a path to supply common (−)
- Sourcing = provides a path to supply source (+)

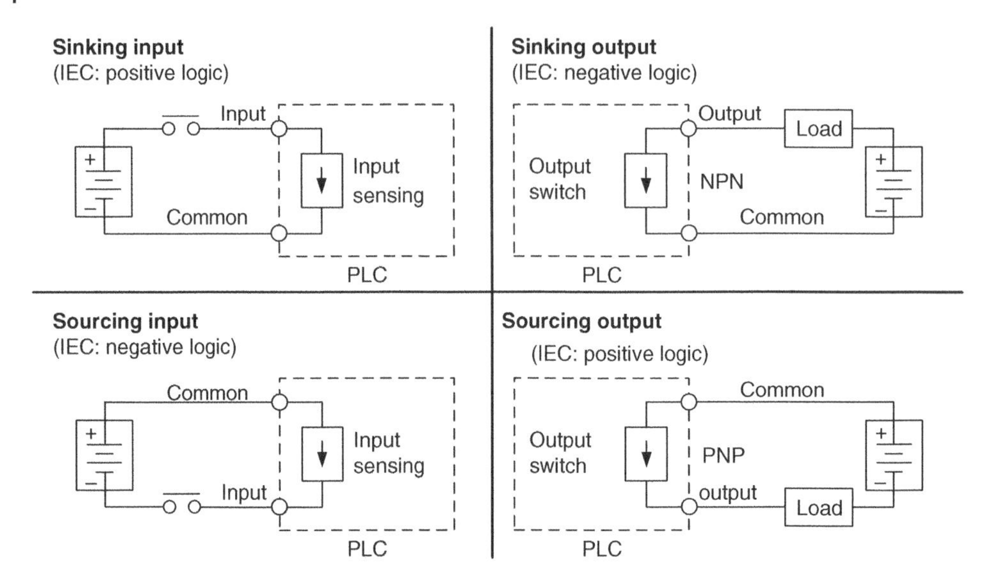

Figure 4.53 Illustration of all four combination.

4.13 Measures for the Avoidance of Systematic Failures

Systematic failures can be avoided with a correct design, engineering, and production of each single product or component used in the SCS and with a correct design, engineering production, **and maintenance** of the machinery SCS.

As detailed in IEC 61508 series, everything starts with a proper activity plan, that includes how modifications to the safety system, once it has been validated, are made. That is organized in a **Functional Safety Plan (FSP)** that is detailed in annex I of IEC 62061 and annex G of ISO 13849-1.

Moreover, the technical aspects to be adopted for a correct approach to the design of the safety system are summarized in two types of safety principles:

- Basic safety principles
- Well-tried safety principles

4.13.1 The Functional Safety Plan

The approach comes from IEC 61508-1 [5], chapter 6, and the aim is to prevent incorrect specification, implementation, or modification issues: in essence, **to avoid the systematic part of the SCS failures**. In other words, it is part of the actions to put in place to reduce the probability of Systematic Failures. The FSP, written by either the component or the machinery manufacturer, depending what product or system we are considering, should:

- Identify the relevant activities linked to the SRP/CS or SCS design process (specification, design, integration, analysis, testing, verification, validation) and detail when they should take place;
- Identify the roles and resources necessary for carrying out and reviewing each of these activities;
- Identify the procedures for release, configuration, documentation, and modification of hardware and software design;
- Establish a validation plan.

Once the machinery has been installed, the manufacturer should indicate, in the instruction manual, that modifications with an impact on the SCS should initiate a return to an appropriate design phase for its hardware and/or for its software, including the validation of the modified safety system. The **Management of Changes** is a key aspect of the FSP.

4.13.2 Basic Safety Principles

They represent good engineering practices that, combined with specific solutions, reduce the probability of Systematic failures.

4.13.2.1 Application of Good Engineering Practices

They could be defined as the "*Proper selection, combination, arrangements, assembly and installation of components/system by applying manufacturer's application notes, e.g. catalogue sheets, installation instructions, specifications, and use of good engineering practice in similar components/systems.*"

Let's consider **some practical examples.**

Inside industrial control panels, there are cables and contactors protected by circuit breakers or fuses. Protecting a 1.5 mm^2 (16 AWG in North America) cable with a 16 A (15 A in the United States and Canada) circuit breaker is normally a non-correct engineering practice that can cause a systematic failure.

Once I was told, by an engineer who designed a control panel with a 10 A contactor protected by a 100 A motor protection (self-protected combination motor controller, in north American language), that the solution was acceptable, since the contactor status was monitored. His idea was: "if the contactor gets stuck, I immediately notice it and I can take immediate counter measures."

The monitoring of a Final Element is required in Architecture 1oo2D of IEC 62061, for example, or in Category 4 of ISO 13849-1. That is needed in order to claim a certain level of Diagnostic Coverage (DC). However, if I do not properly protect the contactor (the final element) from short circuit or even overload, I cannot even claim that I designed a Safety Function, since it lacks a Basic Safety Principle.

4.13.2.2 Use of De-energization Principles

When using the de-energization principle, "*The safe state is obtained by the release of energy to all relevant devices*". This principle is not to be followed when loss of energy would create a hazard, e.g. the release of workpiece caused by loss of clamping force.

That is one of the first aspects an engineer learns: **a Safety control system must be effective in case a wire is cut**. If the wire is normally energized, in case it is interrupted for any reason, the system detects it and places the machine into a safe state. If the loop is normally de-energized, the wire interruption would not be detected and the safety function would be lost, even if two (or more) channels in parallel are used.

4.13.2.3 Correct Protective Bonding (Electrical Basic Safety Principle)

By this Basic Safety Principle it is meant that "*One side of the control circuit, one terminal of the operating coil of each electromagnetic operated device, or one terminal of another electrical device are connected to the protective bonding circuit.*"

This is the concept explained in Section 4.12.4.

Further examples of Basic Safety Principles are indicated in ISO 13849-2, annexes A.1, B.1, C.1, and D.1.

4.13.3 Well-Tried Safety Principles

They are engineering principles that are effective in the design of SCS, to avoid or reduce the probability of failures that can influence the performance of a safety function. Hereafter are some examples taken from ISO 13849-2.

4.13.3.1 Positively Mechanically Linked Contacts

> *Use of positively mechanically linked contacts for, e.g. monitoring function in Category 2, 3, and 4 systems (see EN 50205, IEC 60947-4-1:2001, annex F, IEC 60947-5-1:2003 + A1:2009, annex L).*

This is a synonym of Direct Opening Action, described in § 4.12.1.

4.13.3.2 Fault Avoidance in Cables

> *To avoid short circuits between two adjacent conductors, either*
> - *use cable with shielding connected to the protective bonding circuit on each separate conductor, or*
> - *in flat cables, use one earthed conductor between each signal conductor.*

Normally, shielding is used with analog signals, to avoid interferences by the environment. Digital signals do not require shielding in this respect. However, in case a single connection is used, as shown in Figure 4.54, inside or outside of the control panel, in a Category 3 or 4 SCS, shielding that particular connection is a way to exclude a possible short circuit between cables.

Moreover, as explained in § 4.11.3, provided some proper considerations are made, a fault exclusion on the single cable can be done, even without the use of a shielded cable.

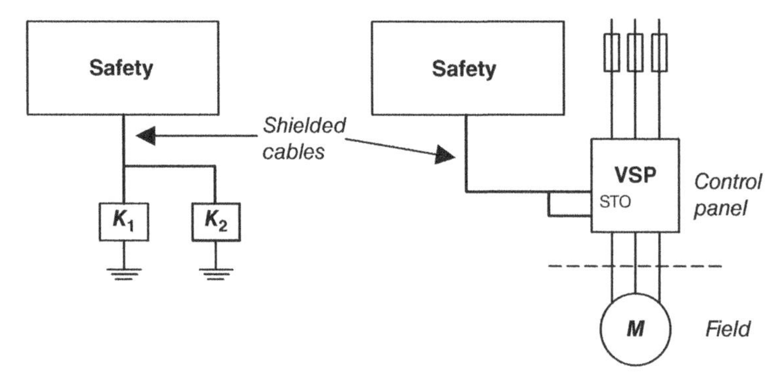

Figure 4.54 Example of the use of shielded cabling for digital signals.

4.14 Fault Masking

4.14.1 Introduction to the Methodology

In § 3.3.6, indications of possible values of DC are given. No indication is present how to deal with the case of several interlocking devices, with free voltage contacts (FVCs), connected in series: a common practice in Machinery. In [30] annex J a methodology is proposed to calculate the DC of those elements in series.

When interlocking devices with FVCs are connected in series, the detection of the single fault can be masked by the actuation of any other non-faulty interlocking device, through the relative opening/closing of the guard.

> ***[ISO 14119] 3 Terms and definitions***
> ***3.37 Fault Masking.*** *Unintended resetting of faults or preventing the detection of faults in the SRP/CS by operation of parts of the SRP/CS which do not have faults.*

Fault masking is an effect that can occur when several devices share a common mechanism for their diagnostic function. If Fault Masking takes place, the capability of the system to detect faults is limited or completely lost. When calculating the parameters in accordance with ISO 13849-1 or IEC 62061, the effect of fault masking on the DC and therefore on the PL and SIL, must be taken into account. The implication is that the possibility of Fault Masking has a direct impact on the value of DC of the Safety Subsystem, and therefore it has an impact on the PL or SIL of the safety function.

This problem only concerns the series connection of interlocking devices with FVCs. Safety switches with monitored outputs (e.g. an RFID device with OSSD output) can reach PL e even with several of those devices connected in series.

In general, to avoid Fault Masking, these solutions are possible:

- To use additional interlocking devices, individually connected to a monitoring device, in combination with appropriate diagnostic procedures.
- To avoid the series of interlocking devices and use dedicated digital inputs for each interlocking device.
- To use interlocking devices with internal diagnostics and monitored outputs (e.g. most OSSD output safety switches).

ISO 14119 [30] contains a methodology that allows, in case of interlocking devices connected in series, to determine the level of DC of the input subsystem. In that respect, it is important to distinguish whether, for each movable guard, one or two interlocking devices are used. The former situation is defined as **Single Arrangement**, the latter as **Redundant Arrangement.**

The standard defines six types of cabling.

4.14.1.1 Redundant Arrangement with Star Cabling

It is a cabling structure whereby each interlocking device is wired with a single cable to a dedicated terminal block inside the Industrial Control Panel. A graphical representation is shown in Figure 4.55. However, the contacts are connected in series inside the panel.

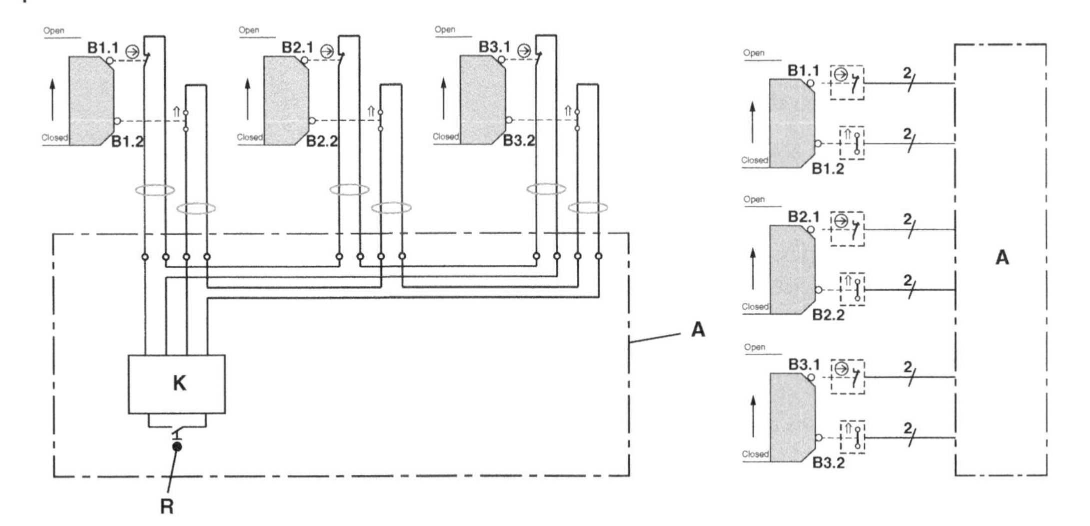

Figure 4.55 Redundant arrangement with star cabling.

Note: the symbol ⇑ means that the contact B1.2 **is shown in its actuated position**: when the movable guard is closed, both contacts B1.1 and B1.2 are closed. When the guard opens, both contacts open (please refer to § 4.12.1.1).

Hereafter the legend for the above and the following drawings:

- **A** is the Electrical Control Panel
- **B1, B2, B3** are three interlocked movable guards with the corresponding Interlocking devices with FVCs
- **K** is the Safety Logic Unit or Logic Solver
- **R** is the manual reset pushbutton

4.14.1.2 Redundant Arrangement with Branch Cabling

It is a cabling structure whereby a single cable from the digital input of the Safety Logic Unit is wired to the first interlocking device and from this interlocking device to the next, and so on, until the last interlocking device and the resulting signal is wired, the same way, back to the Logic Unit. A graphical representation is shown in Figure 4.56.

4.14.1.3 Redundant Arrangement with Loop Cabling

It is a cabling structure whereby a single cable from the electrical cabinet is wired to the first interlocking device and, from this one, to the next, and so on, until the last interlocking device, while the signal return to the electrical cabinet **in a separate cable**. A graphical representation is shown in Figure 4.57.

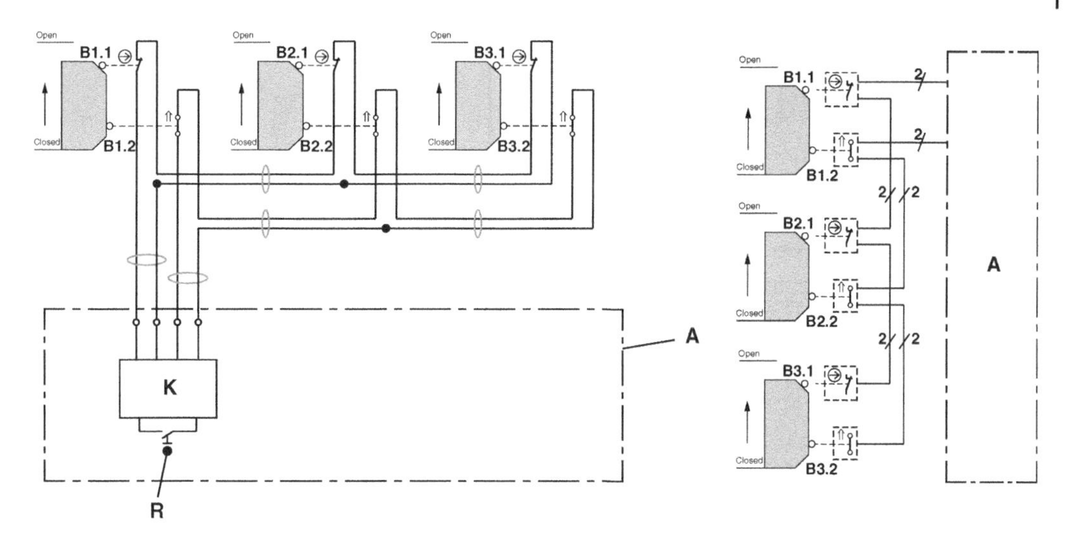

Figure 4.56 Redundant arrangement with branch cabling.

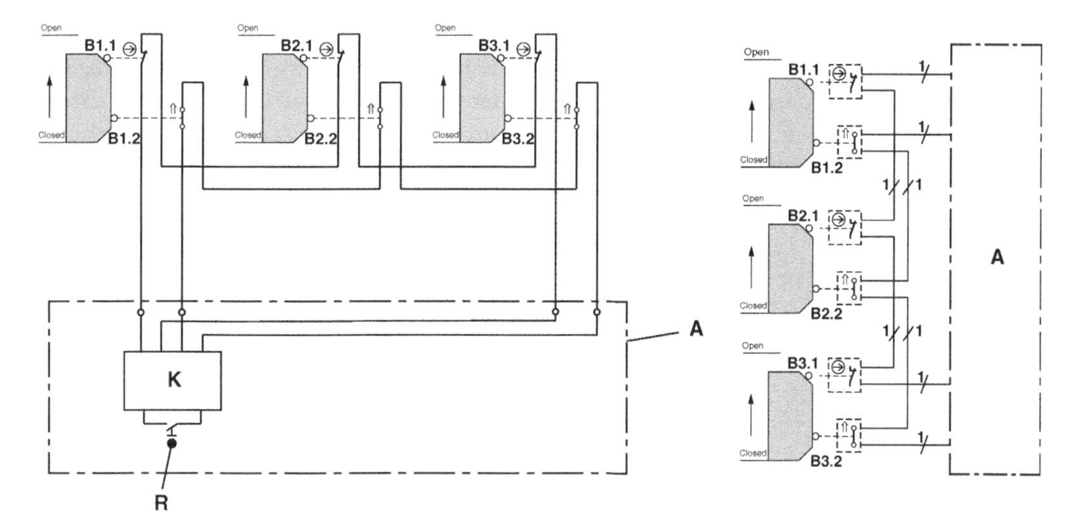

Figure 4.57 Redundant arrangement with loop cabling.

4.14.1.4 Single Arrangement with Star Cabling

It is a cabling structure whereby **each interlocking device has two FVCs**, wired to a dedicated terminal block inside the Industrial Control Panel. Moreover, the contacts are connected in series inside the panel. A graphical representation is shown in Figure 4.58.

Hereafter the legend for the above and the following drawings:

- **A** is the Electrical Control Panel
- **B1, B2, B3** are three interlocked movable guards with the corresponding Interlocking devices with FVCs
- **K** is the Safety Logic Unit or Logic Solver
- **R** is the manual reset Function

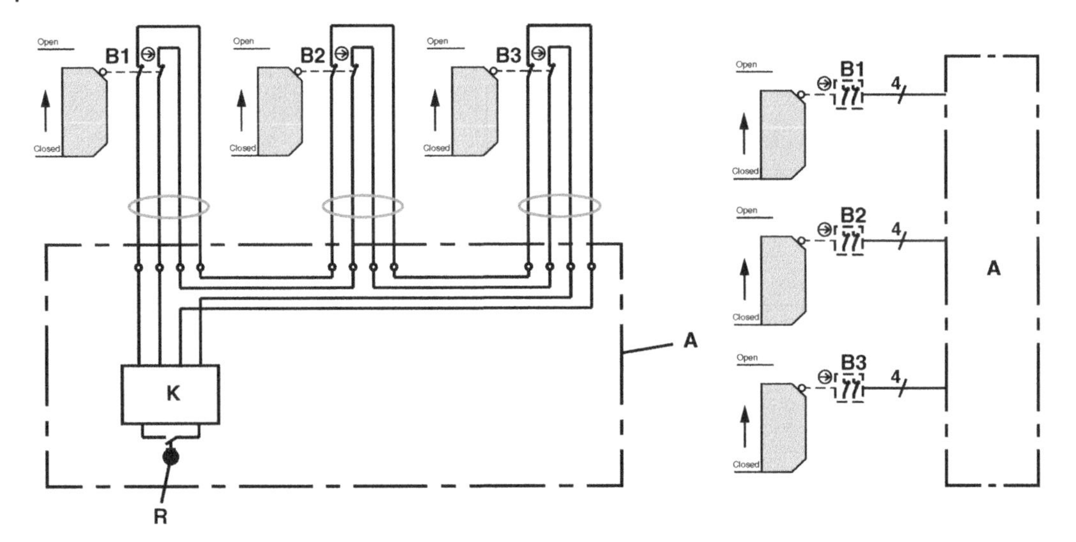

Figure 4.58 Single arrangement with star cabling.

4.14.1.5 Single Arrangement with Branch Cabling

It is a cabling structure whereby each interlocking device has two FVCs. Two cables from the digital input of the Safety Logic Unit are wired to the first interlocking device and from this interlocking device to the next and so on, until the last one and the same way back to the Logic Unit. A graphical representation is shown in Figure 4.59.

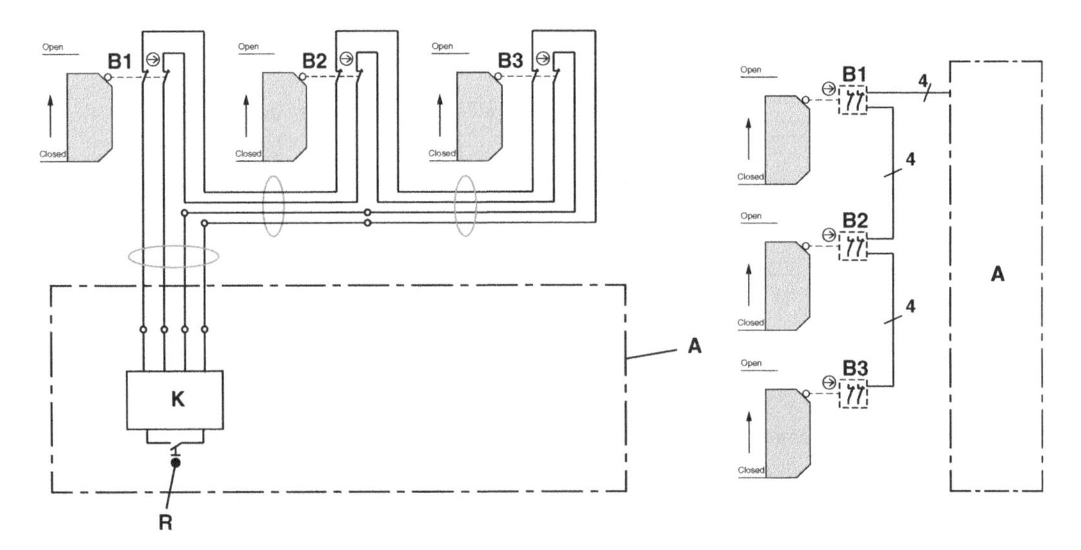

Figure 4.59 Single arrangement with branch cabling.

4.14.1.6 Single Arrangement with Loop Cabling

It is a similar cabling structure whereby two cables from the electrical cabinet are wired to the first interlocking device and from this one to the next and so on, until the last interlocking device; the signals return to the electrical cabinet in a separate cable. A graphical representation is shown in Figure 4.60.

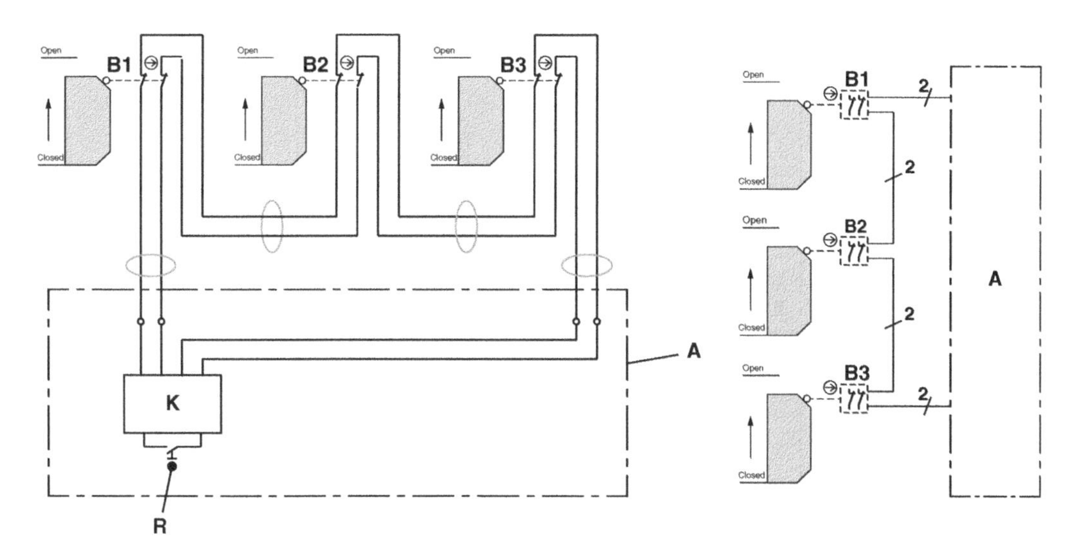

Figure 4.60 Single arrangement with loop cabling.

4.14.2 Fault Masking Example: Unintended Reset

ISO 14119 standard [30] indicates a few examples of possible fault masking. Figure 4.61 shows a *"Cable fault with unintended reset."*

Hereafter the detailed sequence of events shown in the figure.

1. Starting situation with both guards closed.
2. A short circuit with another cable at the same voltage (not indicated), happens to one of the two cables.
3. The **guard B1 is opened** in order to enter the safeguarded area. The Safety Logic Units **detects** there has been **a fault** since the line on the left is de-energized, while the one on the right is not.
4. The **guard B1 is closed,** since the person has left the safeguarded area. The logic unit remains with in error, since it remembers the fault.
5. Before the operator tries to reset the logic, a second one enters the area from a different gate: **guard B2 Opens**. The logic detects the correct opening of the access to the safeguarded area and it "forgets" the input discrepancies detected at Step 3.
6. **Guard B2 is closed and the logic is reset**. The first fault is masked and it remains undetected.
7. **Guard B1 is opened again** and there is a fault on one of the interlock devices. Now, there are two undetected faults and the safety function is lost: therefore, when B1 opens, movements inside the area are not stopped.

The conclusion is that, when interlocking devices with FVCs are connected in series, faults may happen and be masked. The probability depends upon how many entrances to the safeguarded area are present and how frequently they are opened. A masked fault remains and, if a second fault happens, it could compromise the safety function, even if implemented with a redundant architecture.

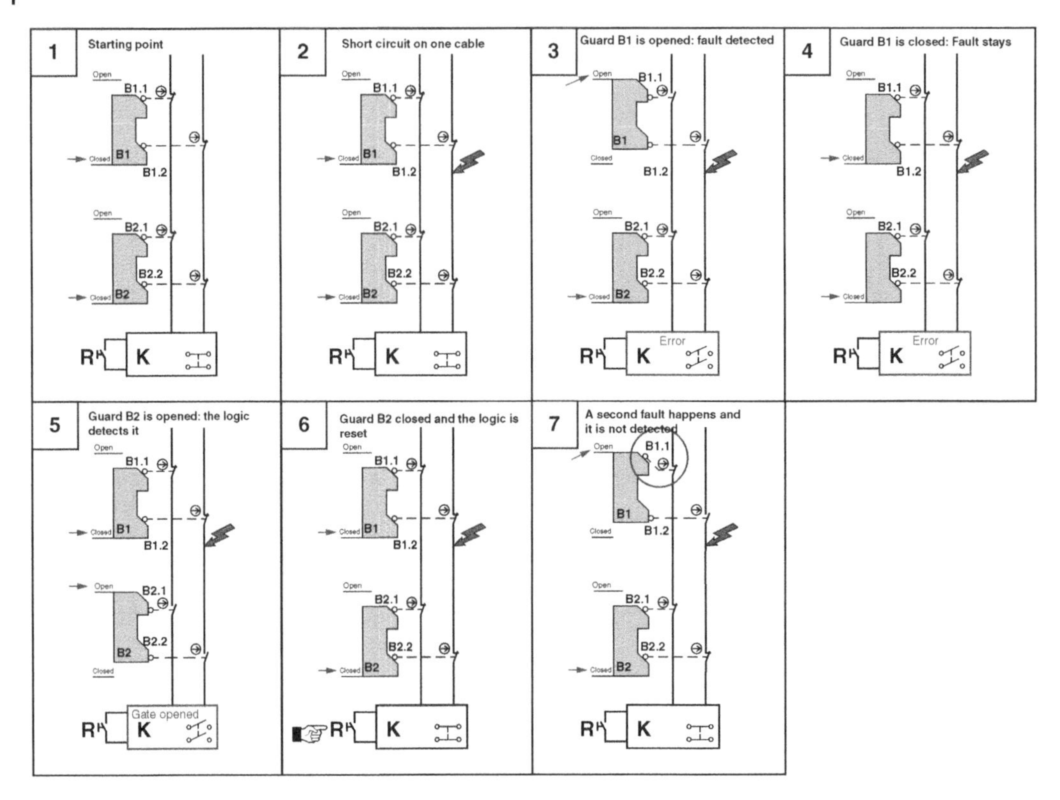

Figure 4.61 Example of a cable fault with unintended reset.

That is the reason why the DC is lower compared with the case when each interlocking device is cabled directly to the safety unit.

4.14.3 Methodology for DC Evaluation

ISO 14119 [30], annex K, describes two possible methods to determine the DC value of the subsystem:

1. **A simplified method**
2. **A regular method**

4.14.3.1 The Simplified Method

With reference to the simplified method, the DC can be estimated directly form Table 4.13.

Data to be considered are:

- Number of frequently used Movable Guards (frequently means more than once per hour).
- Number of additional movable guards (used less than once per hour).

Other info to be considered are:

- Distance between the movable guards
- Accessibility of the movable guards
- Number of operators

Table 4.13 Maximum achievable DC (simplified method).

Number of frequently Used movable guards[a,b]	Number of additional (non-frequently used) Movable guards[c]	Maximum achievable DC[d]
0	2–4	Medium
	5–30	Low
	>30	None
1	1	Medium
	2–4	Low
	≥5	None
>1	≥0	None

[a] If the frequency is higher than once per hour.
[b] If the number of operators capable of opening separate guards exceeds one then the number of frequently used movable guards shall be increased by one level.
[c] The number of additional movable guards may be reduced by one if one of the following conditions are met
- When the minimum distance between any of the guards is more than 5 m or
- When none of the other movable guards is directly reachable from the position where the operator actuates a frequently used movable guard

[d] In any case, if it is foreseeable that fault masking will occur (e.g. multiple movable guards will be open at the same time as part of normal operation or service), then the DC is limited to none.

4.14.3.2 Regular Method

In the regular method, Table 4.14 uses the same parameters of the simplified method, but not to directly determine the maximum achievable DC, but to determine the probability of fault masking (FM).

The maximum achievable DC depends on the fault masking probability level (FM) and the type of cabling used in combination with the switch arrangement and the diagnostic capabilities of the overall system to detect faults.

In ISO 14119 [30], three tables show the maximum reachable DC depending on those parameters.

4.14.3.3 Example

In the example shown in Figure 4.62, there are three interlocked mobile guards to access the safeguarded area (Doors A, B and C).

There is an interlocking device on each movable guard (Single Arrangement). Each interlocking device has two VFCs that open when the movable guard is opened.

The contacts are connected in series to a logic unit that evaluates both channels.

The interlocking devices can be **cabled in Branch or Star.**

Guard A is opened regularly (once four hours).

The other guards (B and C) are normally seldom opened (once a week).

One operator normally handles the robot cell.

Table 4.14 Maximum achievable DC (regular method).

Number of frequently Used movable guards[a,b]	Number of additional movable guards[c]	Fault masking probability level (FM)[d]
0	2–4	1
	5–30	2
	>30	3
1	1	1
	2–4	2
	≥5	3
>1	≥0	3

[a] If the frequency is higher than once per hour.
[b] If the number of operators capable of opening separate guards exceeds one then the number of f frequently used movable guards shall be increased by one level.
[c] The number of additional movable guards may be reduced by one if one of the following conditions are met
- When the minimum distance between any of the guards is more than 5 m or
- When none of the other movable guards is directly reachable from the position where the operator actuates a frequently used movable guard

[d] In any case, if it is foreseeable that fault masking will occur (e.g. multiple movable guards will be open at the same time as part of normal operation or service), then the fault masking probability level (FM) is 3.

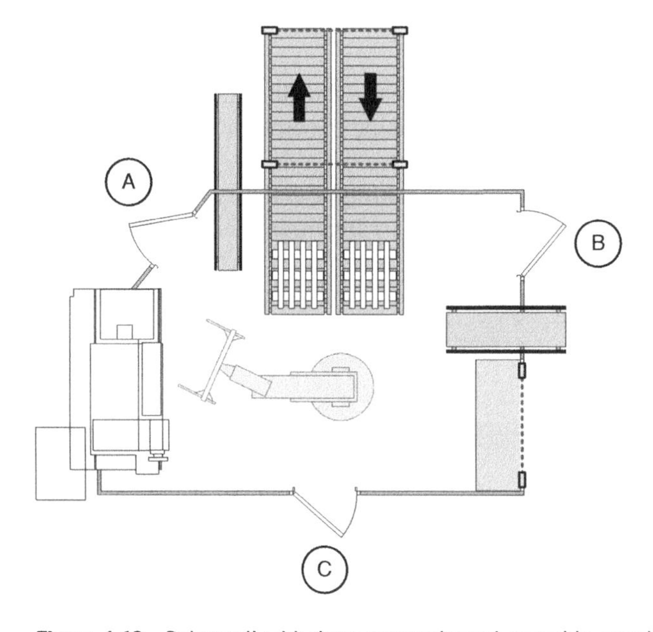

Figure 4.62 Robot cell with three access through movable guards.

Using the **Simplified method**, the maximum DC level that can be obtained is "**Low**" as shown in Table 4.15.

Using the **Regular method,** the resulting Fault Masking probability level is $FM = 2$.

According to Table 4.16, for dynamic signals connected in Branch/Star method, the maximum achievable DC is **"Medium"**.

Table 4.15 Simplified method table applied to the example.

Number of frequently Used movable guards[a,b]	Number of additional movable guards[c]	Maximum achievable DC[d]
0	2–4	Medium
	5–30	Low
	>30	None
1	1	Medium
	2–4	Low
	≥5	None
>1	≥0	None

[a] If the frequency is higher than once per hour.
[b] If the number of operators capable of opening separate guards exceeds one then the number of f frequently used movable guards shall be increased by one level.
[c] The number of additional movable guards may be reduced by one if one of the following conditions are met

- When the minimum distance between any of the guards is more than 5 m or
- When none of the other movable guards is directly reachable from the position where the operator actuates a frequently used movable guard

[d] In any case, if it is foreseeable that fault masking will occur (e.g. multiple movable guards will be open at the same time as part of normal operation or service), then the fault masking probability level (FM) is 3.

Table 4.16 Regular method table applied to the example.

Protected multicore cable with or without positive (+U) voltage wire					
			Maximum achievable DC		
Arrangement	Cabling	Signal evaluation of redundant channels with	*FM* = 3	*FM* = 2	*FM* = 1
Single arrangement	Branch/star	Same polarity (+U/+U)	Medium	Medium	Medium
		Inverse polarity (+U/GND)	None	Low	Medium
		Dynamic signals	Medium	Medium	Medium
	Loop	Same polarity (+U/+U)	Medium	Medium	Medium
		Inverse polarity (+U/GND)	None	Low	Medium
		Dynamic signals	Medium	Medium	Medium

As it can be seen from the example, with the regular method the maximum achievable DC is “Medium”, while the simplified method limits the DC to low. The reason is that the simplified method does not consider how interlocking devices are wired and does not evaluate their signals.

5

Design and Evaluation of Safety Functions

5.1 Subsystems, Subsystem Elements, and Channels

5.1.1 Subsystems

Each safety function is performed by either an SCS (IEC 62061) or an SRP/CS (ISO 13849-1), and **it consists of one or several subsystems**. The concept of a Safety-related control System composed of a series of subsystems is coming from IEC 62061 and, in general, the Reliability theory. ISO 13849-1 is coming from the concept of categories applicable to the entire Safety-related control System. However, the new edition of the ISO standard adopts fully the concept of subsystems, and it clarifies that the categories are applicable to subsystems only.

Also on this aspect, ISO 13849-1 aligns with IEC 62061. The IEC standard indicates the Reliability of a safety function with a level of SIL, with no mention of architectures; the same is done now by the ISO standard. Therefore, with the new edition of ISO 13849-1, when specifying the Reliability level required by a safety function, the correct statement is, for example, **PL d and not PL d, Category 3. The category is only applicable to sub-functions, and it is a mean to calculate the PFH_D of the subsystem**. I could have a Safety-related control System whereby the input subsystem is a Category 1 and the output a Category 4.

5.1.2 Subsystem Element and Channel

In both standards, the term subsystem is used within a strongly defined hierarchy of terminology: **subsystem is the first level subdivision of a system**. The way to identify a **subsystem** is that its failure will result in the loss of the whole safety function. Here is the definition:

> ***[ISO 13849-1] 3.1 Terms and definitions***
> ***3.1.45 Subsystem.*** *Entity which results from a first-level decomposition of an SRP/CS and whose dangerous failure results in a dangerous failure of a safety function.*

The parts resulting from subdivisions of a subsystem are called **subsystem elements**.

> ***[IEC 62061] 3.2 Terms and definitions***
> ***3.2.6 Subsystem Element.*** *Part of a subsystem, comprising a single component or any group of components.*

Functional Safety of Machinery: How to Apply ISO 13849-1 and IEC 62061, First Edition. Marco Tacchini.

A subsystem element can comprise hardware or a combination of hardware and software; software-only components are not considered subsystem elements. SISTEMA [60], the software developed by IFA, defines a subsystem element as a **block**.

Failure in a (subsystem) element does not affect the safety function. **An element** is made of one or more components. A **Component** is therefore a part of a (subsystem) element which cannot be physically divided into smaller parts without losing its particular function. An example is any electrical component, such as a resistor, that has specific electrical characteristics, and that may be connected to other electrical components to form an element.

Let's consider a safety-related control system made of:

- an Interlocking device;
- a Safety controller; and
- two Solenoid Valves that act to the same cylinder.

The Interlocking device is the Input Subsystem and its representation, as part of a safety function, is a sub-function. **The two solenoid valves** are the Output subsystem, and each Valve is a Subsystem Element.

Each of the two valves represents a channel within the subsystem, and this **output subsystem** is a Category 3 or 4, provided a proper Diagnostic is present.

> ***[ISO 13849-1] 3.1 Terms and definitions***
> ***3.1.47 Channel.*** *Element or group of elements that independently implement a safety function or a part of it.*

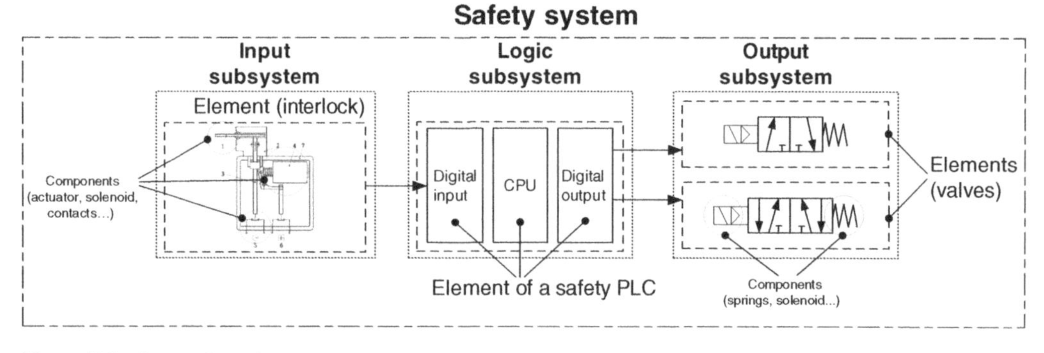

Figure 5.1 Parts of a safety system.

Again, a failure of the subsystem compromises the Safety System, while a failure in one of the channels keeps the Safety System in place.

In Figure 5.2, the Safety-related Control System in Figure 5.1 is shown in a Safety-related Block Diagram representation.

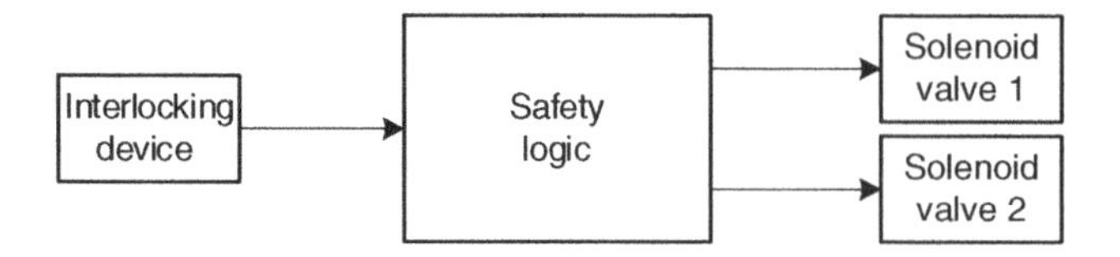

Figure 5.2 The RBD representation of the safety system shown in Figure 5.1.

The two Solenoid Valves are the **output subsystem** of the SCS or of the SPR/CS; each Valve is a **subsystem element**. ISO 13849-1 gives a more practical definition of a subsystem.

One subsystem can be part of several safety functions. For example, a contactor can be used to de-energize a motor either in the event of the detection of a person in a safeguarded area or in the event of the opening of an interlocked movable guard.

5.1.3 Decomposition of a Safety Function

A safety function can be decomposed into sub-functions that are allocated to subsystems. A dangerous failure of any subsystem results in the loss of the safety function.

Figure 5.3 provides an example of decomposition, starting with detection and evaluation of an "initiating event" (e.g. an Emergency pushbutton) and is ending with an output causing a safe reaction of a "machine actuator" (e.g. motor or cylinder).

Safety Function 2 and 3 may refer to a robot cell provided with an access gate (Interlocking device) and an opening protected by an AOPD. Both safety functions trigger the same event: for example, both the Robot and a blade inside the cell are stopped.

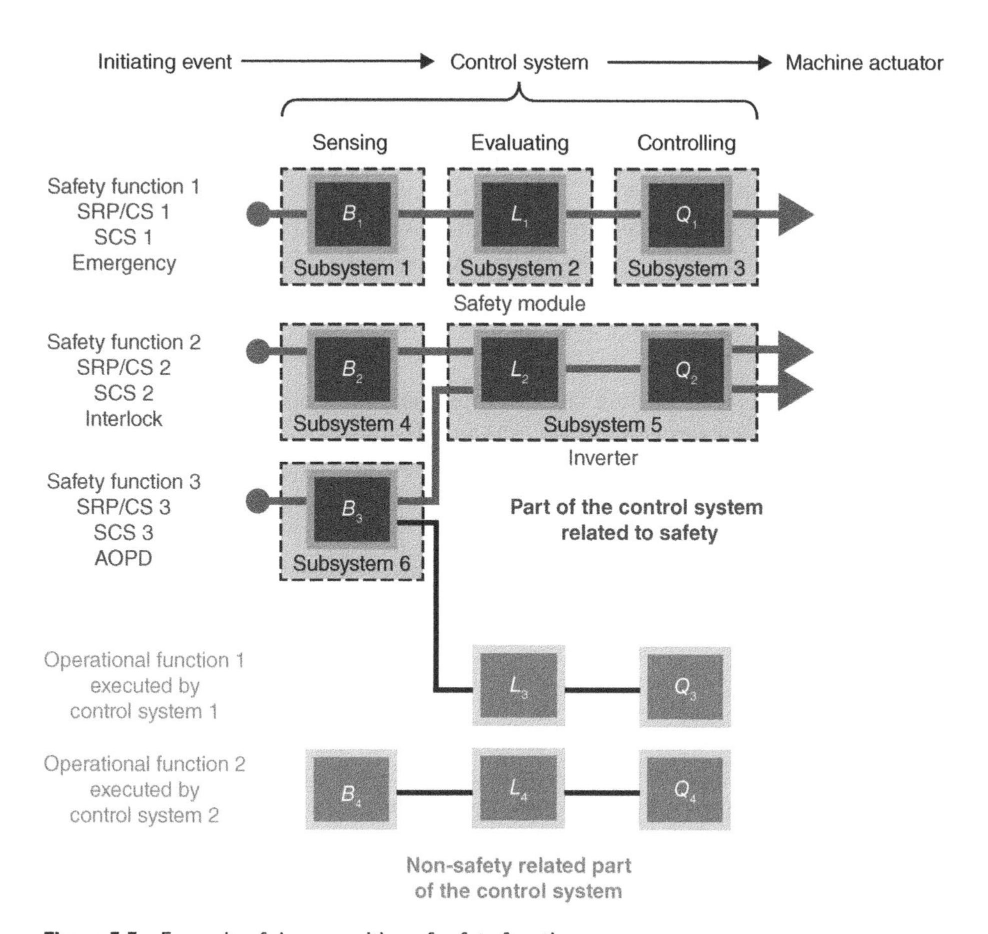

Figure 5.3 Example of decomposition of safety functions.

The AOPD input subsystem may also be shared with the General Purpose PLC for specific automation tasks.

5.1.4 Definition of Device Types

In high demand mode of operations, devices can be used in a safety-related control system only in case they are provided with reliability data. ISO 13849-1 distinguishes four types of devices. Some can be used directly as an SRP/CS or subsystem element in a safety function, because the manufacturer has already developed the device for this specific application: it is the case for devices Type 1 and devices Type 4. Other devices are defined and assessed as an SRP/CS or subsystem element only, through the user's design process: these are devices Type 2 and devices Type 3.

5.1.4.1 Device Type 1

A device Type 1 has the highest integration level. They are normally Pre-designed safety systems with integrated diagnostics. **This type is SIL or PL classified, in line with the intended use**. Devices of this type are developed in accordance with safety standards like IEC 61508 series or IEC 61131-6. Examples are safety light curtains, safe drives/drive functions, some types of Interlocking Devices, safety relays, and safety PLCs.

5.1.4.2 Device Type 2

Devices of this type are not necessarily developed in accordance with safety standards; however, this does not exclude application in accordance with ISO 13849–1 or IEC 62061. Examples are **non-safety-related electronics**, e.g. operational amplifier, proximity switch, pressure sensor, hydraulic valve, and General Purpose PLCs.

Additional information like circuit structure, diagnostic coverage (DC), and consideration of common cause failure (CCF) are needed in order for the user to assess a safety function using these types of devices.

5.1.4.3 Device Type 3

Type 3 devices are components with a failure mode that depends on the operating cycles: they are represented by the classical **Electromechanical component**!

Additional information like the number of operations, number of activations, circuit structure, DC, and consideration of CCF are needed in order for the user to assess a safety function using these types of devices.

Devices of this type are not necessarily developed in accordance with safety standards; however, this does not exclude application in accordance with ISO 13849-1 or IEC 62061. Examples are power and auxiliary contactors (§ 4.12.2), switches, pneumatic valves, and interlocking devices.

5.1.4.4 Device Type 4

Device Type 4 is a special case of device Type 1. This type has **no random failures** which lead to a dangerous fault: that means the probability of a dangerous fault occurring is negligible ($PFH_D \approx 0$). For components of this type, either one of the following applies for each potential fault:

- fault exclusion is in accordance with ISO 13849-1 or IEC 62061; or
- fault always leads to a safe condition.

Table 5.1 summarizes the different reliability parameters for each type of device.

Table 5.1 Characteristic values of device types.

<table>
<tr><th rowspan="2">Characteristic value</th><th colspan="4">Device type</th><th rowspan="2">Comments</th></tr>
<tr><th>1</th><th>2</th><th>3</th><th>4</th></tr>
<tr><td>PL</td><td rowspan="2">X</td><td></td><td></td><td></td><td>ISO 13849-1</td></tr>
<tr><td>SIL</td><td></td><td></td><td></td><td>IEC 62061</td></tr>
<tr><td>PFH_D</td><td>X</td><td></td><td></td><td></td><td>ISO 13849-1 and IEC 62061</td></tr>
<tr><td>Category</td><td rowspan="2">X</td><td rowspan="2">X</td><td rowspan="2">X</td><td></td><td rowspan="2">ISO 13849-1 and IEC 62061
One of the characteristic values is required</td></tr>
<tr><td>HFT</td><td></td></tr>
<tr><td>$MTTF_D$</td><td></td><td rowspan="4">X</td><td></td><td></td><td rowspan="4">ISO 13849-1 and IEC 62061
One of the characteristic values is required</td></tr>
<tr><td>λ_D</td><td></td><td></td><td></td></tr>
<tr><td>$MTTF$</td><td></td><td></td><td></td></tr>
<tr><td>$MTBF$</td><td></td><td></td><td></td></tr>
<tr><td>B_{10D}</td><td></td><td></td><td rowspan="2">X</td><td></td><td rowspan="2">ISO 13849-1 and IEC 62061
One of the characteristic values is required</td></tr>
<tr><td>B_{10}</td><td></td><td></td><td></td></tr>
<tr><td>RDF</td><td></td><td rowspan="2">O[a]</td><td rowspan="2">O[a]</td><td></td><td>ISO 13849-1</td></tr>
<tr><td>SFF</td><td></td><td></td><td>IEC 62061[a]</td></tr>
<tr><td>T_{10d}</td><td rowspan="2">X</td><td rowspan="2">X</td><td></td><td rowspan="2">X</td><td rowspan="2">ISO 13849-1 and IEC 62061
Exactly one of the characteristic values is required</td></tr>
<tr><td>T_M</td><td></td></tr>
</table>

X, mandatory; O, optional.
[a] If there is no determined safety value from the manufacturer ($MTTF_D$ or B_{10D}).

5.1.4.5 Implication for General Purpose PLCs

From the above classification, it is clear that a microprocessor-based component, for example a digital input card of General Purpose PLC, **can be provided with an MTBF value, but that does not mean it can be used as the only component in a safety application**. A microprocessor-based component, suitable for safety applications, is now classified as a Type 1 component and, in general, in order to be suitable to be used in a Safety Application, it has to be provided with the indicated characteristic values. The reason is that complex electronic components, like General Purpose PLCs, **are not considered as well-tried components** and, therefore, cannot be used in Category 1 subsystems.

> ***[ISO 13849-1] 6.1.11 Well-tried component***
> *[...] Complex components (e.g. PLC, microprocessor, and application-specific integrated circuit) shall not be considered as equivalent to well-tried.*

However, **they may be used in all other Categories if they fulfill all requirements of the standard**, including the basic safety principles and, for Categories 2, 3, and 4, also the well-tried safety principles. So they should, for example, be designed to withstand the expected environmental conditions and used according to the manufacturer's instructions and application notes.

Paragraph 7.3 of ISO 13849-1 also clarifies that components, for which the safety-related embedded software (SRESW) requirements are not fulfilled, for example a General Purpose PLCs, may be used in safety-related applications under the following conditions:

- the subsystem is limited to either PL a or PL b, and it uses a Category B, 2 or 3; or
- the subsystem is limited to PL c with Category 2 or PL d with Category 3, and it is necessary to fulfill the diversity requirements of the CCF, where both channels use diverse technologies, designs, or physical principles.

Moreover, the associated hardware and its safety-related application software (SRASW) should be assessed with respect to the requirements of ISO 13849-1, especially annex F (Measures against Common Cause Failures).

5.2 Well-Tried Components

The concept is defined in both ISO 13849-1 and IEC 62061. Those types of components are compulsory for Category 1 (ISO 13849-1) and in both Basic Subsystem Architecture A and B (IEC 62061).

In all other Categories and Architectures, non-well-tried components can be used as well, provided they have Reliability data. Well-tried components are needed in those two cases since the Safety-related Control System has a single channel and no Diagnostic. Here is the definition.

> ***[ISO 13849-1] 3.1 Terms and definitions***
> ***3.1.50 Well-Tried Component****. Component successfully used in safety-related applications.*

Note: This book lists several definitions given by various standards. Please bear in mind that, in general, in order to understand the meaning of a definition stated in the standards, the whole text of the standard has to be read, since the definition does not list "applicable requirements." For example, in the case of the definition of the well-tried components of ISO 13849-1, the requirements are stated in § 6.1.11 of the standard.

A **well-tried component** for safety-related applications is a component that shall be

- either widely used in the past with documented successful results in similar applications. This aspect opens the door to the "proven in use" concept of IEC 61508-2 or
- listed in the informative annexes A–D of ISO 13849-2:2012; or
- made, verified, and validated using principles that demonstrate its suitability and reliability for safety-related applications according to relevant product and application standards.

It remains as a key guideline that a particular component can be considered as well-tried only if suitable for the specific application, considering for example the environmental influences.

A switch with positive mode actuation is a well-tried component if it complies with IEC 60947-5-1, annex K. That means if a manufacturer has a new switch in his product range and it complies with the above standard, it is a well-tried component from the first unit produced. However, if a manufacturer has a **new component whose category is not mentioned in the annexes of ISO 13849-2** and he wants it to be defined as well-tried, he needs to use it extensively first.

Despite opening the door to the Proven in Use concept, both standards make it clear that **complex electronic components** (e.g. PLC, microprocessor, application-specific integrated circuit) **cannot be considered as equivalent to "well-tried."** For all the above reasons, the concept of well-tried components is not the same as Proven in Use according to IEC 61508 or Prior Use according to IEC 61511-1.

There are PLC manufacturers that have input and output digital modules with reliability data, but only if used in a redundant (1oo2) configuration. The reason is that in a single configuration, it would be a 1oo1 safety subsystem or basic subsystem architecture A according to IEC 62061. In which case, only well-tried components can be used: again, complex electronics cannot be considered well-tried.

5.2.1 List of Well-Tried Components

ISO 13849-2 annex A.3 and D.3 give some examples of well-tried components. We also add a proposal, initiated by IFA [72] and discussed in a Task Group of ISO TC 199 WG8, of additional examples of well-tried components. Please bear in mind that a well-tried component for some applications could be inappropriate for other applications.

5.2.1.1 Mechanical Systems

Table 5.2 shows some examples of well-tried mechanical components.

Table 5.2 List of well-tried mechanical components.

Well-tried component	Conditions for "well-tried"	Standard or specification
Screw	All factors influencing the screw connection and the application are to be considered. See table A.2.	Mechanical jointing such as screws, nuts, washers, rivets, pins, bolts, etc. is standardized.
Spring	See table A.2, "Use of well-tried spring."	Technical specifications for spring steels and other special applications are given in ISO 4960.
Cam	All factors influencing the cam arrangement (e.g. part of an interlocking device) are to be considered. See table A.2.	See ISO 14119 (interlocking devices).
Break-pin	All factors influencing the application are to be considered. See table A.2.	
Brake or locking unit (e.g. pneumatically, hydraulically, or electrically released)	The component fulfills the basic and well-tried safety principles with regard to the intended application; Use of the component according to the specification of the manufacturer.	

5.2.1.2 Pneumatic Systems

Table 5.3 shows a possible list of well-tried pneumatic components.

Table 5.3 List of well-tried pneumatic components.

<table>
<tr><th>Well-tried component</th><th>Conditions for "well-tried"</th><th>Standard or specification</th></tr>
<tr><td>Directional-control valve</td><td rowspan="5">The component fulfills the basic and well-tried safety principles with regard to the intended application.

On-board electronics, if available, do not participate in executing the safety function.[a]
Use of the component according to the specification of the manufacturer.</td><td rowspan="2"></td></tr>
<tr><td>Continuous control valve with safe shut-off function</td></tr>
<tr><td>Stop (shut-off) valve – e.g. gate valve, AND-valve, shuttle valve, non-return (check) valve</td><td rowspan="3"></td></tr>
<tr><td>Flow valve</td></tr>
<tr><td>Pressure valve</td></tr>
</table>

[a] Serial communication units for valve manifolds are complex and cannot be assumed to be well-tried.
If shutting-off the power supply puts the valves into safe state and if a fault exclusion for cross-circuit connections between power supply and control bus (see annex D) can be made, and if for the installed valves, the conditions for "well-tried" are fulfilled, the manifold can be considered a well-tried component.

5.2.1.3 Hydraulic Systems

Table 5.4 shows examples of well-tried Hydraulic components.

Table 5.4 List of well-tried hydraulic components.

Well-tried component	Conditions for "well-tried"	Standard or specification
Directional-control valve	The component fulfills the basic and well-tried safety principles with regard to the intended application.	
Continuous control valve		
Stop (shut-off) valve (e.g. gate valve, non-return (check) valve, counterbalance valve)	On-board electronics, if available, does not participate in executing the safety function;	
Flow valve	Use of the component according to the specification of the manufacturer.	
Pressure valve		

5.2.1.4 Electrical Systems

Table 5.5 shows examples of well-tried electromechanical components.

Table 5.5 List of well-tried electrical components.

Well-tried component	Conditions for "well-tried"	Standard or specification
Switch with positive mode actuation (direct opening action), e.g.: - pushbutton; - position switch; - cam-operated selector switch, e.g. for mode of operation		IEC 60947-5-1:2003, annex K
Emergency stop device		ISO 13850 IEC 60947-5-5
Fuse		IEC 60269-1
Circuit-breaker		IEC 60947-2
Switches, disconnectors		IEC 60947-3
Differential circuit-breaker/ RCD (residual current device)		IEC 60947-2:2006, annex B
Main contactor 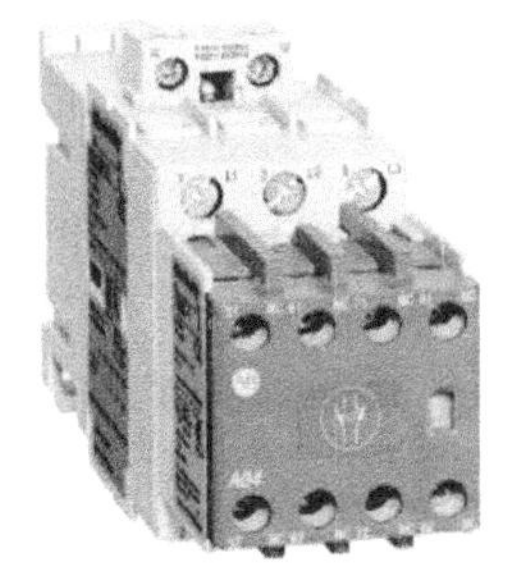	Only well-tried if a) other influences are taken into account, e.g. vibration, b) failure is avoided by appropriate methods, e.g. over-dimensioning (see table D.2), c) The current to the load is limited by the thermal protection device, and	IEC 60947-4-1

Table 5.5 (Continued)

Well-tried component	Conditions for "well-tried"	Standard or specification
	d) The circuits are protected by a protection device against overload. **Note: Fault exclusion is not possible.**	
Control and protective switching device or equipment (CPS)		IEC 60947-6-2
Auxiliary contactor (e.g. contactor relay) 	Only well-tried if a) other influences are taken into account, e.g. vibration, b) there is positively energized action, c) failure is avoided by appropriate methods, e.g. over dimensioning (see table D.2), d) The current in the contacts is limited by a fuse or circuit-breaker to avoid the welding of the contacts, And e) Contacts are positively mechanically guided when used for monitoring. **Note: Fault exclusion is not possible.**	EN 50205 IEC 60947-5-1 IEC 60947-4-1:2001, annex F
Relay  RCG RDG	Only well-tried if a) other influences are taken into account, e.g. vibration, b) positively energized action, c) failure avoided by appropriate methods, e.g. over-dimensioning (see table D.2), and d) the current in the contacts is limited by fuse or circuit-breaker to avoid the welding of the contacts. **Note: Fault exclusion is not possible.**	IEC 61810-3
Transformer		IEC 61558
Cable	Cabling external to enclosure should be protected against mechanical damage (including, e.g. vibration or bending).	IEC 60204-1
Plug and socket		According to an electrical standard relevant for the intended application. For interlocking, see also ISO 14119.

(Continued)

Table 5.5 (Continued)

Well-tried component	Conditions for "well-tried"	Standard or specification
Temperature switch		For the electrical side, see EN 60730-1
Pressure switch		For the electrical side, see IEC 60947-5-1. For the pressure side, see annexes B and C.
Solenoid for valve		

Again, bear in mind that a well-tried component **is not to be confused with a proven in-use component**. As shown in this paragraph, well-tried component is a "well known" component, or better, **a "simple" component having predictable failure modes.** Finally, please notice that the following components cannot be considered as well-tried components: Inductances, Resistors, resistor networks, potentiometers, Capacitors, Discrete semiconductors (e.g. diodes, Zener diodes, transistors, triacs, thyristors, voltage regulators, quartz crystal, phototransistors, light-emitting diodes [LEDs]) and Optocouplers.

5.3 Proven in Use and Prior Use Devices

5.3.1 Proven in Use

The idea of proven in use is coming from the process industry, whereby a component does not have Reliability data; however, it is widely used in that specific application, and the manufacturer has gathered details of its failures in a structured way. Here is the definition:

> ***[IEC 61508-4] 3.8 Confirmation of safety measures***
> ***3.8.18 Proven in Use****. Demonstration, based on an analysis of operational experience for a specific configuration of an element, that the likelihood of dangerous systematic faults is low enough so that every safety function that uses the element achieves its required safety integrity level.*

Proven in use is based on the manufacturer's design basis (e.g. temperature limit, vibration limit, corrosion limit, desired maintenance support) **for his device**. That definition does not consider that the user, in its process environment, has the capability of gathering Reliability data of specific components. That is the reason why IEC 61511-1, in the 2015 edition, defined "**Prior**

Use" components. It deals with the device's installed performance, within a process sector application, in a specific operating environment, which is often different from the manufacturer's design basis.

5.3.2 Prior Use Devices

The concept was introduced in the **second edition of IEC 61511-1** to allow a process owner to use components without appropriate failure rates provided by the component manufacturer.

> ***[IEC 61511-1] 3.2 Terms and definitions***
> ***3.2.51 Prior Use.*** *Documented assessment by a user that a device is suitable for use in a SIS and can meet the required functional and safety integrity requirements, based on previous operating experience in similar operating environments.*

A prior use evaluation involves gathering information concerning the device performance in a similar operating environment. It demonstrates the functionality and integrity of the installed device, including the process interfaces, full device boundary, communications, and utilities. The main intent of the prior use evaluation is to gather evidence that the dangerous systematic failures have been reduced to a sufficiently low level compared to the required safety integrity. **Prior use data** may contribute to a database for the calculation of hardware failure rates.

5.3.3 Prior Use vs Proven in Use

In order to summarize the proven in use and prior use concepts, we may state that **Proven In Use** criteria are based on application data returned to the manufacturer, while **Prior use** deals with the performance of installed devices within a given user application in a specific operating environment. The term **Prior use** was created specifically for the process sector to include operational experience with devices. **Proven in use** is based primarily on manufacturer experience and returns of faulty devices from the field.

5.4 Use of Process Control Systems as Protection Layers

One of the major differences between high and low demand mode safety standards is the use of Automation (General Purpose) PLCs or Distributed Control Systems to perform safety functions. IEC 61511-2 clearly recognizes that possibility (annex A).

> ***[IEC 61511-2] Annex A. 9.3 Guidance to "Requirements on the Basic Process Control System as a protection layer"***
> ***A.9.3.2*** *Risk reduction of ≤10 may be claimed from a BPCS protection layer without the need to comply with IEC 61511-1. This allows the BPCS to be used for some risk reduction without the need to implement BPCS protection layers to the requirements of IEC 61511-1.*
> *The risk reduction claim of ≤10 for the BPCS protection layer should be justified by consideration of the risk reduction capability of the BPCS (**determined by Reliability analysis or prior use data**) and the procedures used for configuration, modification and operation and maintenance.*

The process environment is considered more organized and structured, and that allows the implementation of certain rigor like the one stated again in annex A of IEC 61511-2.

> *When allocating risk reduction to a BPCS protection layer, it is important to ensure that access security and change management are provided. Administrative controls should be used to control access to and modification of the protection layer within the BPCS. Bypassing a BPCS protection layer (e.g., placing BPCS function in manual) should require approval and compensating measures should be in place prior to bypassing to ensure the required risk reduction is maintained. Means should be provided to validate the functionality of the protection layer after changes are made to the BPCS that could affect the operation of the BPCS protection layer.*

On the other hand, the exclusive use of Automation PLCs, or logic controllers, to perform safety applications **is not recognized as suitable for Machinery**: please refer also to § 5.1.4.5.

5.5 Information for Use

The machinery manufacturer should include, in the instruction for use, information about the Safety System and how to properly maintain it.

In the previous edition of ISO 13849-1, it was required to list all safety functions with their level of Reliability. Specific information was required on both the category and performance level of the SRP/CS, as follows:

- the Category, B, 1, 2, 3, or 4;
- the performance level, a, b, c, d, or e.

The ISO team (ISO Technical Committee No. 199, Working Group 8) realized that information was not really relevant for the end user. That is the reason why the new editions of both ISO and IEC standards distinguish between:

- information for the SRP/CS integrator (ISO 13849-1) or Information for use given by the manufacturer of subsystems (IEC 62061); and
- information for the user (ISO 13849-1) or Information for use given by the SCS integrator (IEC 62061)

The latter being the normal situation for a Machinery manufacturer who has to write the Information for use of his machine. **What information shall he give**?

Both standards give a long and exhaustive item list. Hereafter just a few important concepts to remember.

5.5.1 Span of Control

The term was first used in ISO 11161 in 2007; it was adopted in 2015 by ISO 13850, and it will be more and more used in future International Technical Standards dealing with Machinery Safety. Hereafter is its latest definition:

[ISO 11161] [23] ***3 Terms and definitions***
3.8 Span-of-Control. *Predetermined portion of the machinery under control of a specific device or safety function.*

***Note 1 to entry**: A protective device could initiate a stop of a machine or a portion of a machine. For example, an emergency stop pushbutton could cause a local stop or global stop (see ISO 13850).*

The starting point is that a Machinery or a production line may have several safety sensors or devices like interlocking devices, Sensitive Protective Equipment, Emergency pushbuttons, or hold-to-run controls. Each of these components may stop part of the machine; that can be perfectly acceptable, provided a proper risk assessment was performed.

It is important, though, to clearly show to the user the area where each safety sensor operates. That can be done by indicating the **Span of Control in the instruction manual** and/or with signs and labels on the machine.

5.5.2 Information for the Machinery Manufacturer

The information which is important for the correct integration of SRP/CS shall be given to the integrator. This should include:

- limits (e.g. environmental conditions) appropriate information to ensure the continued justification of the fault exclusions;
- descriptions of the interfaces with the SRP/CS and protective devices;
- response time;
- assumed operating limits (e.g. demand frequency);
- control modes and reset;
- maintenance checklists;
- how to access and replace the parts of the SRP/CS;
- means for easy and safe troubleshooting;
- mission time or T_{10D} or Useful lifetime.

Specific information for each safety function on categories and performance level shall be provided:

- the categories of the subsystems forming the SRP/CS;
- the performance level, a, b, c, d, or e;
- the PFH_D value for SRP/CS or the SCS, if relevant for subsystem(s).

5.5.3 Information for the User

The user does not need to know the level of reliability of each safety function. He needs information for proper maintenance:

- setting;
- teaching/programming;
- process /tool changeover;
- cleaning;
- preventive and corrective maintenance;
- troubleshooting/fault finding;

- nature and frequency of inspections and safety functions;
- instructions relating to maintenance operations;
- drawings/diagrams enabling maintenance personnel to perform their tasks;
- information about replacement of components at or before the T_{10D} period ends.

5.6 Safety Software Development

5.6.1 Limited and Full Variability Language

Nowadays, more and more Safety functions make use of Programmable Logic Controllers. In other term, the "Logic" in an I–L–O architecture is implemented within a Safety PLCs or Safety Programmable Module.

A Safety PLC needs to be programmed by the user with the specific Safety Function logic. In other terms, **the machinery manufacturer** writes the Application Software. ISO 13849-1 calls it Safety-related application software (**SRASW**):

> ***[IEC 61508-4] 3.2 Equipment and devices***
> ***3.2.7 Application Software*** *(application data or configuration data). Part of the software of a programmable electronic system that specifies the functions that perform a task related to the EUC rather than the functioning of, and services provided by the programmable device itself.*

That is different from the **System Software**, which is written by the **PLC manufacturer.** ISO 13849-1 calls it Safety-related embedded software (**SRESW**):

> ***[IEC 61508-4] 3.2 Equipment and devices***
> ***3.2.6 System Software.*** *Part of the software of a PE system that relates to the functioning of, and services provided by, the programmable device itself, as opposed to the application software that specifies the functions that perform a task related to the safety of the EUC.*

Normally, in a Safety PLC, the system software is embedded:

> ***[IEC 62061] 3.2 Terms and definitions***
> ***3.2.60 Embedded Software****. Software, supplied as part of a pre-designed subsystem, that is not intended to be modified and that relates to the functioning of, and services provided by, the SCS or subsystem, as opposed to the application software.*

Those are the key concepts to have in mind when we want to understand the software part of a Safety Function. ISO 13849-1 has similar definitions.

In case of a safeguarded space with two access gates, where each access has an interlocking device, physically, each interlocking device enters the Safety PLC installed in the control panel. The output of the safety PLC stops three motors. The first interlocking device should stop only one of the motors, and the second stops all of them, for example. To implement the two safety functions, a Safety PLC is used. It comes with Embedded software, and the manufacturer has to program the application software.

One way to classify the Application Software is between Limited and Full Variability Language:

Limited Variability Language (LVL). The programmer has a limited number of "Building Blocks" to use. **Examples** of LVL are given in IEC 61131-3 [19]. They include:

- **Ladder Diagram (LD).** It is a graphical language consisting of a series of input symbols (representing behavior similar to devices such as normally open and normally closed contacts) interconnected by lines (to indicate the flow of current) to output symbols (representing behavior similar to relays), as shown in Figure 5.4.
- **Function Block Diagram (FBD).** in addition to Boolean operators, as shown in Figure 5.4, it allows the use of more complex functions such as data transfer file, block transfer read/write, shift register, and sequencer instructions.

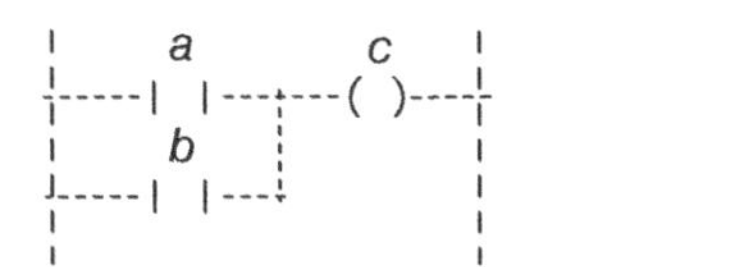


```
   +--------+
a--| >=1    |---c
b--|        |
   +--------+
```

«Wired –OR» in LD language **Boolean «OR» in FBD language**

Figure 5.4 Example of a Boolean "OR."

- **Sequential Function Chart (SFC).** A graphical representation of a sequential program consisting of interconnected steps, actions, and directed links with transition conditions. The SFC elements provide a means of partitioning a programmable controller program organization unit into a set of steps and transitions interconnected by directed links. Associated with each step is a set of actions, and each transition is associated with a transition condition.
- **Boolean Algebra**. It is a low-level language based on Boolean operators such as AND, OR, and NOT with the ability to add some mnemonic instructions.

Full Variability Language (FVL). It gives the programmer a wide range of possibilities. FVL is found in general-purpose computers. **Examples** are instruction lists (IL) and structured text (ST); C, Pascal, C++, Java, SQL, ADA, MATLAB, and Simulink.

Here are the Definitions.

> ***[IEC 62061] 3.2 Terms and definitions***
> ***3.2.62 Limited Variability Language*** *(LVL). Type of language that provides the capability to combine predefined, application specific, library functions to implement the safety requirements specifications.*
>
> ***[IEC 62061] 3.2 Terms and definitions***
> ***3.2.61 Full Variability Language*** *(FVL). Type of language that provides the capability to implement a wide variety of functions and applications.*

One advantage of a LVL is that it forbids unsafe constructions (e.g. nesting loops).

5.6.2 The V-Model

The V-model is a graphical representation of a system development lifecycle. The V-model summarizes the main steps to be taken in conjunction with the corresponding deliverables within the computerized system validation framework or project life cycle development. It describes the activities to be performed and the results that have to be produced during software development **in order to reach the correct level of Software Systematic Capability**.

In other words, the development of safety-related embedded or application software shall primarily consider the avoidance of faults during the software lifecycle by adopting the principles contained in the V-model.

Software development according to a V-model was originally recommended by IEC 61508-3, and it is shown in Figure 5.5.

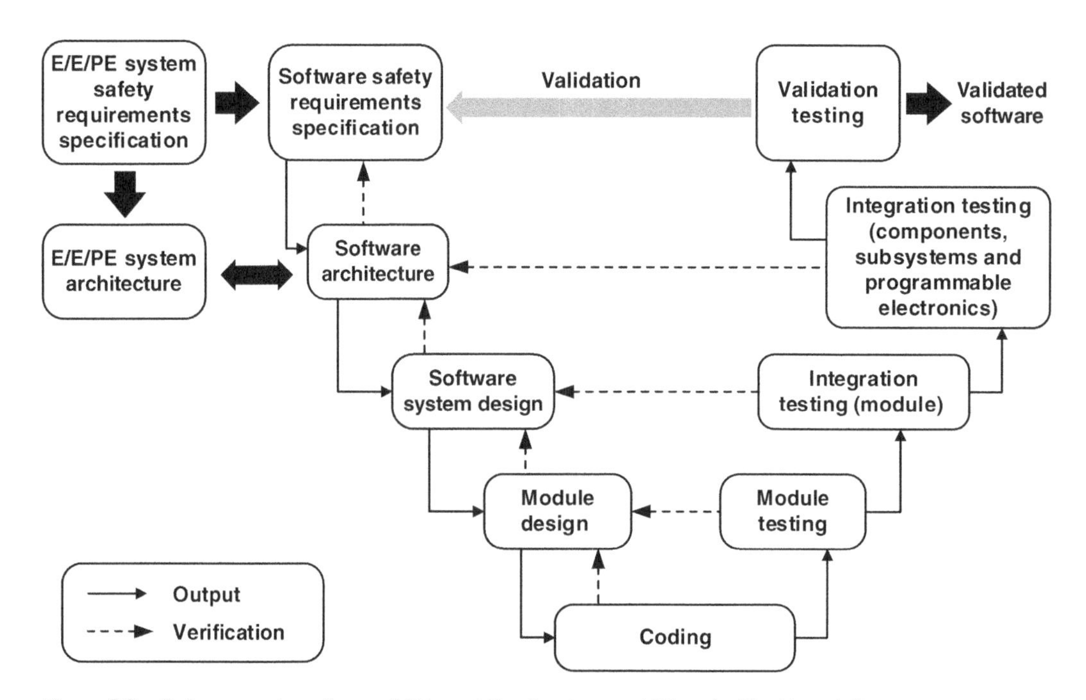

Figure 5.5 Software systematic capability and the development lifecycle (the V-model).

The left side of the "V" starts with the E/E/PE **safety requirement specifications**; those are translated in a certain **Software design,** and finally, it is written in a certain **code**. The right side of the "V" represents **integration of parts and their validation**. Each software module is tested and verified against its requirements. Finally, the whole code is validated against the intended use.

Two important terms are used in the model: Verification and Validation: **Verification is against the requirements** and **validation is against the real world** or the user's needs.

> ***[IEC 61508-4] 3.8 Confirmation of safety measures***
> ***3.8.1 Verification.*** *Confirmation by examination and provision of objective evidence that* ***the requirements have been fulfilled****.*
> ***3.8.2 Validation.*** *Confirmation by examination and provision of objective evidence that the* ***particular requirements for a specific intended use*** *are fulfilled.*

5.6.3 Software Classifications According to IEC 62061

IEC 62061 defines three different situations in relation to the hardware/software combination that the machinery manufacturer may have. They are summarized in Table 5.6.

Table 5.6 Different levels of application software.

<table>
<tr><th>SW level</th><th>Main principle</th><th>Subprinciple</th><th>Example</th></tr>
<tr><td>1</td><td>Platform (combination of hardware and software) pre-designed according to IEC 61508, or other functional safety standards linked to IEC 61508 e.g. IEC 61131-6.
Application software making use of a limited variability language (LVL).</td><td rowspan="2">Application software complying with IEC 62061.</td><td>Safety PLC with LVL or Safety programmable relay</td></tr>
<tr><td>2</td><td rowspan="2">Platform (combination of hardware and software) pre-designed according to IEC 61508, or other functional safety standards linked to IEC 61508 e.g. IEC 61131-6.
Application software making use of another language than a limited variability language (LVL).</td><td>Safety PLC with FVL complying with IEC 62061</td></tr>
<tr><td>3</td><td>Application software complying with IEC 61508-3</td><td>Safety PLC with FVL according IEC 61508</td></tr>
</table>

Software Level 1: the typical application for a machinery manufacturer: a Safety PLC with Limited Variable language. With this type of applications, there is no limitation to the maximum SIL that can be reached: each Safety Function **may reach SIL 3**.

Software Level 2: the machinery manufacturer is given the possibility by the PLC manufacturer to program its Safety PLC (certified to IEC 61508 or IEC 61131-6) using a Full Variability Language. However, the program developed in FVL only complies with the requirements of IEC 62061. That is possible; however, the **maximum SIL reachable is SIL 2**.

Software Level 3: Similar to Level 2, but the FVL application software is compliant with the criteria detailed in IEC 61508-3. In this case, **maximum SIL reachable is SIL 3**.

Please consider that despite IEC 62061 allows the use of an FVL, the design of complex electronic subsystems is not within the scope of the standard.

5.6.3.1 Software Level 1

Software level 1 is of reduced complexity due to the use of **pre-designed** safety-related hardware and software modules. For that reason, the simplified V-model in Figure 5.6 can be used.

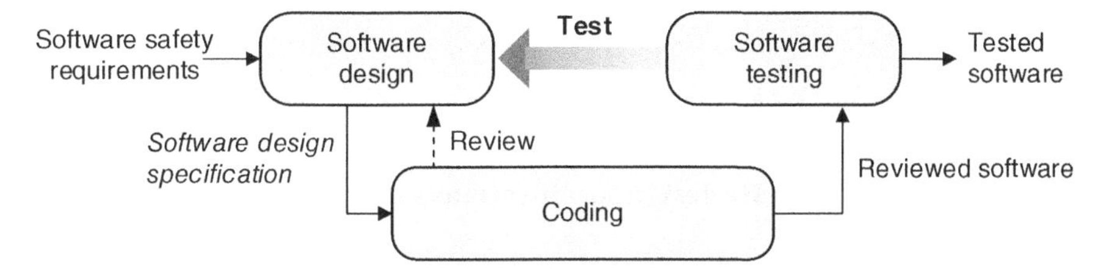

Figure 5.6 V-model for software level 1.

The design of customized or self-created software modules can be necessary (e.g. in the case of the library modules provided by the component manufacturer being inadequate or not suitable). The design of software modules customized by the designer, still using LVL software, is an additional activity that shall be carried out according to the V-model in Figure 5.7. The result of the latter figure is an input to the coding of Figure 5.7.

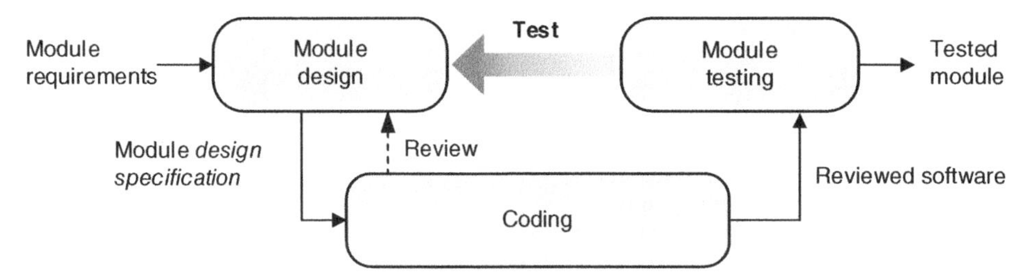

Figure 5.7 V-Model for software Level 1.

You may notice that the language used to describe the V-Model by IEC 62061 (similar to the one used in ISO 13849-1) is different than the original one used by IEC 61508-3 (Figure 5.5).

In both cases though, the V-model is a static model used to structure the software design into small parts. The left side represents requirements, the things to achieve where the output of each phase is reviewed, and the right side details the testing of the software.

The Review (Verification according to IEC 61508-3) means to check the output of a phase in the V-model against the requirements of the input of the same phase. The Test is a synonym for Validation.

5.6.3.2 Software Safety Requirements for Level 1

Where software is to be used in any part of an SCS implementing a Safety Function, a software safety requirements specification shall be developed, documented, and managed throughout the lifecycle of the SCS.

The following information shall be considered:

a) specification of the safety function(s);
b) configuration or architecture of the SCS (e.g. hardware architecture, wiring diagram, safety-related inputs and outputs);
c) response time requirements;
d) operator interfaces and controls, such as switches, joysticks, mode selectors, dials, touch-sensitive control devices, keypads, etc.;
e) relevant modes of operation of the machine;
f) requirements on diagnostics for hardware including the characteristics of sensors, final actuators, etc.;
g) effects of mechanical tolerances, e.g. of sensors and/or their sensing counterparts;
h) coding guidelines.

5.6.3.3 Software Design Specifications for Level 1

From the Safety requirements, the **Software design specifications** shall be derived. The following shall be specified:

a) the logic of the safety functions, including safety-related inputs and outputs and proper diagnostics on detected faults. Possible methods include, but are not limited to, cause and effect table, written description, or function blocks;

b) test cases that include:
   - the specific input value(s) for which the test is carried out and the expected test results including pass/fail criteria;
   - fault insertion or injection(s);

For simple functions, the test cases can be given implicitly by the specification of the safety function.

a) diagnostic functions for input devices, such as sensing elements and switches, and final control elements, such as solenoids, relays, or contactors;
b) functions that enable the machine to achieve or maintain a safe state;
c) functions related to the detection, annunciation, and handling of faults;
d) functions related to the periodic testing of SCS(s) on-line and off-line;
e) functions that prevent unauthorized modification of the SCS (e.g. password);
f) interfaces to non-SCS;
g) safety function response time.

5.6.3.4 Software Testing for Level 1

The main goal of software testing is to ensure that the functionality, as detailed in the software design specification, is achieved. The main output of software testing is a test report with test cases and test results allowing an assessment of the test coverage.

When pre-designed input cards or software modules, which incorporate failure detection and reaction, are utilized (e.g. discrepancy of input signals or feedback contact of output), **then the test of those failure detection and reaction is not necessary**. In that case, only the integration of these input cards or software modules in accordance with the manufacturer's specification shall be tested.

5.6.3.5 Validation of Safety-Related Software

The validation of software shall include:

- the specified functional behavior and performance criteria (e.g. timing performance) of the software when executed on the target hardware;
- verification that the software measures are sufficient for the specified required SIL of the safety function; and
- measures and activities taken during software development to avoid systematic software faults.

In the case of small programs, an analysis of the program by means of reviews or walk-through of control flow, procedures, etc. using the software documentation (control flow chart, source code of modules or blocks, I/O and variable allocation lists, cross-reference lists) **can be sufficient.**

5.6.4 Software Safety Requirements According to ISO 13849-1

The guidelines given by the ISO standard are applicable to both the development of safety-related embedded (SRESW) or application (SRASW) software. Their scope is the avoidance of faults during software development, and the objective is to have readable, understandable, testable, and maintainable software.

Annex N of ISO 13849-1 details the measures to apply to SRASW developed either with LVL or with FVL.

In case FVL software is used, the full V-Model shown in Figure 5.8 has to be followed.

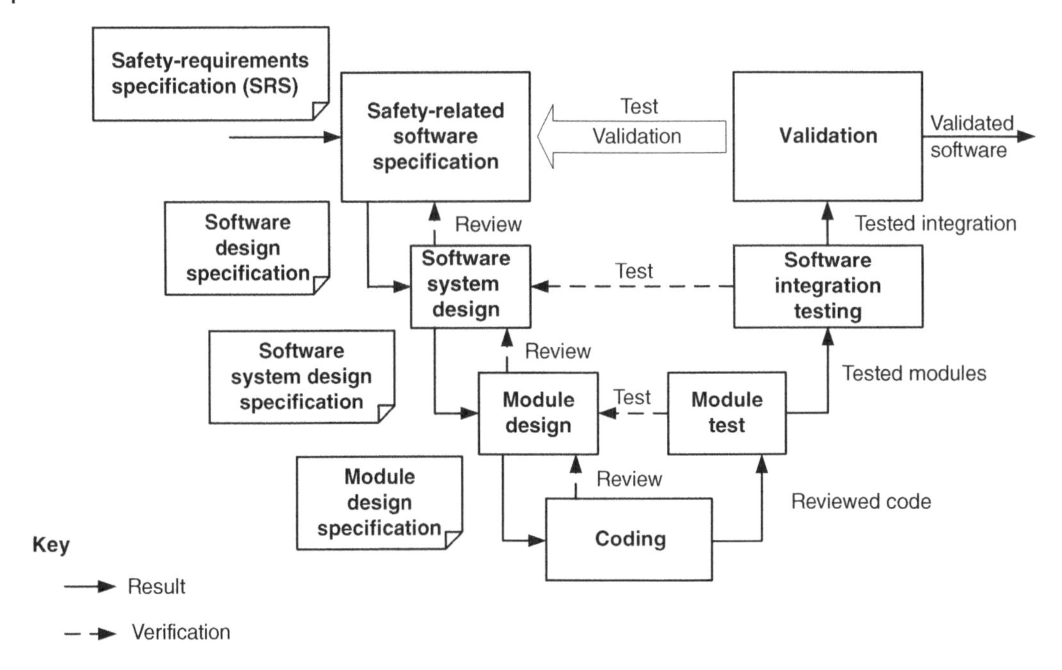

Figure 5.8 The V-model according to ISO 13849-1.

If pre-assessed safety-related hardware and software modules are used, in combination with LVL, **a simplified software lifecycle shown in Figure 5.9 is applicable**. Typically, this applies to the use of module-based programming in LVL, that only requires simple interconnections to be configured, which limits the inputs and outputs to a pre-defined set of values, including a combination of modules.

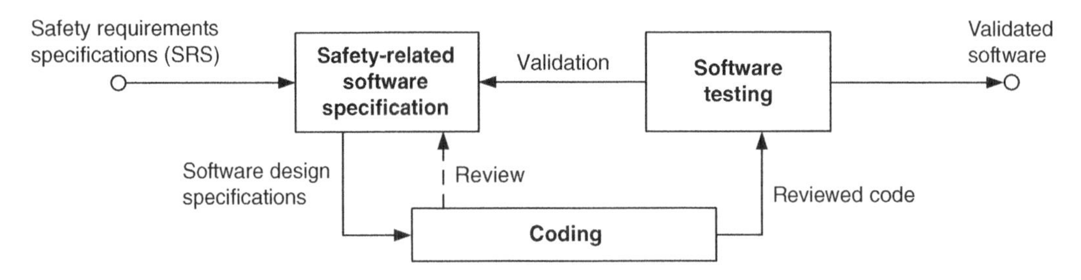

Figure 5.9 Simplified V-model in case LVL is used.

5.6.4.1 Requirements When SRASW is Developed with LVL

SRASW written in LVL and complying with the following requirements does not have any limitation to the reliability level that the Safety Function can achieve.

In general, the following **basic measures** shall be applied:

- development lifecycle with verification and validation activities;
- documentation of specification and design;
- modular and structured programming;
- functional testing;
- appropriate development activities after modifications.

In case the Safety subsystem has a reliability level between PLr c and PLr e, the following additional measures are applicable. Compared to the ISO standard, only a part of the requirements are listed.

a) The software design specification shall be reviewed (see also annex J), made available to every person involved in the lifecycle and shall contain the description of:
   - safety functions with required PL and associated operating modes;
   - performance criteria, e.g. reaction times;
   - communication interfaces;
   - detection and control of hardware failures to achieve the required DC and fault reaction.
b) Whenever reasonable and practicable, safety-related FB libraries provided by the tool manufacturer shall be used.
c) The code shall be readable, understandable, and testable: symbolic variables (instead of explicit hardware addresses) should be used.
d) Testing:
   - the appropriate validation method is black-box testing of functional behavior and performance criteria (e.g. timing performance);
   - for PL d or e, test case execution from boundary value analysis is recommended;
   - test planning is recommended and should include test cases with completion criteria and required tools;
   - I/O testing shall ensure that safety-related signals are correctly used within SRASW.

5.6.4.2 Software-Based Manual Parameterization

There are situations whereby the software module only requires parametrization. For example, a converter with integrated sub-functions can be parameterized via a PC-based configuration tool for setting the upper speed limit parameter. In another case, to establish the detection zone of a laser scanner, parameters such as angle and distance can be configured following the manufacturer's safety documentation and the machine risk assessment.

The objective of the requirements for software-based manual parameterization is to guarantee that the safety-related parameters specified for a safety function or a sub-function are correctly transferred into the hardware performing the safety function or a sub-function. Different methods can be applied to set such parameters; even dip switch-based parameterization can be used to set or change safety-related parameters. However, PC-based tools with dedicated parameterization software, commonly called configuration or parameterization tools, are becoming more prevalent.

Software-based manual parameterization shall use a dedicated tool provided by the manufacturer or supplier of the SRP/CS or the related subsystem(s). This tool shall have its own identification (name, version). The SRP/CS or the related subsystem(s) and the parameterization tool shall have the capability to prevent unauthorized modification, for example by using a dedicated password. Parameterization while the machine is running, shall be permitted only if it does not cause an unsafe state.

The following requirements shall be satisfied:

- The design of the software-based manual parameterization shall be considered as a **safety-related aspect of SRP/CS design** that is described in a safety requirements specification.
- The SRP/CS or subsystem shall provide means to check the data plausibility, e.g. checks of data limits, format, and/or logic input values.
- The integrity of all data used for parameterization shall be maintained.

5.7 Low Demand Mode Applications in Machinery

5.7.1 How to Understand if a Safety System is in High or in Low Demand Mode

5.7.1.1 Milling Machine

Usually, machinery safety systems operate in high demand mode of operation.

The access to the working area of a milling machine, shown in Figure 5.10, may happen once every 15 minutes because of loading/unloading activities.

Every 15 minutes the operator opens the interlocked movable guard to exchange the piece to be milled. Every 15 minutes the interlocked guard is opened, and all the movements inside the machine are stopped: that is the Safety Function we are considering.

Every 15 minutes contactor coils are de-energized, and the control diagnostics verify that each one of them has opened.

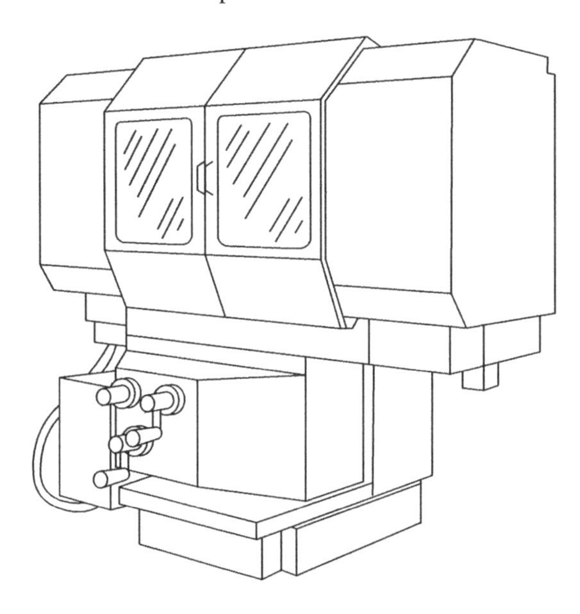

Figure 5.10 Milling machine with safeguards.

For this type of application, the approach of ISO 13849-1 and IEC 62061 fits well with the reliability assumptions of the components used.

Sensors and Actuators are often electromechanical components, and the event of activating them regularly is within the spirit of a high demand mode Safety-related control System. When the interlocked door is opened, the control system can detect a failure: in other words, each component can be tested every time it is used. Of course, only detectable failures (λ_{DD}) can be checked, but that is not an issue since, as described in the next chapters, that aspect is taken into account in the calculation of the PFH_D of each safety function.

5.7.1.2 Industrial Furnaces

Sometimes, equipment falling under the Machinery Directive contains Safety-related control Systems in low demand mode as well. As an example, let's consider an **industrial furnace**.

Normally, the Gas Train (Figure 5.11) has a **low gas pressure safety loop** that causes the closure of **two shut-off valves**. The low-pressure input subsystem is represented by two low-pressure switches (PSL), and normally, they are installed before the cut-off valves and after the manual on/off valve.

The furnace can also be equipped with a **high-temperature input subsystem** (Figure 5.12): two thermocouples controlled by a dedicated electronic module, and the manufacturer, during the risk assessment, has decided that it is a Safety Function.

If the gas pressure is not very reliable or if the furnace is switched off every week by closing the manual valve, the pressure switches work in high demand mode since, every week, they detect a gas low pressure, and the two valves are consequently closed: in that way no gas can reach the burner, and therefore the flame goes off: the system **is in a safe state**. In this case, also the two on/off valves work in high demand since they close because of a demand upon the safety function (gas low pressure).

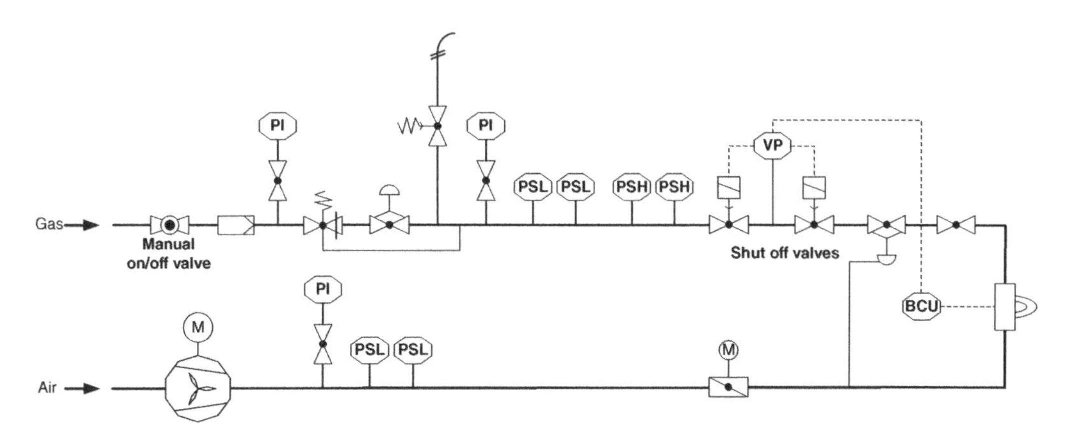

Figure 5.11 Schematics of a gas train.

However, **the temperature input subsystem works in low demand mode** since, probably, during the furnace lifetime, it will never intervenes because the control system manages to keep the maximum temperature within a certain range.

The high-temperature subsystem must be evaluated according to IEC 61511-1, while both the low-pressure subsystem and the output subsystem can be analyzed using either IEC 62061 or ISO 13849-1.

For the high-pressure safety function, we need to understand if it works in high or in low demand mode. The question to be asked is: "while the furnace is working, how often the gas pressure reaches the high threshold?". If the gas line is reliable, probably very rarely, even less than once in a year. In that case, the high-pressure Safety subsystem is working in low demand mode, and IEC 61511-1 should be used instead of ISO 13849-1 or IEC 62061 to assess its reliability.

5.7.2 Subsystems in Both High and Low Demand Mode

Electromechanical components are provided with a B_{10D} value that allows the calculation of an $\mathrm{MTTF_D}$ or λ_D, based upon the number of operations. B_{10D} values are calculated having a certain number of components under test and by switching on and off the components several times per hour; please refer to Chapter 1 for more details. That means B_{10D} values should **only be used when the component is part of a high demand mode subsystem**. Moreover, considerations should be made when the number of operations is less than once per month.

In high demand mode, the more a component is used, the less reliable the subsystem system will be, since the high demand mode standards take into consideration the component Fatigue: the more it is used, the higher the Fatigue.

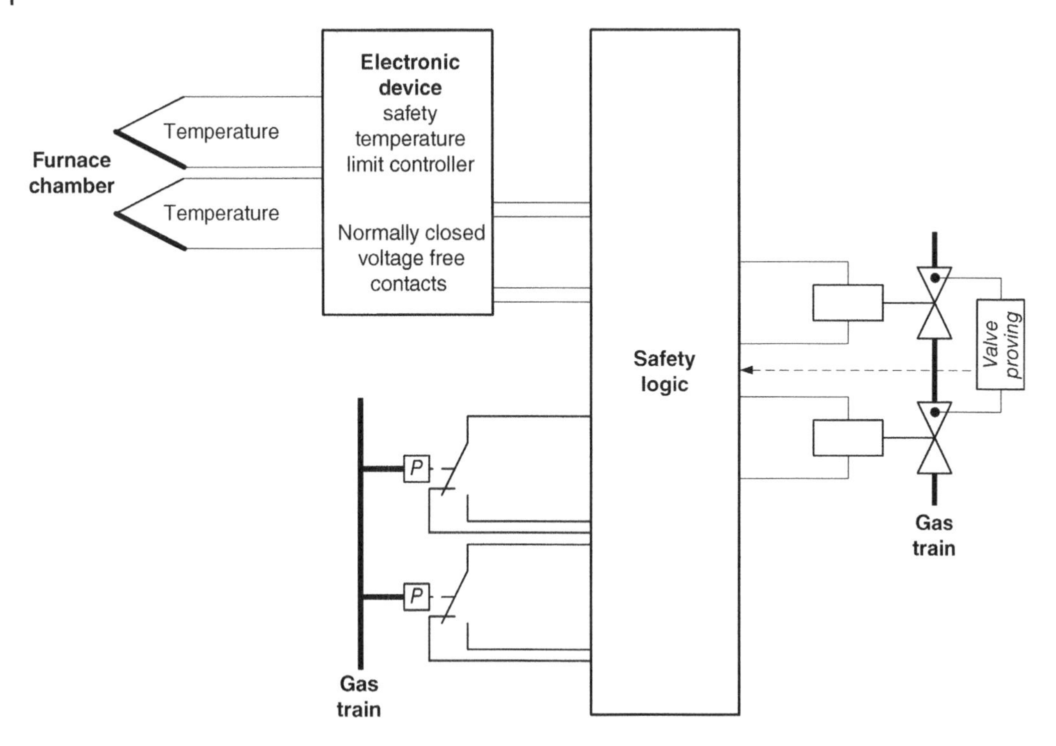

Figure 5.12 Example of a mixed high and low demand mode safety control system.

Example

A contactor has a $B_{10D} = 1\,400\,000$.

Please refer to Equation 3.5.2.

If used every minute over 16 h/day, 220 days/year, it has an $MTTF_D = 66$ years

If used every 10 minutes over 16 h/day, 220 days/year, it has an $MTTF_D = 663$ years

If used once per hour over 16 h/day, 220 days/year, it has an $MTTF_D = 3977$ years

For electromechanical components, the high demand mode of operation works well for the weekly actuation of safety components. When the time is longer, other considerations should be made: if a contactor stays closed for several months, what guarantees that, when de-energized, it will open?

That is the reason why, **when a subsystem is used in high demand mode**, the minimum number of demand upon the safety function has to be guaranteed. The way in which the standards tackle the issue is by requiring a manual activation of the safety function, depending upon the reliability level to be reached.

[IEC 62061] 7.3.3 Fault consideration and fault exclusion

7.3.3.4. *[...] When a functional test is necessary to detect a possible accumulation of faults or a undetected fault before the next demand, it shall be made within the following test intervals:*

- *at least every month for SIL 3;*
- *at least every 12 months for SIL 2.*

[ISO 14119] 9.2 Assessment of faults and fault exclusions
***9.2.1** [...] If for detecting of a fault, a manual test (e.g. opening of a guard) is necessary but frequency of access to the safeguarded area is seldom, the following intervals shall be chosen:*

- *at least every 12 months for PL d with Category 3 or 2 (according to ISO 13849-1) or SIL 2 with HFT (hardware fault tolerance) = 1 (according to IEC 62061:2015).*
- *at least every 1 month for PL e (according to ISO 13849-1:2015) or SIL 3 (according to IEC 62061:2015).*

Similarly, in the Recommendation for Use sheets (RfUs) [61], it is stated:

***Question:** What are the minimum requirements concerning the frequency of tests for failure detection in a safety-related system with two channels with electromechanical outputs (relays or contactors)?*

***Solution:** A functional test (automatic or manual) to detect failures shall be performed within the following test intervals: a) at least every month for PL e with Category 3 or Category 4 (according to EN ISO 13849-1) or SIL 3 with HFT (hardware fault tolerance) = 1 (according to EN 62061); b) at least every 12 months for PL d with Category 3 (according to EN ISO 13849-1) or SIL 2 with HFT (hardware fault tolerance) = 1 (according to EN 62061).*

***Note:** It is recommended that the functional test is initiated by the control system of the machine. If this is not possible, then it is recommended that the control system of the machine reminds the user (e.g. by an appropriate indication at the control panel) to perform a functional test of the safety function. If this is also not possible, an appropriate requirement has to be contained in the instructions for use.*

However, that is required only if the subsystem is designed and used in high demand mode.

For Safety Functions used less than once per year, **the low demand mode approach comes into play**. If a process on/off safety valve is triggered by a dangerous situation once every ten years (expected frequency of demand upon the safety function), the standard IEC 61511-1 [16] is worried that after ten years, when the valve has to close to bring the system to a safe state, it doesn't. Since it is not actuated for the whole 10-year period, there is no way to check if, during this time, it had an internal undetected failure.

That is the reason why the standard requires a **Proof Test (§ 3.7)** at least every year, in our example. The Proof Test shall detect either all the detectable dangerous faults or a part of them; the latter is defined as **Imperfect Proof Test (§ 3.7.3.1)**. **The timing of the proof test depends upon the expected demand upon the safety function,** and it has a direct influence on the reliability of the safety subsystem.

In case of subsystems in low demand mode of operation, the higher the frequency of the test, the higher the Safety Function Reliability.

Example

A valve has a $\lambda_{DU} = 4.3\ 10^{-7}\ (h^{-1})$.

Please refer to Equation 1.10.3.

If used in an $HFT = 0$ subsystem and tested (T_i) every four years, it has $PFD_{avg} = 7.53\ 10^{-3}$.

If used in an $HFT = 0$ subsystem and tested (T_i) every three years, it has $PFD_{avg} = 5.65\ 10^{-3}$.

If used in an $HFT = 0$ subsystem and tested (T_i) every two years, it has $PFD_{avg} = 3.77\ 10^{-3}$.

For this output subsystem, I cannot use B_{10D} values to calculate the failure rates.

5.7.3 How to Address Low Demand Mode in Machinery

When there is a Safety Function that has a Demand Rate (DR) less or equal to once per year, it falls under the low demand mode of operation. IEC 61511-1 offers good guidance; however, if it is possible to force it manually with a frequency higher than once per year, the safety function falls under the high demand mode of operations and, for example, IEC 62061 can be used for its analysis.

In general, when the DR is estimated to be low, the machinery manufacturer has two options:

- **A high demand mode** can be assumed by activation of the safety function (that "naturally" would be triggered less that once per year) more frequently than once per year. The manufacturer can specify that obligation in the instruction manual.
- **A low demand mode** has to be considered. Therefore, IEC 61511-1 has to be used for the design of the safety function.

Please refer to Figure 5.13 for a graphical representation of the concept.

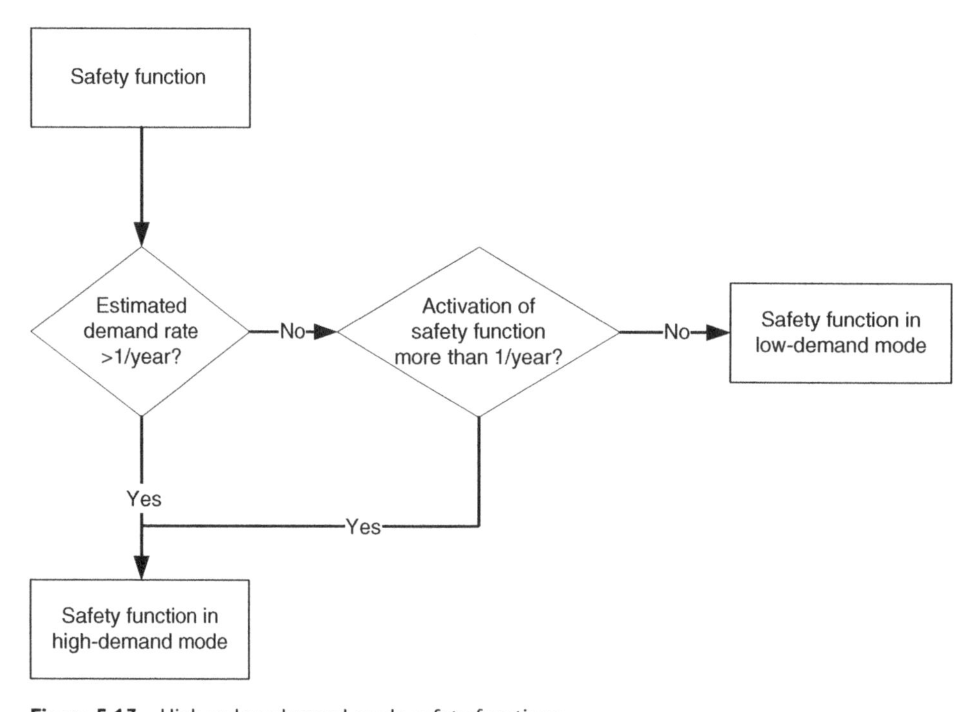

Figure 5.13 High vs low demand mode safety functions.

5.7.4 Subsystems Used in Both High and Low Demand Mode

If a safety function is used in low demand mode of operations, the approach of IEC 61511-1 should be used.

In Machinery, when a low demand mode safety function is present, it often happens that the output subsystem is shared between low and high demand mode safety functions. Let's consider, for example, a high-pressure switch in low demand mode that stops a motor pump. The motor is also part of a high demand mode safety function, linked to an interlocked door. Please refer to Figure 5.14.

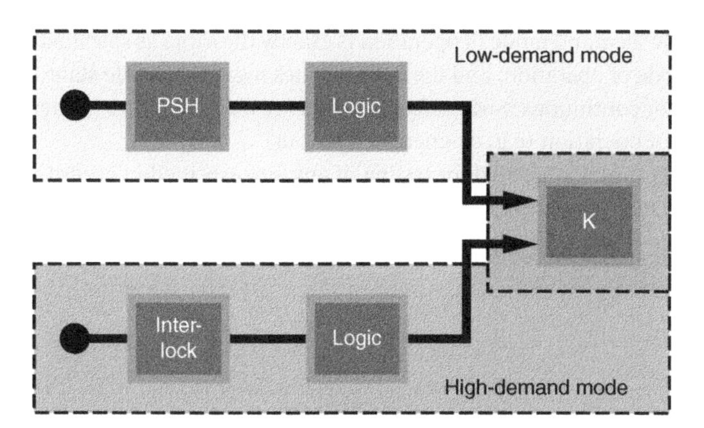

Figure 5.14 A contactor shared between a high and low demand mode safety control system.

Therefore, **the contactor is used in high demand mode, but shared with a low demand mode** Safety Instrumented System. We propose two methods to assess the reliability of these types of **"mixed" safety systems**.

5.7.5 How to Assess "Mixed" Safety Systems: Method 1

5.7.5.1 How to Estimate the Failure Rate of the Shared Subsystem

Let's suppose the contactor has a $B_{10D} = 1.4\ 10^6$, and it is used 30 times/h since the interlocked gate is opened with that frequency.

$$\lambda_D = \frac{0.1 * n_{op}}{B_{10D} * 8760} = \frac{30 \cdot 24\,\text{hours} \cdot 365}{(1.4 \cdot 10^7) \cdot 8760} = 2.14 \cdot 10^{-6}\left(\text{h}^{-1}\right) = 2143\,\text{FIT}$$

A Siemens contactor used **in low demand mode** of operation has a failure rate of 100 FIT, according to its datasheet. The ratio of dangerous failures is <40%. That means, conservatively, $\lambda_D \approx$ 40 FIT. However, if the contactor is shared with a high demand mode system, the failure rate to be used in low demand mode calculations is **2143 FIT** since it is the higher of the two values.

Let's now suppose the contactor has a $B_{10D} = 1.4\ 10^6$ and is used 12 times in a year, triggered only by an emergency stop in high demand mode (in Figure 5.14, the interlocking device is replaced by an emergency stop pushbutton).

$$\lambda_D = \frac{0.1 * n_{op}}{B_{10D} * 8760} = \frac{12}{(1.4 \cdot 10^7) \cdot 8760} = 9.78 \cdot 10^{-11}\left(\text{h}^{-1}\right) = 0.1\,\text{FIT}$$

In this case, if the contactor is shared between a high and low demand mode system, the failure rate to be used in both high and low demand mode calculations is **40 FIT** since it is the higher of the two values.

5.7.5.2 Relationship Between PFD_{avg} and PFH_D

In general, safety functions implemented in high demand or continuous mode can be used in low demand mode. For this case, there is a simplified conservative method to estimate the PFD_{avg} value from the PFH_D.

For an SCS with a specified safety function quantified by a related PFH_D value for high demand or continuous mode of operation, an estimated value for the PFD_{avg} in a low demand mode application can be derived from the PFH_D under certain circumstances. Provided that:

1. The safety function to be used in low demand mode of operation is exactly the same as specified for high demand or continuous mode of operation, and the system states regarded as safe states in the context of the high demand or continuous mode safety function are also safe states in the context of the low demand mode of operation (e.g. de-energized output).
2. Compulsory actuations of the safety function needed for testing, if any, are executed in accordance with the requirements of the manufacturer.

An estimated value for the PFD_{avg} may be derived from the PFH_D value for high demand mode using the following equation:

$$PFD_{avg} = \frac{PFH_D \cdot T_1}{2}$$

where T_1 is the Mission Time of the SCS expressed in hours or the Proof Test interval **in case the latter is smaller**. The indicated PFD_{avg} equation tends to deliver conservative results.

Example

PFD_{avg} calculation.

Let's consider a subsystem having a $PFH_D = 1.5 \cdot 10^{-7}$ (h^{-1}) (SIL 2) and a Mission Time of 20 years. If the safety function is used in low demand mode, with a Proof Test of three years, the subsystem is still in the SIL 2 range.

$$PFD_{avg} = \frac{1.5 \cdot 10^{-7} \cdot 3 \cdot 8760}{2} = 2 \cdot 10^{-3}$$

If the safety function has a Proof Test of 20 years, the subsystem is in SIL 1 in low demand mode.

$$PFD_{avg} = \frac{1.5 \cdot 10^{-7} \cdot 20 \cdot 8760}{2} = 1.3 \cdot 10^{-2}$$

5.7.5.3 Safety Functions 1 with a Shared Subsystem: Method 1

Let's now consider a safety function with two input subsystems, as shown in Figure 5.15:

- A pressure Transmitter working in low demand mode.
- An emergency stop Button working in high demand mode.

The frequency of demand upon the safety functions are the followings:

- Emergency stop: **once a month.**
- The pressure transmitter detects a high pressure **once every 30 years.**

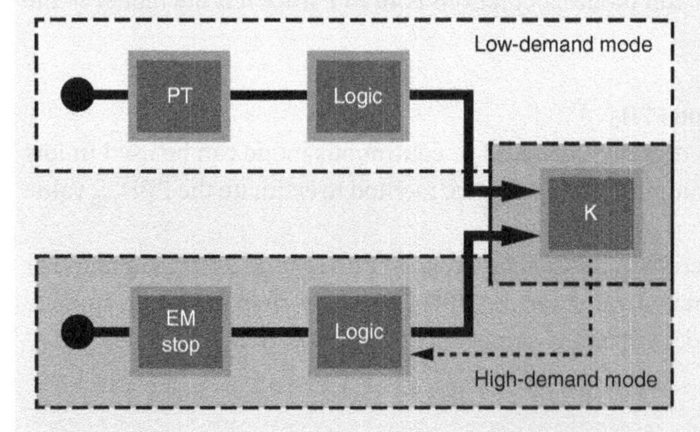

Figure 5.15 SCS in both high and low demand mode.

Let's now show how to calculate the reliability level of the low demand mode Safety System. The contactor has a $B_{10D} = 1.4\ 10^6$ and is used once a month.

$$\lambda_D = \frac{0.1 * n_{op}}{B_{10D} * 8760} = \frac{12}{(1.4 \cdot 10^7) \cdot 8760} = 9.78 \cdot 10^{-11}\ \mathrm{h}^{-1} = 0.1\ \mathrm{FIT}$$

The same contactor, used in low demand mode of applications, has a failure rate of 100 FIT, according to its datasheet. The ratio of dangerous failures is <40%. That means, conservatively, $\lambda_D \approx 40$ FIT. In this case, the failure rate to be used in the low demand mode calculations is 40 FIT **since it is the higher of the two values.**

For the PFD_{avg} calculation we need the $\lambda_{DU} = \lambda_D(1 - DC)$.

If the contactor is monitored, **since it is used in a high demand mode safety subsystem as well**, we assume $DC = 59\%$ for a single channel $\lambda_{DU} = 40$ FIT $(1 - 0.59) =$ **16.4 FIT** $= 1.64 \cdot 10^{-8}\ \mathrm{h}^{-1}$.

Since it is an electromechanical component $SFF \approx DC$ (§ 3.1.3 and § 3.2.3), Table 3.9, valid for a Type A safety-related element or subsystem, limits the maximum reachable SIL to SIL 1. That is due to the low DC.

The pressure transmitter has a $\lambda_{DU} = 9.9 \cdot 10^{-8}\ \mathrm{h}^{-1}$ according to its datasheet.

The safety logic has a $PFD = 10^{-6}$, with a Proof test equal to 20 years, again, according to its datasheet.

Let's suppose a Perfect Proof Test ($PTC = 100\%$) every three years for both the pressure transmitter and the contactor.

For the pressure transmitter, $PFD_{avg} = \dfrac{9.9 \cdot 10^{-8} \cdot 3 \cdot 8760}{2} = 1.3 \cdot 10^{-3}$

For the Safety Logic, $PFD = 10^{-6}$

For the Contactor, $PFD_{avg} = \dfrac{1.64 \cdot 10^{-8} \cdot 3 \cdot 8760}{2} = 2.2 \cdot 10^{-4}$

PFD_{avg_TOT} is obtained by adding all the PFD_{avg} of the elements

$$PFD_{avg_{TOT}} = 1.5 \cdot 10^{-3}$$

The SIL level of the safety system is limited by the Architectural constraint to SIL 1 (due to the contactor) even if the PFD_{avg_TOT} can be associated to a SIL 2 level (§ 2.1.6). In case a Perfect proof test on the contactor is not possible and a Partial Test Coverage (PTC) is given by the component manufacturer, the formula to be used is Equation 3.7.3.1.

5.7.5.4 Safety Functions 2 with a Shared Subsystem: Method 1

Let's now consider a safety function with two input subsystems, as shown in Figure 5.16:

- A pressure Transmitter working in low demand mode.
- An interlocked door working in high demand mode.

The frequency of demand upon the safety functions are the followings:

- Door opens once every hour.
- The pressure transmitter detects a high pressure once every 30 years.

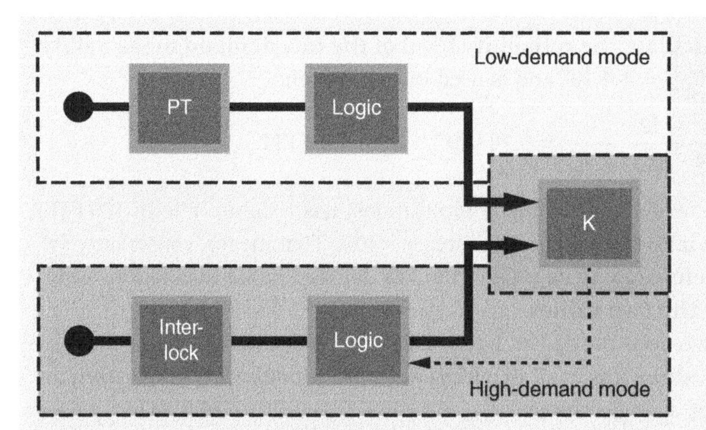

Figure 5.16 SCS in both high and low demand mode.

Let's now show how to calculate the reliability level of the low demand mode Safety System.

The contactor has a $B_{10D} = 1.4\ 10^6$ and is used every hour.

$$\lambda_D = \frac{0.1 * n_{op}}{B_{10D} * 8760} = \frac{24 \cdot 365}{(1.4 \cdot 10^7) \cdot 8760} = 7.14 \cdot 10^{-8}\ \text{h}^{-1} = 71\ \text{FIT}$$

The same contactor, used in low demand mode of applications, has a failure rate of 100 FIT, according to its datasheet. The ratio of dangerous failures is <40%. That means, conservatively, $\lambda_D \approx 40$ FIT. In this case, the failure rate to be used in the low demand mode calculations **is 71 FIT**, since it is the highest of the two values.

For the PFD_{avg} calculation we need the $\lambda_{DU} = \lambda_D(1 - DC)$.

If the contactor is monitored, **since it is used in a high demand mode safety system as well,** we assume $DC = 59\%$ for a single channel $\lambda_{DU} = 71$ FIT $(1 - 0.59) = 29$ FIT $= 2.91 \cdot 10^{-8}\ \text{h}^{-1}$.

Table 3.9, valid for Type A safety-related element or subsystem, limits the maximum reachable SIL to SIL 1. That is due to the low DC, equal to SFF for electromechanical components (§ 3.4 and § 3.2.3).

The pressure transmitter has a $\lambda_{DU} = 9.9 \cdot 10^{-8}\ \text{h}^{-1}$ according to its datasheet.

The safety logic has a $PFD = 10^{-6}$ With a Proof test equal to 20 years, again, according to its datasheet.

Let's suppose a Perfect Proof Test ($PTC = 100\%$) every three years for both the pressure transmitter and the contactor.

For the pressure transmitter, $PFD_{avg} = \dfrac{9.9 \cdot 10^{-8} \cdot 3 \cdot 8760}{2} = 1.3 \cdot 10^{-3}$

For the Safety Logic, $PFD = 10^{-6}$

For the Contactor, $PFD_{avg} = \dfrac{2.91 \cdot 10^{-8} \cdot 3 \cdot 8760}{2} = 3.82 \cdot 10^{-4}$

PFD_{avg_TOT} is obtained by adding all the PFD_{avg} of the elements

$$PFD_{avg_{TOT}} = 1.7 \cdot 10^{-3}$$

The SIL level of the safety system is limited by the Architectural constraint to SIL 1 (due to the contactor) even if the PFD_{avg_TOT} can be associated to SIL 2 level (§ 2.1.6).

5.7.6 How to Assess "Mixed" Safety Systems: Method 2

5.7.6.1 How the Method Works

Let's take, as an example, the "mixed" system shown in Figure 5.12 (industrial Furnace). It can be represented with reliability block diagrams as shown in Figure 5.17.

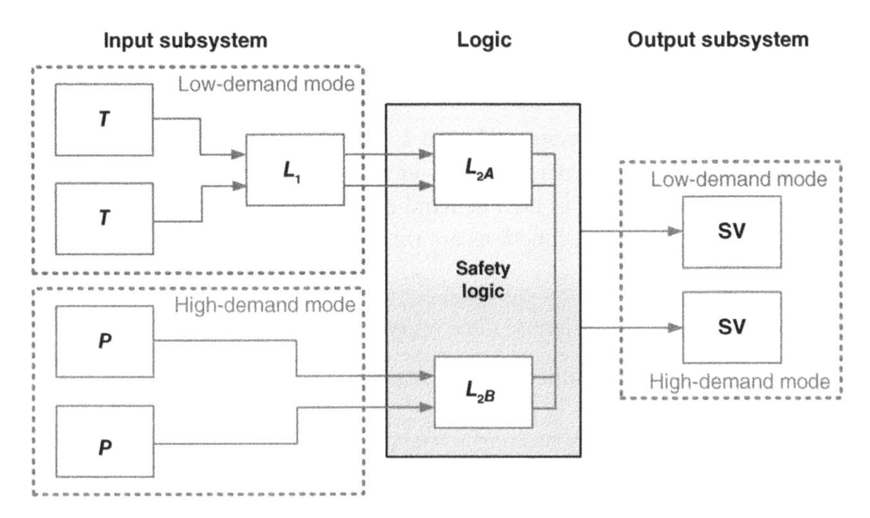

Figure 5.17 SCS in both high and low demand mode.

This simplified method is applicable only if both the high and the low demand safety system have to reach the same reliability level: let's suppose SIL 2.

In that case, the input, logic, and output subsystem is assessed in high demand. If it reaches the required SIL level, the low demand safety subsystem is assessed only as an input.

However, the criteria of the "ratio of probability of failures of each subsystem" is considered, as shown in Figure 5.18.

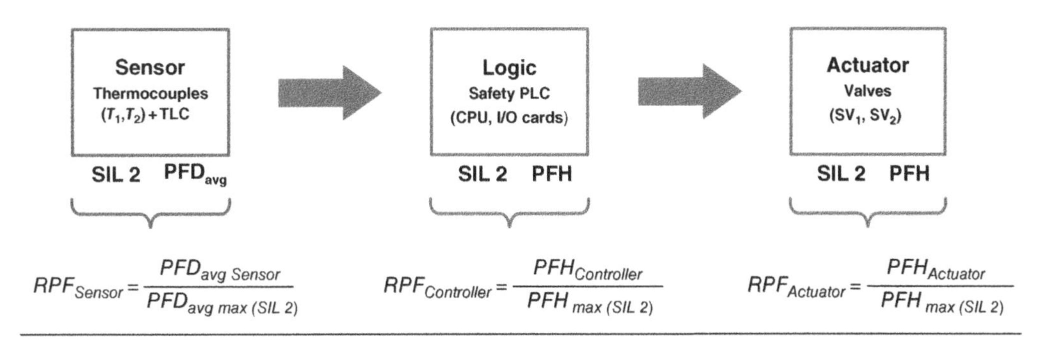

Figure 5.18 Quantitative SIL verification using the approach of the ratio of probability of failures of each subsystem.

Table 5.7 shows the maximum values for PFD_{avg} and PFH_D for respective target SIL.

Table 5.7 $PFD_{avg\ max}$ and PFH_{Dmax} for respective target SIL.

SIL	$PFD_{avg\ max}$	$PFH_{D\ max}$
1	10^{-1}	10^{-5}
2	10^{-2}	10^{-6}
3	10^{-3}	10^{-7}

5.7.6.2 Safety Function 2 with a Shared Subsystem: Method 2

Let's still consider the safety function with two input subsystems made of a pressure Transmitter in low demand mode and an interlocked door in high demand mode, as shown in Figure 5.19.

The frequency of demand upon the safety functions are the followings:

- Door opens once every hour.
- The pressure transmitter detects a high pressure once every 30 years.

Both the high and the low demand safety function must reach SIL 1.

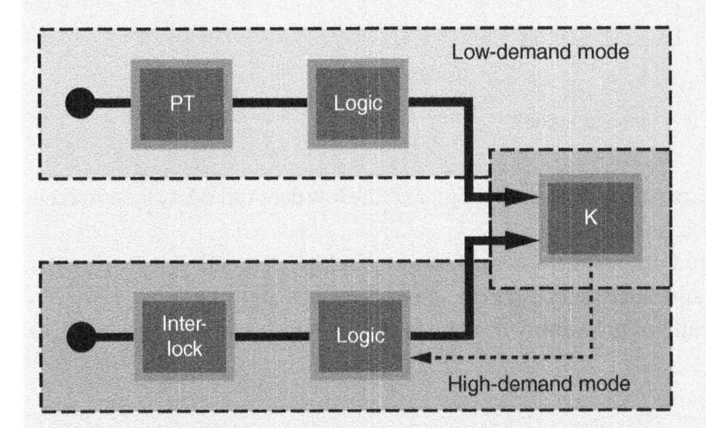

Figure 5.19 SCS in both high and low demand mode.

Let's suppose we obtain the following results for the high demand safety function:

- Interlock Subsystem: $PFH_D = 1.14\ 10^{-6}\ (h^{-1})$; SIL 1
- Logic: $PFH_D = 4.29\ 10^{-8}\ (h^{-1})$; SIL 3
- Output Subsystem (contactor): $PFH_D = 1.14\ 10^{-6}\ (h^{-1})$; SIL 1

Therefore the safety function reaches SIL 1.

For the low demand safety system, we only assess the input subsystem and verify that it reaches SIL 1. Let's suppose we obtain $PDF_{avg} = 2{\cdot}10^{-4}$ and, based upon the architectural constraints, that means SIL 2 ($SFF = 91\%$ and Type B component).

We only have to verify the RPF for the low demand safety system:

- Input subsystem: $RPF = 2{\cdot}10^{-4}/10^{-1} = 2{\cdot}10^{-3}$
- Logic: $RPF = 4.29{\cdot}10^{-8}/10^{-5} = 4.29{\cdot}10^{-3}$
- Output subsystem: $RPF = 1.14{\cdot}10^{-6}/10^{-5} = 1.14{\cdot}10^{-1}$

The sum of the RPF is $\ll 1$, and therefore also, the low demand safety system meets the minimum required level of SIL 1.

6

The Categories of ISO 13849-1

6.1 Introduction

ISO 13849-1 is intended to be used in the design and evaluation of safety-related parts of the control system (SRP/CS) and only the part of the control system that is safety-related falls under the scope of the standard. It applies to SRP/CS for high demand and continuous modes of operation, including their subsystems, regardless of the type of technology and energy used: electrical, hydraulic, pneumatic, or mechanical. **ISO 13849-1 does not apply to low demand mode of operation.**

The ability of safety-related parts of control systems to perform safety functions under foreseeable conditions is indicated by one of five levels, called **performance levels or PL**. Annex A of ISO 13849-1 contains a method that can be used for the determination of the PLr of a safety function performed by the SRP/CS. Annex A of IEC 62061 could also be used as an alternative. In general, any such method will show a variance **because of the subjective nature of the evaluation criteria**.

The required performance level corresponds to the required risk reduction to be provided by the safety function: the greater the contribution to the risk reduction, the higher the required safety performance. The performance levels of safety functions are defined in terms of **Average probability of dangerous failure per hour**. There are five performance levels, ranging from providing a low contribution to risk reduction for **PL a**, to a high contribution to the risk reduction for **PL e**. The defined ranges of probability of a dangerous failure per hour are shown in Table 4.2.

The probability of dangerous failure of a safety function depends on several factors, including hardware and software structure, the extent of fault detection mechanisms (indicated as **diagnostic coverage** or DC), the reliability of components (indicated, for example, with the parameter **Mean time to dangerous failure** or $MTTF_D$) and the influence of common cause failures (CCF).

In order to facilitate the design of an SRP/CS and the assessment of the achieved PL, ISO 13849-1 employs a methodology based on the **categorization of architectures,** with specific design criteria ($MTTF_D$ and DC_{avg},) and **specified behavior under faults conditions**. These architectures are allocated to one of five levels termed Categories B, 1, 2, 3, and 4.

The first edition of ISO 13849-1 was the evolution of EN 954-1, and it was still based upon the so-called **deterministic approach**. Despite the approach from Reliability theory was introduced in ISO 13849-1 with the second edition, the so-called **probabilistic approach**, the five categories defined by EN 954-1 were kept as basic elements of the standard.

One of the differences between EN 954-1 and ISO 13849-1 is that, in the former, the categories were associated to the entire SRP/CS, while in the latter, they are used to represent subsystems. This association is clearly stated in the new edition.

Functional Safety of Machinery: How to Apply ISO 13849-1 and IEC 62061, First Edition. Marco Tacchini.

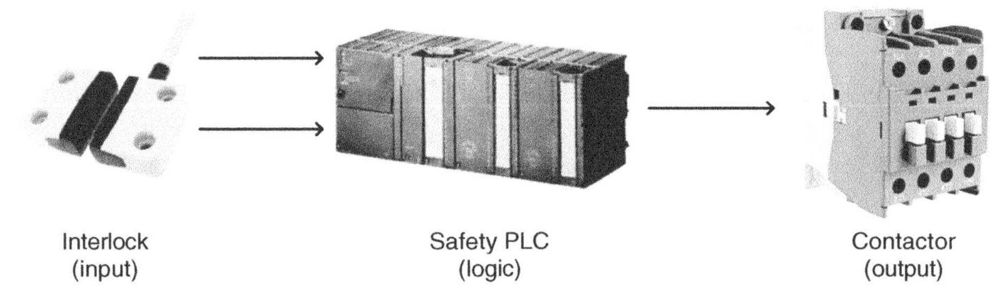

Figure 6.1 An SCP/CS.

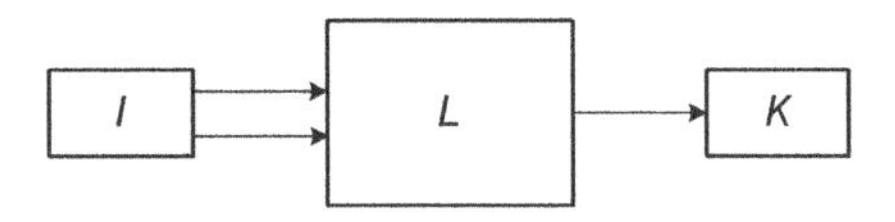

Figure 6.2 The safety function shown with block diagrams.

Let's consider the SRP/CS in Figure 6.1 and its representation in RBD as shown in Figure 6.2. The input subsystem could be a Category 3, while the output is a Category 1; the Safety PLC is usually a Category 4.

In EN 954-1, the Category was the indication of the Reliability level of an SRP/CS. Type C Standards were requiring, for example, an SRP/CS Category 3 or Category 1: that was the common language used. In the new edition of ISO 13849-1, the concept is made clear: **the Category is a way to achieve the Performance level of a subsystem**. Therefore, it is improper to describe an SRP/CS in terms of a Category: a safety system has a PFH_D and a Performance Level (or a SIL, if IEC 62061 is used) but, necessarily, no Category (nor Architecture).

The new edition contains several changes, compared with the previous one.

In Chapter 4 of ISO 13849-1, **a new Figure 2** clarifies that ISO-13849-1, and that is valid for IEC 62061 as well, is applicable only when the risk reduction is achieved with the use of a safety-related control system. The figure is similar to Figure 4.3 of this book.

The term **Safety Requirements Specification**, already present in the 2015 edition, is now fully developed in § 5.2 and in the **annex M** of the standard. Please refer to § 4.9 of the book.

A new chapter 7 was added, to be read together with annex N, dealing with **software development** and how to avoid systematic failures in software. Please refer to § 5.6 of the book.

Probably the major change, with respect to the 2015 edition, is the fact the **Validation** process was moved from ISO 13849-2 to ISO 13849-1 and it was better detailed. Please refer to Chapter 8 of the book.

Among the measures to be adopted to reduce the systematic failures of a Safety System, IEC 61511 has the concept of **Management of Functional Safety**. It recommends the creation of a team that is responsible for carrying out and reviewing each of the SRP/CS safety life-cycle phases. It can be quite a time-consuming activity that may not be completely feasible when dealing with Machinery Safety. In the new edition of ISO 13849-1, the concept has been added in annex G.5. **A functional safety plan** should be drawn up and documented for each SRP/CS design project and should be updated as necessary. Please refer to § 4.13.1 of the book.

6.1.1 Introduction to the Simplified Approach

The numerical quantification of the probability of failure of a subsystem can never be attained exactly, but only by approximation with the aid of statistical methods or other estimations are possible.

Any validated and recognized method can be used for this purpose. Such methods include reliability block diagrams (used in IEC 62061), fault tree analysis, Markov modeling (used in ISO 13849-1) or Petri nets.

However, in general, engineers lacking prior experience in quantification of the probability of failure of safety related control systems require some degree of support. This need was addressed, in ISO 13849-1, by developing a **simplified approach,** also called simplified method which, whilst being based upon sound scientific principles (Markov modeling), describes a simple method for quantification in successive steps.

The starting point of the simplified method is the observation that the majority of safety-related control systems **can be grouped in very small number of basic types**, or to combinations of these basic types.

These types are, at one end of the spectrum, the **single-channel untested system** having components with different reliability level; in the middle of the spectrum, the same type, but enhanced by testing; and at the other end, the **two-channel system** featuring high-quality testing. Systems with more than two channels are rare in machinery.

That was the starting point for the development of **the probabilistic approach of ISO 13849-1**. At the time, it was decided that the five categories of EN 954-1 could cover the majority of SRP/CS used in Machinery and, for that reason, continuity was intentionally assured with the previous standard.

EN 954-1 defined five structures as Categories, ISO 13849-1 supplements the former Category definition with quantitative requirements for the component reliability ($MTTF_D$), the DC of tests (DC_{avg}), and the resistance to CCF. In addition, it maps the Categories to five basic structural types, termed "designated architectures."

Therefore, ISO 13849-1 provides a simplified approach, based upon the definition of five designated architectures, that fulfills specific design criteria and behavior under a fault condition.

6.1.2 Physical and Logical Representation of the Architectures

The Categories are therefore important to achieve a specific PL for a subsystem. However, the standard clarifies that they show a logical representation of the subsystem structure, which may differ from its physical one.

> ***[ISO 13849-1] 6.1.3.2 Designated Architectures – Specification of Categories***
> ***6.1.3.2.1 General.*** *[...] The designated Architectures show a logical representation of the structure of the subsystems for each Category.*
>
> ***Note 1****: For Categories 3 and 4,* ***not all parts are necessarily physically redundant*** *but there are redundant means of assuring that a single fault cannot lead to the loss of the sub-function. Therefore, the technical realization (for example, the circuit diagram) can differ from the logical representation of the architecture.*

Another way to state the same concept is that each one of the five Categories of ISO 13849-1 **describes the required behavior of the subsystem with respect to its resistance to faults**.

Let's consider the guard locking mechanism of an interlocking device. The market offers interlocking devices that can reach **PL e**; they have redundant electrical channels, like two Voltage Free Contacts (VFCs) or two Output Signal Switching Device contacts (OSSD), **but the Guard Locking Mechanism is a single element**. That is not an uncommon solution: the reason is that, in mechanical devices with a single channel Architecture, the detection of faults by the control system may not be possible in certain situations, or its cost would be unjustifiable. However, it is important that all probable faults are evaluated by the interlocking device manufacturer and that **any dangerous failure mode is either eliminated or proven to be technically improbable**. This can be achieved by over-dimensioning critical parts of the device and subsequently testing them. If that is done, the single channel locking mechanism can be used in a redundant Architecture, in our example an interlocking device with guard locking, **since it achieves the relevant Category 4 behavior**.

Just to state the concept in a different way, where mechanical faults are proved to be technically improbable, **continued performance of the safety function in the presence of a single fault is assumed**. Of course, the specific Fault Exclusion can only be justified if the device is used within its manufacturer's specification.

IEC 61508-2 accepts the use of Fault exclusion. Please also refer to § 4.11:

> ***[IEC 61508-2] 7.4.4.1 General requirements***
> ***7.4.4.1.1*** *With respect to the hardware fault tolerance requirements.*
> *[...]*
> *c) when determining the hardware fault tolerance achieved,* ***certain faults may be excluded****, provided that the likelihood of them occurring is very low in relation to the safety integrity requirements of the subsystem. Any such fault exclusions shall be justified and documented (see Note 2).*

6.1.3 The Steps to be Followed

The performance level shall be determined for each subsystem and/or each combination of subsystems that provide a safety function. The PL of the subsystem shall be determined by going through the following aspects:

1) **the architecture**
   a) Decompose the Safety Related Parts of the control system into subsystems (§ 5.1.3)
   b) Assign a category to each subsystem;
   c) Evaluate whether the applicable qualitative requirements of the category are met (Table 6.1), including:
      - basic safety principles (§ 4.13.2)
      - well-tried safety principles (§ 4.13.3)
      - well-tried components (§ 5.2)

   d) Evaluate whether the required behavior under fault conditions is met;
2) **The $MTTF_D$** value for single components (annex C and D of ISO 13849-1);
3) The **DC**, limited in any case by the selected Category (annex E of ISO 13849-1);
4) The **CCF** has to reach at least 65 points (annex F of ISO 13849-1);
5) The effect of the safety-related software design on the operation of the hardware (annex J of ISO 13849-1);
6) The effect of measures against systematic failures (annex G of ISO 13849-1)

Depending upon the subsystem Category, only some of the qualitative requirements are applicable. Table 6.1 shows when the different methodologies used to avoid Systematic Failures and CCF shall be used, depending upon the category employed.

Table 6.1 Applicability of the qualitative requirements.

	B	1	2	3	4
Basic safety principles (§ 4.13.2)	X	X	X	X	X
Well-tried components (§ 5.2)		X			
Well-tried safety principles (§ 4.13.3)		X	X	X	X
CCF (§ 3.6)			X	X	X

When a safety function is designed using one or more subsystems, each subsystem can be designed either using PLs according to ISO 13849-1, or using SILs according to IEC 62061 and or IEC 61508. Subsystems designed according to IEC 61508 may be used but shall be restricted to those designed for high demand or continuous mode that use Route 1_H (§ 3.4.8). They can be combined as indicated in § 6.4.

6.2 The Five Categories

6.2.1 Introduction

Subsystems designed according to ISO 13849-1 should be in accordance with the requirements of **one of the five categories** that are fundamental to achieve a specific Performance Level. The categories describe the required behavior of subsystems in respect of its resistance to faults, based upon the design considerations previously indicated ($MTTF_D$, $DC_{avg,}$ etc.).

Category B is the basic Category where the occurrence of a fault can lead to the loss of the safety function. In **Category 1** an improved resistance to faults is achieved by using high-quality components.

With Categories 2, 3, and 4, higher Reliability of the subsystem is achieved by improving fault tolerance (Category 3 and 4 only) and diagnostic measures. In **Category 2**, since there is no redundancy, higher reliability is achieved by periodically checking that the safety function is performed without faults (DC). In **Categories 3 and 4**, the DC works together with Redundant channels, so that a single fault will not lead to the loss of the safety function.

In Category 4 and whenever reasonably practicable in Category 3, such faults should be detected.

The 5 Categories are represented by specific safety-related block diagrams, each one meeting the requirements of the Category. The Markov modeling used by IFA [72] engineers only considered those 5 Architectures; it is possible to deviate from them, but that implies to go through a new modeling.

For each subsystem, the maximum value of $MTTF_D$ for each channel is limited to 100 years. For Category 4 subsystems only, the maximum value of $MTTF_D$ for each channel is limited to 2500 years.

6.2.2 Category B

Subsystems of Category B must use **Basic Safety Principles,** detailed in ISO 13849-2 [14], and shall be designed according to relevant standards. This should guarantee that they can withstand expected operating stresses and influences of the processed material, like detergents (thanks for example, to the use of stainless steel) or other relevant external influences, like mechanical vibrations.

The Reliability block diagram of Category B is shown in Figure 6.3. It is shown as having an Input a Logic and an Output however, each category is applicable to a subsystem: either an Input subsystem or an output subsystem, for example. Therefore, do not be confused by the figure: it shows a subsystem and not necessarily a Safety-related control system.

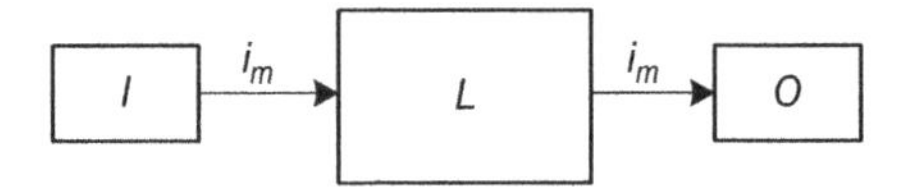

Keys:

- i_m represents the interconnecting means, typically electrical wires
- *I* represents the Input
- *L* represents the Safety Logic; it can be also a wire or a Safety Module (non-programmable) or a Programmable Logic.
- *O* represents the output; it can be a contactor or a solenoid valve

Figure 6.3 Designated architecture for Category B and 1.

With this category, there is no need of DC and CCF considerations are not relevant; the maximum PL achievable is **PL b**.

Being a single channel, a single fault can lead to the loss of the safety sub-function.

6.2.3 Category 1

In Category 1, the same requirements as those for Category B apply; moreover, well-tried safety principles should be followed. Additionally, Category 1 is the only one requiring the use of **well-tried components**. The Safety-related Block Diagram is the same as for Category B. There is no DC; CCF considerations are not relevant and **the maximum PL achievable is PL c**.

A fault can lead to the loss of the safety function; however, the $MTTF_D$ of a single channel in Category 1 is higher than in Category B; consequently, the loss of the safety function is less likely.

6.2.3.1 Example of a Category 1 Input Subsystem: Interlocking Device

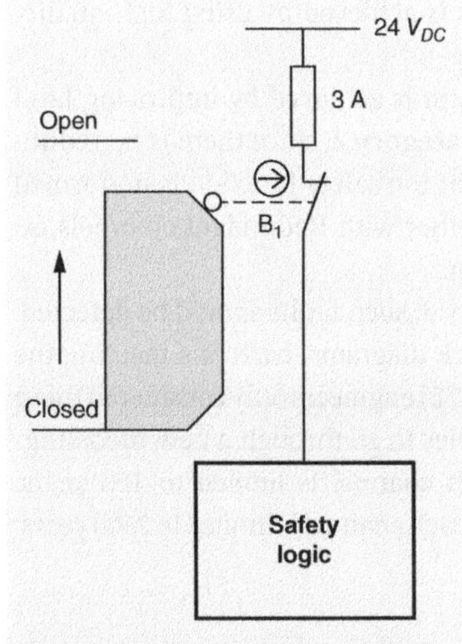

Figure 6.4 Category 1 input subsystem.

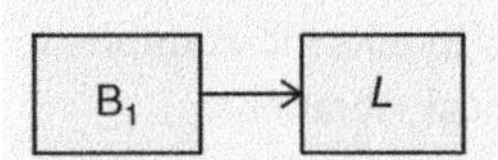

Figure 6.5 Category 1 input subsystem, represented as a RBD.

Let's consider an electromechanical interlocking device, connected to a Safety Logic. When the door is opened, the interlocking device output system (a Voltage Free Contact) opens and the input to the safety logic is de-energized. The **circuit structure** is shown in Figure 6.4, while in Figure 6.5 the same input subsystem is represented as a **Safety-related Block Diagram**. The interlocking device has a $B_{10D} = 20 \cdot 10^6$ and it is supposed to open **twice per hour**.

Looking at the manufacturer's datasheet, its **Mission Time is 20 years** and it should be protected by a max. 4 A fuse type gG, installed on the 24 V_{dc} line. That is important to avoid **systematic failures**: if the interlocking device output system is not properly protected from short circuits, all Reliability calculations have no real meaning. That is also the case for the maximum ambient temperature of 80 °C stated by the manufacturer, or the maximum impulse voltage U_{imp} of 2.5 kV that the component can withstand. Those are just examples: in general, being a Category 1 subsystem, **basic and well-tried safety principles must be applied.**

Moreover, being a Category 1 subsystem, **the components used should be well-tried** and it is the case.

The interlocking device is a "*Switch with positive mode actuation*" that complies with IEC 60947-5-1 (§ 5.2.1.4) and therefore, it is a well-tried component.

Let's now focus on the **probability of random failures**.

- **The first step** is to calculate the **$MTTF_D$**. We assume the machine is working 240 days/year and 8 h/day. Using the formula from Chapter 1:

$$n_{op} = 2 \cdot 8 \cdot 240 = 3\,840\ \frac{1}{\text{year}}$$

$$MTTF_D = \frac{B_{10D}}{0.1 \cdot n_{op}} = \frac{20 \cdot 10^6}{0.1 \cdot 3\,840} = 52\,083\ \text{years}$$

However, since we are in Category 1, the subsystem $MTTF_D$ has to be limited to 100 years.

- Moreover, being a Category 1, there is no diagnostic. That means we assume a $DC < 60\%$.
- The last step is to refer to table K.1 of ISO 13849-1, same as Table 6.3 in the book where, for Category 1 and $MTTF_D = 100$ years, the $\boldsymbol{PFH_D = 1.14 \cdot 10^{-6}\ h^{-1}}$, which corresponds to PL c. Finally, we verify that T_{10D} is higher than the interlocking device mission time: that means the interlocking device can be used up to its mission time of 20 years.

$$T_{10D} = \frac{B_{10D}}{n_{op}} = \frac{20 \cdot 10^6}{3840} = 5\,208\ \text{years}$$

6.2.4 Category 2

In Category 2, both Basic and Well-tried safety principles must be followed.

It is a **single channel architecture** with the monitoring of each subsystem done, in its most general form, by an external unit called **Test Equipment**. In case a fault is detected, the TE signals it to the "outside world" thanks to an output: the OTE.

Hereafter how the Safety-related block diagram looks like.

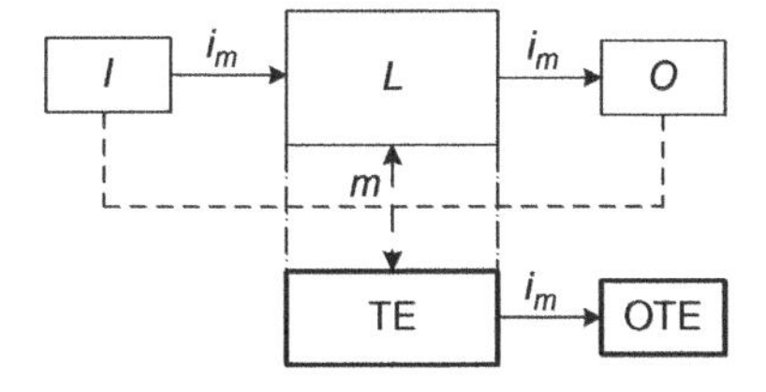

Keys:

- *i_m* represents the interconnecting means, typically electrical wires
- ***I*** represents the Input
- ***L*** represents the Safety Logic; it can be also a wire or a Safety Module (non-programmable) or a Programmable Logic.
- ***O*** represents the output; it can be a contactor or a solenoid valve, for example
- ***m*** represents the monitoring done by the Test Equipment (TE)
- **OTE** is the output of the test equipment.

Figure 6.6 Category 2 architecture.

Compared to a Category 1, you can notice the presence of a **Test Channel,** made of a Test Equipment **or** TE and of its output, the OTE or output of the Test Equipment.

Using a Category 2 architecture, all performance levels, except PL e, can be achieved.

Also, for this category, Figure 6.6 shows a subsystem and not necessarily a Safety-related control system. If we take the example of Figure 6.14, the input subsystem (emergency stop push button) is a Category 4 Subsystem, while **the output subsystem is indeed a Category 2, whereby**:

- The functional channel, I–L–O in Figure 6.14, is the Contactor K_{P1}
- The Test Equipment is the Automation PLC
- The OTE is the contactor K_G.

The fact that a PL d can be achieved, a certain level of DC (at least Low) needs to be present: the functional channel needs to be tested at suitable intervals by the test equipment. The check of the safety function shall be performed prior to the initiation of a hazardous situation, for example:

- Prior to the start of a new cycle and/or,
- Prior to the start of other movements and/or,
- Immediately upon a demand of the safety function and/or,
- Periodically during operations, if the risk assessment and the kind of operation shows that it is necessary.

Any check of the safety function allows its operation, if no fault is detected, or it generates an output (OTE), if a fault is detected.

It is important to highlight that, in case of PL d, the OTE **must initiate a safe state,** which is maintained until the fault is cleared. On the other hand, in case of PL c, a safe state is not required and **it would be enough to provide a warning**.

> ***[ISO 13849-1] 6.1.3.2 Designated Architectures – Specification of Categories***
> ***6.1.3.2.4 Category 2.***
> *[...] For **PLr d** the output (OTE) shall initiate a **safe state** that is maintained until the fault is cleared. For PLr up to and including **PLr c**, whenever practicable the output (OTE) shall initiate a safe state that is maintained until the fault is cleared. When this is not practicable (e.g. welding of the contact in the final switching device) it may be sufficient for the output of the test equipment OTE to provide a warning.*

The DC_{avg} of the functional channel shall be at least Low. The $MTTF_D$ of the Functional channel shall be low-to-high, depending on the required performance level (PL_r).

In any case, measures against **CCF** are applicable.

Of all the five categories, Category 2 is probably the most difficult one to understand: let's try to clarify its usage and understand the reasons of its limitations. For that, we need to go through its Markov Modelling.

6.2.5 Markov Modelling of Category 2

Back in the early 2000, when IEC 61508 series showed its probabilistic approach to functional Safety, a team from IFA [72] reacted by making a Markov modeling of each category of EN 954-1.

Category 2 can be represented by the Safety-related Block Diagrams shown in Figure 6.7. **F** represents the functional Channel (I–L–O) and **M** represents the Test Equipment (TE + OTE).

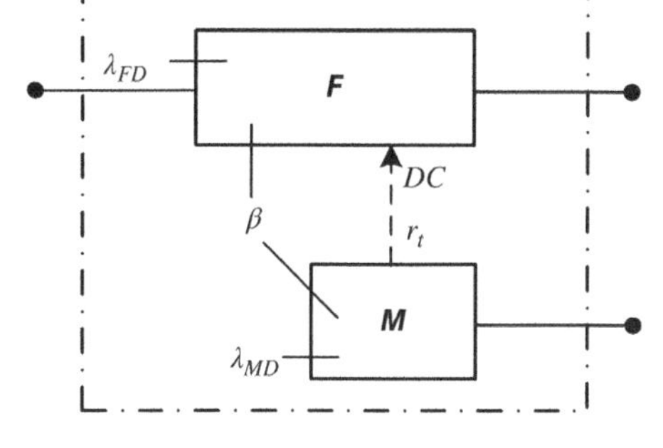

Figure 6.7 Category 2 architecture.

- λ_{FD} is the dangerous failure rate of the **Functional Channel F**.
- λ_{MD} is the dangerous failure rate of the **Test Channel M.**
- β is the **Common Cause factor** that influences at the same time the functional channel F and test channel M;
- r_t is the **Test rate** of the functional channel; in other terms, how often the Functional Channel is tested by the Test Channel
- r_d is the **Demand rate** of the safety function: how often the Safety Function is required.

6.2.5.1 The OK State

Figure 6.8 shows the starting Markov model for a Category 2 Architecture.

As you can see, it is a repairable system since from both the **Operating Inhibition** state and the **Hazardous Event** state, the system is brought back to the **OK** state. As already explained in Chapter 1, that is a different approach from IEC 62061. However, for the same safety-related control function, the difference in PFH_D in the two cases is small, since what is important is that **the system is considered the ultimate safety barrier**. That means, for failures that cause the loss of the overall safety function, the overall safety-related system repair has a negligeable influence on the PFH_D value since, in both cases, the safety system fails, before, in each case, being repaired.

One final word on the subject. A safety-related control system in high demand mode, during its mission time of typically 20 years, may statistically reach the hazardous event several times, especially when the PFH_D is high. Imagine, for example, a **PL a** low-risk applications with $PFH_D = 6 \cdot 10^{-5}\ h^{-1}$. The system is statistically facing a hazardous event 20 [years]·365 [days]· 24 hours·$6 \cdot 10^{-5}$ [h^{-1}] = 10.5 times during 20 years mission time and, each time, the system is normally repaired.

Let' now analyze and simplify the Markov Graph for a Category 2 (or a 1oo1D Subsystem). The machine is in the OK state when it is working normally and all Safety-related control Systems are vigilant and not affected by any fault.

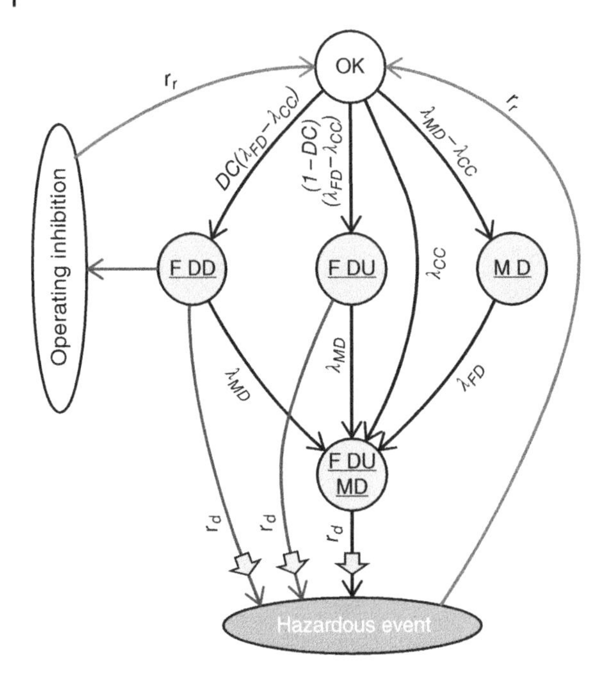

Figure 6.8 State transition model for a Category 2 architecture (1oo1D).

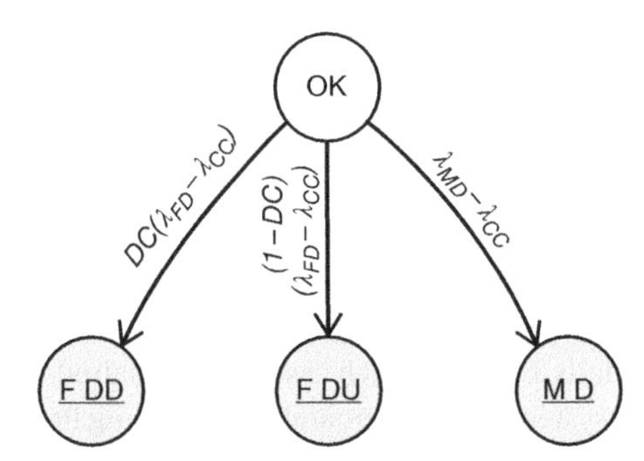

Figure 6.9 Transition from the OK to the failure states.

6.2.5.2 From the OK State to the Failure State

The SRP/CS we are analyzing is a 1oo1D architecture, and therefore a failure can happen either to the Functional Channel **F**, or to the Monitoring Channel **M**. A dangerous failure in the Functional Channel can either be Detected or Undetected, that is the way IEC 61508 series reasons.

Therefore, from the OK state, the system can move to three possible states, as shown in Figure 6.9:

- **F DD**: The system has a failure that is detected
- **F DU**: The system has an undetected failure
- **M D**: The monitoring Channel has a failure

Being λ_{FD} the dangerous failure rate, its detected part is:

$$\lambda_{FDD} = DC \cdot \lambda_{FD}$$

But we need to take into consideration the **CCF** between the Functional Channel **F** and the Monitoring channel **M**. β is the common cause factor and $\boldsymbol{\lambda_{CC}}$ **is** the Common Cause Failure Rate. Therefore, the probability that the SRP/CS moves from the **OK** state to **F DD** is linked to the following Failure Rate:

$$\lambda_{FDD} = DC \cdot (\lambda_{FD} - \lambda_{CC})$$

In case the Monitoring channel cannot detect the failure, the system moves to a so-called Dangerous Undetected State, indicated as <u>**F DU**</u> in the drawing. The probability that the Safety system moves from the <u>**OK**</u> state to <u>**F DU**</u> is linked to the following Failure Rate:

$$\lambda_{FDU} = (1 - DC) \cdot (\lambda_{FD} - \lambda_{CC})$$

The probability of a failure of the Monitoring Channel is linked to the following Failure Rate:

$$\lambda_M = (\lambda_{MD} - \lambda_{CC})$$

The status is defined as <u>**M D**</u>

6.2.5.3 From the Failure State to the Hazardous Event

In the Markov Model of Category 2 Architecture, a "temporary state" called <u>**F DU + MD**</u> is defined, but just for the system completeness. The transition to this state is either from the <u>**F DD**</u> or from the <u>**F DU**</u> in case a failure of the Monitoring Channel happens (**λ_{MD}**). Alternatively, from the <u>**M D**</u> state, in case of a failure in the Functional Channel (**λ_{FD}**).

Please consider that all the transitions described so far are **Exponentially Distributed**: they may happen randomly.

Now, what is important, is the transition to the <u>**Hazardous State**</u>. The SRP/CS enters this state when:

- It has a Failure (just one!) <u>**and**</u>
- There has been a demand upon the safety Function (**r_d**)

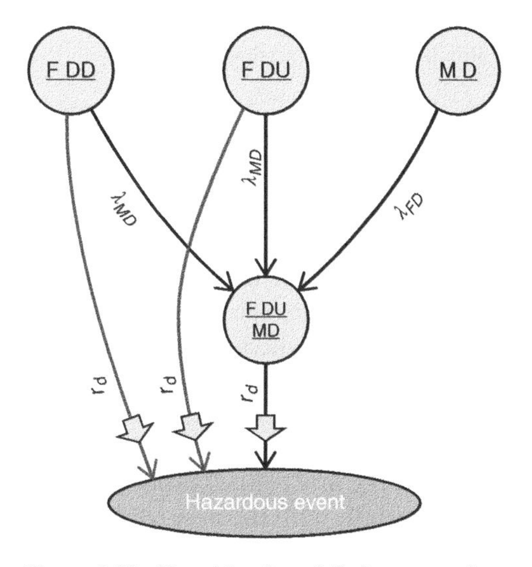

Figure 6.10 Transition from failed states to the hazardous event functional channel.

As an example, the access gate to a robot cell was opened, and the interlocking device has failed to open the electrical contact: **the consequence is that a person is inside the area and the Robot is still running.**

At the end, what is important is the probability that the system moves from the OK state to the Hazardous State; that probability is given by the sum of the probabilities of 3 transitions, described with a big arrow in Figure 6.10:

- Transition from the <u>**F DD**</u> state to the <u>**Hazardous State or Event**</u>
- Transition from the <u>**F DU**</u> state to the <u>**Hazardous State**</u>
- Transition from the <u>**F DU + MD**</u> state to the <u>**Hazardous State**</u>

Please consider that all these transitions are Uniformly Distributed and the frequency is $\mathbf{r_d}$: the frequency of the request upon the safety function.

6.2.5.4 Other States in the Transition Model

One other state displayed in the Markov Model is the **Operation Inhibition** one. The **OK** system moves to this state in two steps:

1. A detected failures in the SRP/CS has happened: transition from the **OK** state to **F DD**.
2. The Monitoring Channel tests the Functional Channel before a request upon the safety function and **it brings the system to a safe state**. In practice, the safety system stops the machine or the dangerous movement. In our example of the Robot Cell, the safety system detects a failure when the person enters the robot cell, the Robot is still running and the safety system stops it, before a person reaches the robot.

Finally, the last transitions are:

- From the **Hazardous Event** to the **OK** State or
- From the **Operation Inhibition** to the **OK** State

That happens when the system is repaired.

6.2.5.5 The Simplified Graph of the Markov Modelling

The Graph can be simplified, using conservative assumptions, to demonstrate that the PFH_D values of a Category 2 Architectures, indicated in table K.1 of ISO 13849-1, can be calculated using formulas. **Those are comparable to the formulas used in IEC 62061**.

The purpose of this exercise, that was prepared in 2017 by some mathematicians from IFA [72], had exactly that scope: to demonstrate that the reliability level of the same architecture, analyzed with either IEC 62061 or ISO 13849-1 is essentially the same. **That is again another way to show that, especially with the new edition of IEC 62061, the two standards are aligned on many aspects, as ever it has been.**

The simplified model is shown in Figure 6.11. The steps from the full to the simplified model are not detailed in the book.

From the graph, the instantaneous PFH_D value for the 1oo1D Architecture can be calculated as follows:

$$PFH_D(t) = \lambda_B \cdot \frac{r_d}{r_t + r_d} \cdot p_{OK}(t) + [(1 - DC)(\lambda_{FD} - \lambda_{CC}) + \lambda_A + \lambda_{CC}] \cdot p_{OK}(t) + \lambda_{FD} \cdot p_{MD}(t)$$

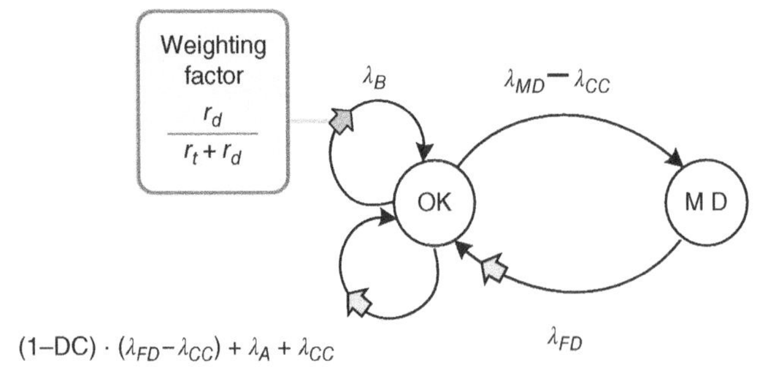

Figure 6.11 Simplified Markov model for calculation of PFH_D in 1oo1D architecture.

When properly integrated,

$$PFH_D = \frac{1}{T_M}\int_0^{T_M} PFH_D(t)\cdot dt$$

where T_M is the Mission Time.

The PFH_D of a 1oo1D Architecture is:

$$PFH_D = \lambda_{FD} - DC\,\frac{r_t}{r_t + r_d}\cdot\frac{(\lambda_{FD}-\lambda_{CC})\left[\lambda_{FD}\,(\lambda_{FD}+\lambda_{MD}-\lambda_{CC})T_M + (\lambda_{MD}-\lambda_{CC})\left(1-e^{-(\lambda_{FD}+\lambda_{MD}-\lambda_{CC})T_M}\right)\right]}{(\lambda_{FD}+\lambda_{MD}-\lambda_{CC})^2\,T_M}$$

6.2.5.6 The Importance of the Time-Optimal Testing

There is a problem with the above formula and it is the term called TRTE: **Time-related Test Efficiency.**

$$TRTE = \frac{1}{1 + (r_d/r_t)}$$

The reason why the term is important is because it shows that, in Category 2, the Reliability of the Safety Function depends upon the ratio between the demand rate and the test rate. That is something not desirable in high-demand mode of operation, while it is the rule in Low demand mode of operation (Proof Test). Bottom line: in high-demand mode of operation, we do not want that the Reliability of a Safety function depends upon how often the subsystem is tested vis a vis how often it is used.

A way to restate the issue is the following. We are in a single channel with a diagnostic function. That means, in principle, the reliability should be higher compared to a 1oo1 system, however **that is true only if we can detect a failure before there is a request upon the safety function**. If the system is not able to do so, the 1oo1D system is equivalent to a 1oo1. Since we do not know when there will be a failure (random variable with an exponential distribution curve), the **test rate has to be much higher than the Demand Rate in order to be relatively sure that the Monitoring Channel has a chance to detect a failure before a demand is made upon the safety function**.

If the monitoring function has a high chance of detecting a failure before a demand upon the safety function happens, the reliability of the SRP/CS will be independent from the ratio r_d/r_t. In other terms, it is important that the TRTE be as close as possible to 1. That is achieved when the test rate is 100 times the demand rate: that is called "**time-optimal testing**":

$$r_t \geq 100 \cdot r_d$$

That concept is also stated in IEC 61508-2, § 7.4.4.1.4.

In the new edition of the ISO 13849-1, it is indicated that, in case the ratio is only 25 times, the Monitoring function is still considered effective and the PFH_D has just to be increased by 10%.

An alternative to time-optimal testing is that the test of the safety function happens together with its demand and, in case a failure is detected, the overall time **to bring the machine to a safe state** is shorter than the time to reach the hazard.

6.2.5.7 1oo1D in Case of Time-Optimal Testing

In case of time-optimal testing, the formula becomes:

$$PFH_D = \lambda_{FD} - DC\cdot\frac{(\lambda_{FD}-\lambda_{CC})\left[\lambda_{FD}\,(\lambda_{FD}+\lambda_{MD}-\lambda_{CC})T_M + (\lambda_{MD}-\lambda_{CC})\left(1-e^{-(\lambda_{FD}+\lambda_{MD}-\lambda_{CC})T_M}\right)\right]}{(\lambda_{FD}+\lambda_{MD}-\lambda_{CC})^2\,T_M}$$

If $|X| \ll 1$, the exponential function can be replaced, without notable loss of precision, by its quadratic approximation:

$$e^x \approx 1 + x + \frac{1}{2}x^2$$

Substitution of the exponential function, in the PFH_D Equation, by the above approximation yields:

$$PFH_D = \lambda_{FD} - DC(\lambda_{FD} - \lambda_{CC})\left[1 - \frac{1}{2}(\lambda_{MD} - \lambda_{CC})T_M\right]$$

λ_{CC} takes into consideration the common cause failure of both the Functional Channel (FD) and the Test Channel (MD), and it can be estimated using the following equation:

$$\lambda_{CC} = \beta \cdot \min(\lambda_{FD}; \lambda_{MD})$$

The reason why this exercise was done is not only to understand better the Markov Modelling behind Category 2 subsystems; it was also done to show that, despite ISO 13849-1 and IEC 62061 use different modeling, the difference in PFH_D between the two approaches is negligible. The same exercise was done for Category 3 and 4 and compared to architecture 1oo2D, with similar results.

In ISO 13849-1, the values of PFH_D are listed in table K.1. However, the models behind those numbers can be simplified, on the safe side, and represented with formulas, as shown in this paragraph for a 1oo1D architecture. **Despite ISO 13849-1 uses the Markov chains and it assumes the systems as repairable, its simplified equations are very similar to those of IEC 62061, which uses reliability block diagrams and it assumes the systems as non-repairable.**

6.2.6 Conditions for the Correct Implementation of a Category 2 Subsystem

The Safety-related Control System has to be done in such a way that any check of the safety function shall either:

- allow operations if no faults are detected, or
- generate an output (OTE: output of the test equipment) which initiates appropriate control actions, if a fault is detected.

In case we want to reach PL d, the output (OTE) **must initiate a safe state** which is maintained until the fault is cleared.

If the risk assessment requires a PL c, or lower levels, whenever practicable, the output (OTE) shall initiate a safe state; otherwise, **it is enough to provide a warning**.

The Figure 6.12 summarises all possible conditions related to a correct implementation of a Category 2 subsystem.

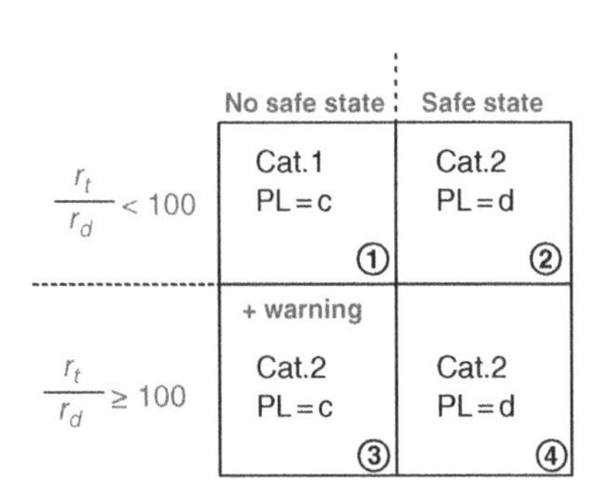

Figure 6.12 Summary of Category 2 conditions.

- **Situation 1.** No Optimal testing is possible; moreover, the Safety subsystem cannot reach a safe state in case a fault is detected. We cannot claim a Category 2 Architecture but only a Category 1, if well-tried components are used.

- **Situation 2.** No Optimal testing is present; however, in case a fault is detected, we can bring the system to a **safe state**. We can claim a Category 2 and we can reach a maximum of PL d, if all other conditions exist. No need to use well-tried components.
- **Situation 3.** We have the time-optimal testing but the system can only generate an alarm (**warning**), in case a fault is detected: we can claim a Category 2 for the SRP/CS but we can reach PL c, at best.
- **Situation 4.** We have both optimal testing and, in case a fault is detected, we can bring the system to a **safe state**. We can claim Category 2 and we can reach PL d, if all other conditions exist.

There is a final condition to be mentioned. In principle, the Monitoring channel is not required to have reliability data. In reality, we need to give it a minimum value. The reason why the test channel needs a certain level of Reliability is because, despite its Reliability does not affect the Safety Function, it should not reasonably fail.

The ISO standard states that **the $\mathbf{MTTF_D}$ of the test channel shall not be lower than half the $\mathbf{MTTF_D}$ value of the functional channel.**

How to achieve it? Let's consider the output subsystem of the functional channel made by a contactor that is de-energized twice per hour (mainly because of it being part of both the automation and the safety system).

Assuming a $B_{10D} = 1\,300\,000$ and 8760 hours of functioning per year

$$MTTF_D = \frac{1\,300\,000}{0.1 \cdot 17\,520} = 742\,\text{years}$$

Let's suppose the monitoring system is made with a general purpose (automation) PLC, with an MTBF (that is the parameter usually provided) of 500 years. That corresponds, worst case, to a $MTTF_D$ of 500 years (please refer to Chapter 1). The Automation System has an acceptable reliability level since its $MTTF_D$ (500 years) is greater than (742/2 =) 371 years.

6.2.7 Examples of Category 2 Circuits

6.2.7.1 Example of Category 2 – PL c

Please consider the electrical Architecture shown in Figure 6.13: an emergency stop button triggers the stop of a motor. The motor contactor (or magnetic motor controller, in North American

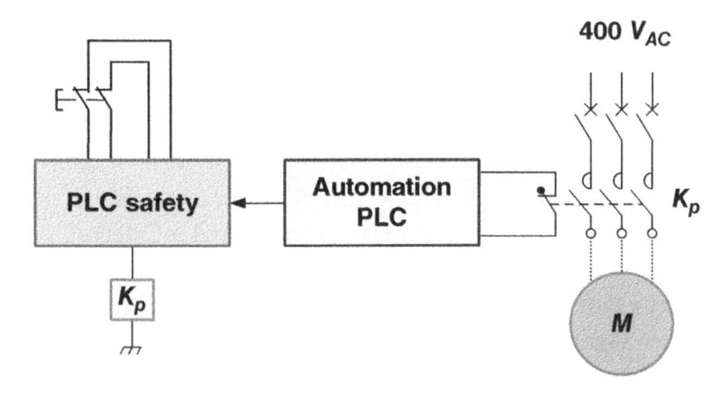

Figure 6.13 Simplified electrical diagram.

language) is monitored by a general purpose PLC. **The output subsystem could be classified as a Category 2,** because the Automation PLC triggers the safety system to switch on and off the motor 100 times more often than the activation frequency of the emergency pushbutton. Therefore, the output subsystem is indeed a Category 2 with a time-optimal testing, however its Reliability is limited to PL c since, in case of a failure of the contactor, there is no way to bring the system to a safe state.

6.2.7.2 Example of Category 2 – PL d

Let's now consider the electrical Architecture shown in Figure 6.14. This time, in case the safety logic de-energizes the contactor K_{P1} and if it does not detect its opening (status contact S_{KP1}), it de-energizes the coil of a second contactor K_G. **Please notice that K_G does not need to be monitored.**

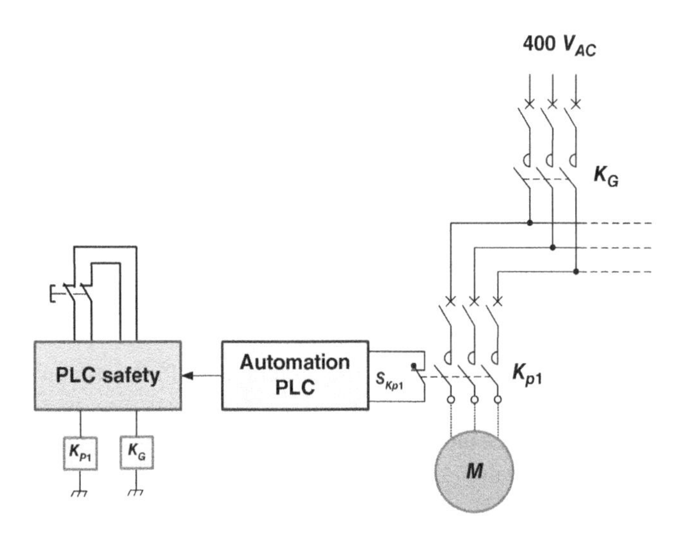

Figure 6.14 Simplified electrical diagram.

This is a Category 2 output subsystem that can reach PL d, since the way it is designed "*testing is performed immediately when a demand is made upon the safety function* (has K_{P1} opened?), ***and** the overall time for detection of the fault and for bringing the machine into a non-hazardous state is shorter than the time to reach the hazard* (if it hasn't, K_G is immediately de-energised)."

Please notice that K_{P1} is the contactor that works with the higher frequency, and therefore it represents the Functional Channel; K_G represents the output of the TE.

Let's suppose a different scenario: K_G is working with the higher frequency (it would be the Functional Channel) and it is not monitored; K_{P1} is monitored but working with the lower frequency (it is this time the OTE): the system cannot be considered a Category 2 since K_G is not monitored!

The Safety-related Block Diagram of the electrical circuit is the one shown in Figure 6.15: K_{P1} represents the "functional channel."

In case the monitoring and the activation of the second channel were done by the safety module, its block diagram would be the one in Figure 6.16.

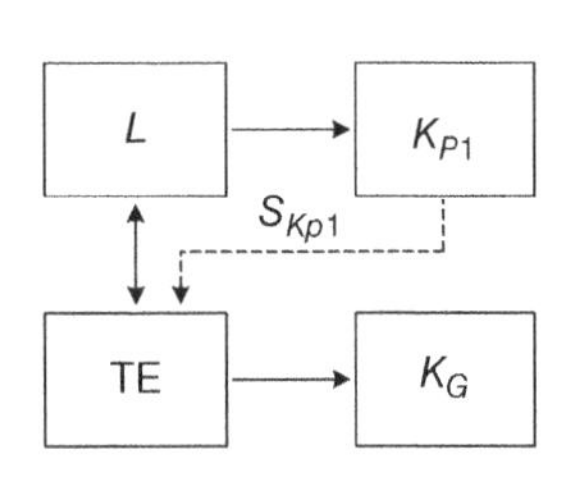

Figure 6.15 Safety-related block diagram for Figure 6.14.

A Final consideration: **it is still a Category 2 in case the OTE (K_G) is activated every time the Final element (K_{P1}) is activated**. Of course, provided the $MTTF_D$ of the OTE is compatible with the frequency of the output element.

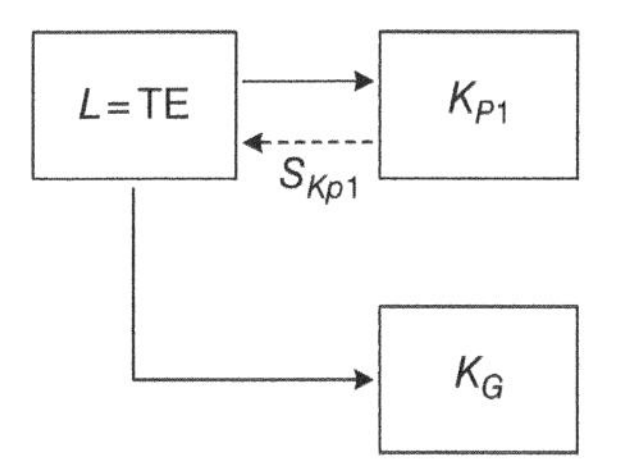

Figure 6.16 Alternative Category 2 architecture.

6.2.7.3 Example of a Category 2 with Undervoltage Coil

In this example, a motor is safely stopped by de-energizing the contactor K_P, whose mirror contact (§ 4.12.2.1) is properly monitored. In case the safety logic (or Safety PLC) detects that the contactor did not open, it de-energizes an undervoltage coil that opens the Circuit Breaker that protects that part of the electrical equipment of the machine (usually that is the Machine Supply Circuit Disconnecting Means). We focus on the output subsystem. The electrical circuit is shown in Figure 6.17.

The Safety-related **Block Diagram of the output subsystem** is shown in Figure 6.18.

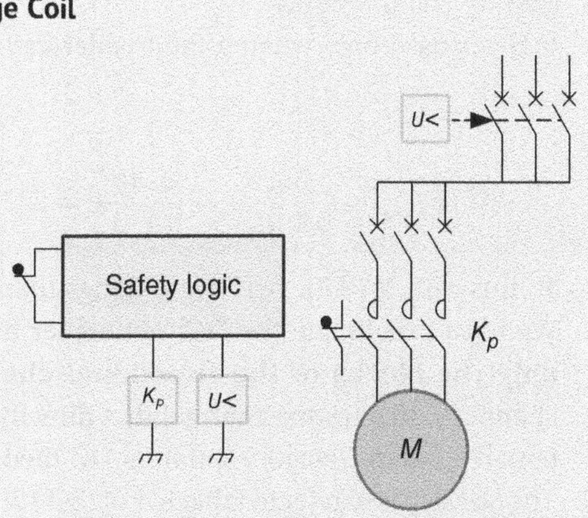

Figure 6.17 Category 2 output subsystem.

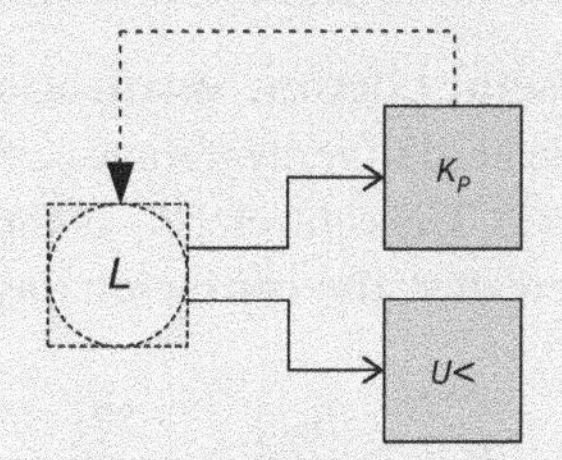

Figure 6.18 RBD of a Category 2 output subsystem.

- **Safety data:**
  K_P has $B_{10} = 1{\cdot}10^6$ with 73% as percentage of dangerous failures (RDF according to par. 1.12.1).
  $U<$ is the so-called Undervoltage Coil with $B_{10D} = 4{\cdot}10^5$ according to table C.1 of ISO 13849-1.
  Looking at the manufacturer datasheet, the Mission Time of both the contactor and the undervoltage coil is 20 years.
- **Usage frequency:**
  K_P is supposed to open **twice per minute, considering 8 hours per day, 240 days per year**, mainly driven by an automation PLC, not shown in the drawing but that communicates with the Safety Logic.
  $U<$ is supposed to open **once per month** (triggered by the detection of a fault, or manually triggered).
- **Avoidance of Systematic Failures**
  Since we claim Category 2, **basic and well-tried safety principles are applied**. We also verified that enough measures have been applied to prevent **CCF** (score $CCF > 65$).

Let's now focus on the calculation of **probability of random failures**.

1. **The first step** is to calculate the $\mathbf{MTTF_D}$ of the contactor K_P:

$$n_{op} = 120 \cdot 8 \cdot 240 = 230\,400\ \frac{1}{\text{year}}$$

$$B_{10D} = \frac{1 \cdot 10^6}{0.73} = 1.37 \cdot 10^6$$

$$MTTF_{D_{KP}} = \frac{B_{10D}}{0.1 \cdot n_{op}} = \frac{1.37 \cdot 10^6}{0.1 \cdot 230\,400} = 59\ \text{years}$$

2. Now, we have to check that the **$\mathbf{MTTF_D}$ value of the test channel is not lower than half the $\mathbf{MTTF_D}$ value of the functional channel**. In this case, functional channel and testing channel have the Safety PLC in common (for both LOGIC and TEST EQUIPMENT). This is the reason why we compare $MTTF_D$ of contactor (functional channel) and undervoltage coil (testing channel) only.
   U< is used only when a fault is detected, once per month:

$$n_{op} = 12$$

$$MTTF_{D_{U<}} = \frac{B_{10D}}{0.1 \cdot n_{op}} = \frac{4 \cdot 10^5}{0.1 \cdot 12} = 333\,333\ \text{years}$$

   In this case $MTTF_{D_{U<}}$ is even higher than $MTTF_{D_{KP}}$, so the condition is fulfilled;
3. We have to estimate the **DC:** remember that the calculation of **$\mathbf{DC_{avg}}$ takes into account only the blocks of the functional channel**, according to ISO 13849-1. Being a single channel with a testing channel that directly monitors the status of K_P, we assume the highest possible DC in Category 2 that is DC medium ($90\% \le DC < 99\%$);
4. The last step is to refer to table K.1 of ISO 13849-1 (same as Table 6.2) where, for Category 2 with medium DC and $MTTF_D = 56$ years (Since in table K.1, $MTTF_D = 59$ is not present, we have to choose the first conservative value), the $\boldsymbol{PFH_D = 5.10 \cdot 10^{-7}\ h^{-1}}$, which corresponds to PL d.

- **Useful Lifetime verification**
  We need to verify if some components have to be replaced before their mission time expires (§ 1.12). The contactor K_P is the most critical element in this respect. As shown in the formula hereafter, **the contactor has to be replaced approximately every six years.**

$$T_{10D} = \frac{B_{10D}}{n_{op}} = \frac{1.37 \cdot 10^6}{230\,400} = 5.9\ \text{years}$$

6.2.8 Category 3

In Category 3, both Basic and Well-tried safety principles must be used. Each Category 3 subsystem should be designed so that **a single fault does not lead to the loss of the safety function**.

Moreover, whenever reasonably practicable, a single fault shall be detected at or before the next demand upon the safety function. The RBD is shown in Figure 6.19.

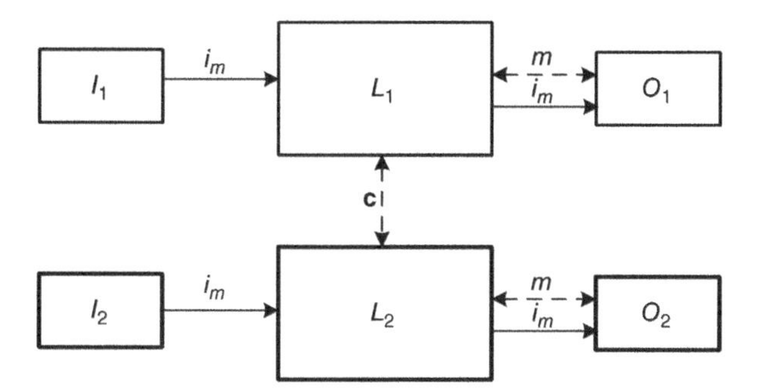

Keys:

- i_m represent the interconnecting means, typically, electrical wires
- I_1 **and** I_2 represent the Inputs (for example two interlocking devices)
- ***L*** represents the Safety Logic; usually a Safety Module (non programmable) or a Programmable Logic.
- ***O*** represents the output; it can be a contactor or a solenoid valve, for example
- **c** is the cross monitoring
- ***m*** **dashed lines** represent reasonably practicable fault detection

Figure 6.19 Category 3 architecture.

The DC_{avg} of each subsystem shall be at least low. The $MTTF_D$ of each redundant channel shall be low-to-high, depending upon the required performance level (PL_r).

Measures against **CCF** are applied.

In Category 3, the requirement of single-fault detection does not mean that all faults will be detected. Consequently, the accumulation of undetected faults can lead to a hazardous situation at the machine.

The subsystem behavior of this Category is therefore characterized by:

- Continued performance of the safety function in the presence of a single fault.
- Detection of some, but not all, faults.
- Possible loss of the safety function, due to accumulation of undetected faults.

6.2.8.1 Diagnostic Coverage in Category 3

We see applications where the DC is done externally to the Safety Logic, or better, to the Functional Channel. In that case, it has to be justified that the simplified approach (annex K) is still applicable. Determination of PFH_D based on other modeling techniques is also possible. Figure 6.20 shows how the Safety-related Block Diagram would look like.

The issue is that the PFH_D values of table K.1 do not correctly represent the Reliability level of that SRP/CS. An updated Markov Modelling should be done. However, we estimate the difference in PFH_D would not be significant, provided at least a Low MTTF can be claimed for the Test Equipment. The requirement for the MTTF of the Test Equipment is therefore less than what is required in Category 2, since in Category 3 and 4 we have a redundant subsystem.

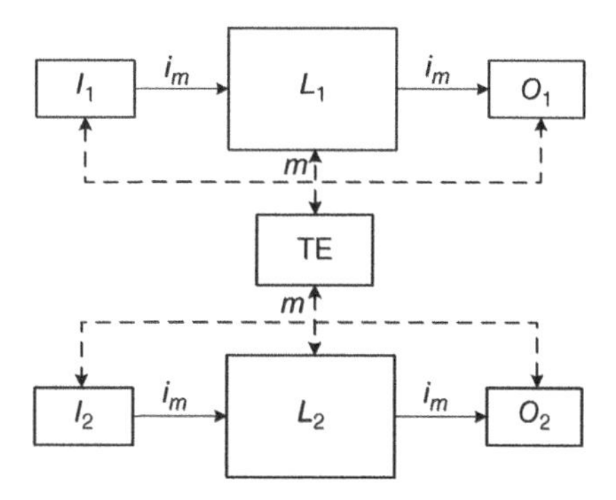

Figure 6.20 Category 3 Architecture with external test equipment.

One of the conditions would also be that, in case a fault is detected in one of the two channels (for example, the welding of the contacts in a contactor), **the safe state** (the other contactor is supposed to open) is maintained until the fault is cleared. For that reason, despite it is possible to monitor the contactor status in an Automation PLC, a signal must be sent to the Safety Logic that blocks any reset until the fault is cleared (contactor K_R in Figure 6.21). To state it in another way: the Diagnostic can go through a general purpose PLC, but the reaction has to be Safe. Which also means that one single fault is not able to lead to the loss of the safety function.

In the IFA report [53] there are a few examples where a general purpose PLC is used for diagnostics in a Category 3 or 4 Subsystem and the simplified approach is kept.

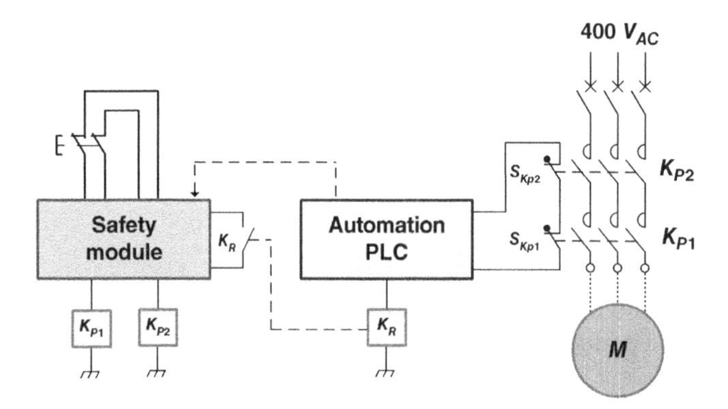

Figure 6.21 Category 3 Architecture with external test equipment.

6.2.8.2 Example of Category 3 for Input Subsystem: Interlocking Device

In this example, a Type 2 Interlocking Device (§ 4.11.2.1), mounted on a movable gate that gives access to a safeguarded area, is controlled by a Safety PLC. In case the guard is opened, the Safety Logic has to detect it and take appropriate actions. We focus on the input subsystem.

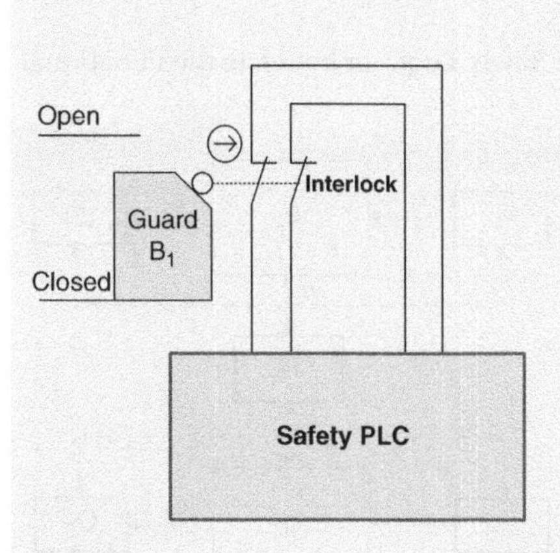

Figure 6.22 Interlocking device input subsystem.

For the electrical circuit shown in Figure 6.22, the Safety-related **Block Diagram input subsystem** is shown in Figure 6.23.

- **Safety data:**
  Interlocking device B_1 has $B_{10D} = 1{\cdot}10^6$ and a Mission Time of 20 years.
- **Usage frequency:**
  The interlocking device is supposed to open **10 times per hour**.
- **Avoidance of Systematic Failures**
  Since we can claim Category 3 (double channels with monitoring), **basic and well-tried safety principles are applied**. We also verified that enough measures have been applied to prevent **CCF** (score $CCF > 65$).

- **Fault Exclusion:**
  Considering the way the interlocking device was installed, we consider negligible the probability of breakage of the actuator (§ 4.11.2.3). Therefore we apply the fault exclusion to the mechanical part of the Interlocking Device: that is the meaning of the FE element in Figure 6.23. In Figure 6.24, we show another possible and equivalent representation.

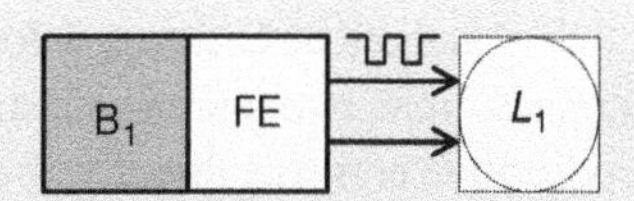

Figure 6.23 Input subsystem represented as RBD.

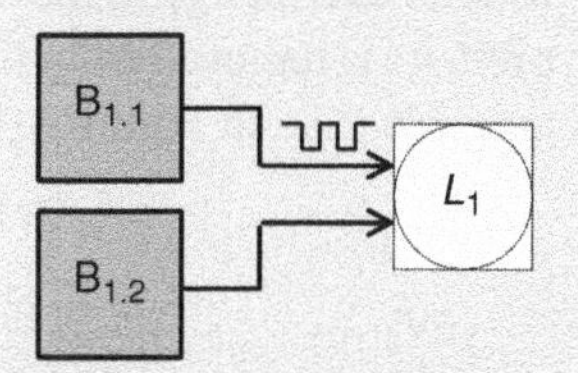

Figure 6.24 Another way to represent the input subsystem as RBD.

Let's now focus on the calculation of **probability of random failures**:

1. **The first step** is to calculate the **$MTTF_D$** of the interlocking device. We assume the machine is working 365 days per year and 16 hours per day

$$n_{op} = 10 \cdot 16 \cdot 365 = 58\,400\ \frac{1}{\text{year}}$$

$$MTTF_{D\,B_1} = \frac{B_{10D}}{0.1 \cdot n_{op}} = \frac{1 \cdot 10^6}{0.1 \cdot 58\,400} = 171\,\text{years}$$

   Being the subsystem in Category 3, $MTTF_D$ must be limited to 100 years;
2. **As a second step**, we estimate the **DC.** The cross monitoring of inputs can only be performed when the guard is opened; therefore, that cannot be defined as "*Cross monitoring of input signals with dynamic test*" since it is not an automatic dynamic test, even if the trigger is present. That can be defined simply as a "*cross monitoring of inputs without dynamic test.*" The corresponding DC can vary "*e.g. depending on how often a signal change is done by the application.*" Since the gate is opened at least once per hour, the maximum DC value that can be reached is 99% (Note 1 of Table 3.6 in § 3.3.6). Since we have the trigger, we assume the highest possible DC in the Category: $DC = 98\%$ (medium).
3. The last step is to refer to table K.1 of ISO 13849-1 (same as Table 6.3) where, for Category 3 with medium DC and $MTTF_D = 100$ years, the $\boldsymbol{PFH_D = 4.29{\cdot}10^{-8}\,h^{-1}}$, which corresponds to PL e, however, **since we did a fault exclusion** on the mechanical part of the interlocking device (§ 4.11.2.3), the final result is:

$$PFH_D = 4.29 \cdot 10^{-8}\,\text{h}^{-1};\ \text{PL d}$$

- **Useful Lifetime verification**
  The concept is present in both IEC 62061 (§ 3.7.4) and ISO 13849-1 (operating lifetime in annex C § C.4.2). We need to verify if the interlocking device has to be replaced before its mission time expires. As shown in the formula hereafter, it has to be replaced after 17 years: it cannot be used for the whole mission time.

$$T_{10D} = \frac{B_{10D}}{n_{op}} = \frac{1 \cdot 10^6}{58\,400} = 17\,\text{years}$$

6.2.8.3 Example of Category 3 for Output Subsystem: Pneumatic Actuator

In this example, two pneumatic solenoid valves control the movement of a cylinder. The general purpose PLC controls directly the 5/2 valve, while the Safety module controls directly the 3/2 cut-off valve. In case a dangerous situation is detected, like the opening of the movable guard of the previous example, the Safety Logic de-energizes both the 3/2 cut-off valve and contactor K_{P1}. The latter cuts energy to the Digital output card that normally drives the 5/2 solenoid valve. When pressure is removed from the circuit (3/2 de-energized), the PSL contact closes; when 5/2 is de-energized (like in Figure 6.25) the status ZSL closes. We focus on the output subsystem.

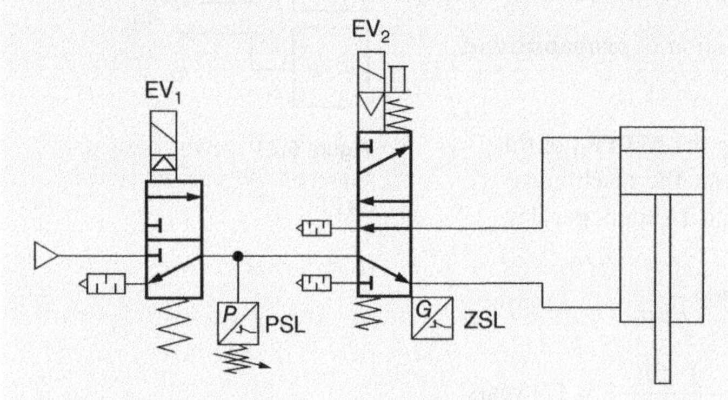

Figure 6.25 Category 3 output subsystem.

For the electrical circuit shown in Figure 6.26, the Safety-related **Block Diagram output subsystem** is shown in Figure 6.27.

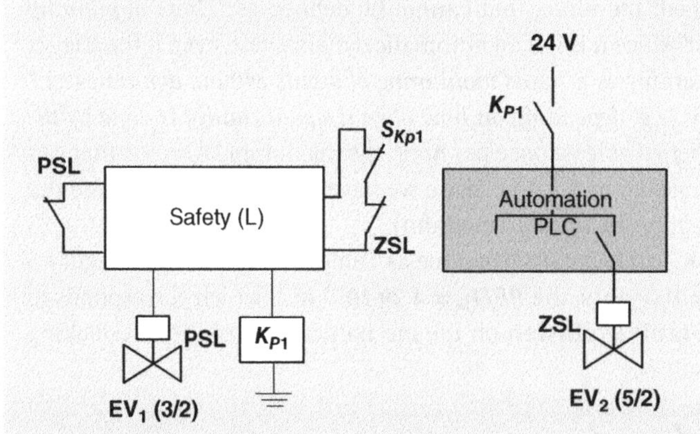

Figure 6.26 General Purpose PLC and the Safety Logic.

- **Safety data:**
  K_{P1} has $B_{10} = 1{\cdot}10^6$ with 73% as percentage of dangerous failures (RDF according to § 1.12.1);
  EV1 and EV2 have $B_{10D} = 20{\cdot}10^6$;
  Looking at the manufacturer datasheet, the Mission Times of both the contactor and the solenoid valves are 20 years.
- **Usage Frequency:**
  EV2 Solenoid Valve switches **twice per minute** (driven by the automation PLC). EV1 and K_{P1} de-energize four times per hour. We assume the machine is working 240 days per year and 8 hours per day.

- **Avoidance of Systematic Failures**
  Since we can claim Category 3 (double channels with monitoring), **basic and well-tried safety principles are applied**. We also verified that enough measures have been applied to prevent **CCF** (score $CCF > 65$).
  EV1 is monitored **indirectly with a** pressure switch (PSL), while EV2 is directly monitored with a status switch (ZSL). Contactor K_{P1} is monitored with its Mechanically Linked Contact S_{KP1} (4.12.2.2).

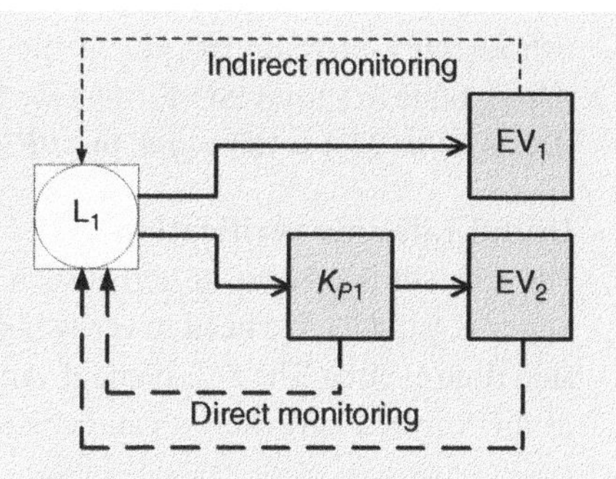

Figure 6.27 output subsystem represented as RBD.

Let's now focus on the calculation of **probability of random failures**:

1. **The first step** is to calculate the **MTTF$_D$** of the two channels. For the 5/2 solenoid valve EV2:

$$n_{op} = 120 \cdot 8 \cdot 240 = 230\,400\ \frac{1}{\text{year}}$$

$$MTTF_{D_{EV2}} = \frac{B_{10D}}{0.1 \cdot n_{op}} = \frac{20 \cdot 10^6}{0.1 \cdot 230\,400} = 868\ \text{years}$$

For its contactor K_{P1}

$$n_{op} = 4 \cdot 8 \cdot 240 = 7680\ \frac{1}{\text{year}}$$

$$MTTF_{D_{KP1}} = \frac{B_{10D}}{0.1 \cdot n_{op}} = \frac{1.37 \cdot 10^6}{0.1 \cdot 7680} = 1\,783\ \text{years}$$

The MTTF$_D$ of the K_{P1} + EV2 channel is

$$MTTF_{D\,KP1+EV2} = \frac{1}{(1/1783) + (1/868)} = 584\ \text{years} \rightarrow 100\ \text{years}$$

The limitation to 100 years is due to the fact we are in Category 3 (§ 6.2.1).
The MTTFD of the second channel (EV1) is:

$$MTTF_{D\,EV1} = \frac{B_{10D}}{0.1 \cdot n_{op}} = \frac{20 \cdot 10^6}{0.1 \cdot 7680} = 26\,042\ \text{years} \rightarrow 100\ \text{years}$$

2. As a second step, we estimate the **DC**.
   The EV2 channel has both a direct monitoring of the Contactor K_{P1} and of the Valve EV2: that means DC 99%. Please refer to table E.1 (same as Table 3.7 in § 3.3.6): *direct monitoring (e.g. electrical position monitoring of control valves, monitoring of electromechanical devices by mechanically linked contact elements)*. Total DC average of the EV2 channel can be calculated according to § 3.3.7

$$DC_{avg-KP1+EV2} = \frac{(DC_{KP1}/MTTF_{KP1}) + (DC_{EV2}/MTTF_{EV2})}{(1/MTTF_{KP1}) + (1/MTTF_{EV2})} = \frac{(99\%/1783) + (99\%/868)}{(1/1783) + (1/868)} = \mathbf{99\%}$$

   EV1 channel has an indirect monitoring of the valve: that means DC 90%. Please refer to table E.1 – Output Device – *indirect monitoring (e.g. monitoring by pressure switch, electrical position monitoring of machine actuators)*.
   The total DC average of the two channels is the following

$$DC_{avg} = \frac{\frac{DC_{avg-KP1+EV2}}{MTTF_{D-KP1+EV2}} + \frac{DC_{EV1}}{MTTF_{D-EV1}}}{\frac{1}{MTTF_{D-KP1+EV2}} + \frac{1}{MTTF_{D-EV1}}} = \frac{\frac{99\%}{584} + \frac{90\%}{26042}}{\frac{1}{584} + \frac{1}{26042}} = \mathbf{98.8\%}$$

It results a Medium DC, this is the reason why the output subsystem is a Category 3.

3. Using table K.1 from ISO 13849-1 we have the following results: for Category 3 with average DC and $MTTF_D = 100$ years, the $\boldsymbol{PFH_D = 4.29{\cdot}10^{-8}\ h^{-1}}$, which corresponds to PL e.

- **Useful Lifetime verification**

  The concept is present in both IEC 62061 (§ 3.7.4) and ISO 13849-1 (operating life time in annex C § C.4.2). We need to verify if some components have to be replaced before their mission time expires. The 5/2 solenoid valve is the most critical element in this respect. As shown in the formula hereafter, it can be used for 20 years without problems.

$$T_{10D} = \frac{B_{10D}}{n_{op}} = \frac{20 \cdot 10^6}{230\,400} = 87\ \text{years}$$

6.2.9 Category 4

In Category 4, both Basic and Well-tried safety principles must be used. Each Category 4 subsystem should be designed so that a single fault does not lead to the loss of the safety function.

Moreover, the single fault must be detected at or before the next demand upon the safety function. When this detection is not possible, an accumulation of undetected faults **should not lead to the loss of the safety function**. The RBD is shown in Figure 6.28.

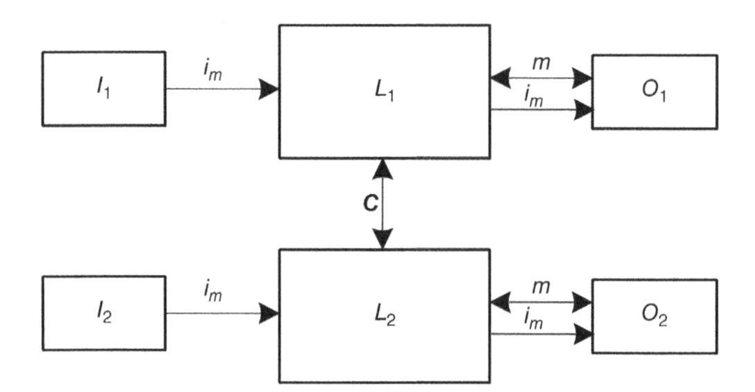

Figure 6.28 Category 4 architecture.

Solid lines for monitoring represent a DC that is higher than in Category 3.

The DC_{avg} of each subsystem shall be at high. The $MTTF_D$ of each redundant channel shall be high and measures against **CCF** shall be applied.

The subsystem behavior of this Category is therefore characterized by:

- Continued performance of the safety function in the presence of a single fault.
- Detection of faults in time to prevent the loss of the safety function.
- The accumulation of undetected faults is taken into account.

6.2.9.1 Category 4 When the Demand Rate is Relatively Low

High demand mode Safety-related Control Systems or SRP/CS may have a relatively low demand rate: for example once a week or once a month. In that case, the assumptions discussed in Chapter 2 may not be so "solid." Therefore, despite it is not stated in ISO 13849-1, we recommend to follow the approach indicated in § 5.7.2 of this book.

6.2.9.2 Example of a Category 4 Input Subsystem: Emergency Stop

In this example, an Emergency stop device, is controlled by a Safety module. In case the button is pressed the Safety Logic takes appropriate actions. We focus on the input subsystem.

For the electrical circuit shown in Figure 6.29, the Safety-related **Block Diagram the input subsystem** is shown in Figure 6.30.

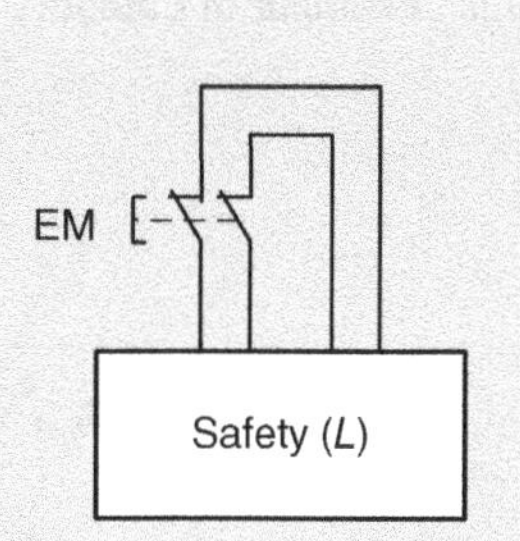

Figure 6.29 Emergency stop input subsystem Category 4.

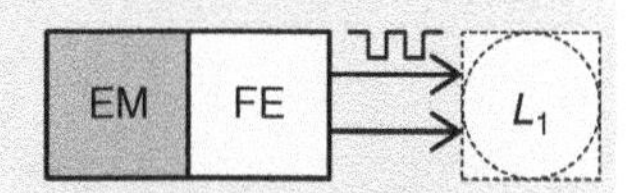

Figure 6.30 RBD of an emergency stop input subsystem Category 4.

- **Safety data:**
  The Emergency Stop has $B_{10D} = 10^5$. The Mission Time is 20 years.
- **Usage frequency:**
  The Emergency stop is supposed to be used **once per month**;
- **Avoidance of Systematic Failures**
  Since we can claim Category 4 (double channels with monitoring), **basic and well-tried safety principles are applied**. We also verified that enough measures have been applied to prevent **CCF** (score $CCF > 65$).
- **Fault Exclusion:**
  The double channel structure is obtained by considering the **double electrical channel** (contacts) and making a **Fault Exclusion for the mechanical part of the device**. In case of the emergency stop with a fault exclusion for the mechanical part of it (e.g. the mechanical link between an actuator and a contact element) it is possible to reach PL e (§ 4.11.2.3).

The input electrical circuit is monitored by the Safety module by means of a **triggered signal (§ 3.2.5).**

Let's now focus on the calculation of **probability of random failures**:

1. **The first step** is to calculate the $\mathbf{MTTF_D}$ of the emergency stop:

$$n_{op} = 12 \frac{1}{\text{year}}$$

$$MTTF_{D_{EM}} = \frac{B_{10D}}{0.1 \cdot n_{op}} = \frac{10^5}{0.1 \cdot 12} = 83\,333 \text{ years}$$

   Being the subsystem in Category 4, MTTF_D must be limited to 2500 years;

2. **As a second step**, we estimate the **DC.** The cross monitoring of inputs can only be performed when the emergency stop is activated; therefore, that cannot be defined as "*Cross monitoring of input signals with dynamic test*" since it is not an automatic dynamic test, even if the trigger is present. That can be defined simply as a "*cross monitoring of inputs without dynamic test.*" The corresponding DC can vary "*e.g. depending on how often a signal change is done by the application.*" Since the emergency stop is opened at least once every month, the maximum DC value that can be reached is 99% (Note 1 of Table 3.6 in § 3.3.6). Since we have the trigger, we assume the highest possible DC in the Category: $DC = 99\%$ (high).
3. The last step is to refer to table K.1 of ISO 13849-1 (same as Table 6.3) where, for Category 4 with high DC and $MTTF_D = 2500$, the $\boldsymbol{PFH_D = 9.06 \cdot 10^{-10}\ h^{-1}}$ **with PL e.** In this case, fault exclusion does not limit the PL (§ 4.11.2.3).

- **Useful Lifetime (operating lifetime) verification**
  We need to verify if the emergency pushbutton has to be replaced before its mission time expires. As shown in the formula hereafter, it can be used for 20 years without problems.

$$T_{10D} = \frac{B_{10D}}{n_{op}} = \frac{10^5}{12} = 8\,333 \text{ years}$$

6.2.9.3 Example of Category 4 for Output Subsystems: Electric Motor

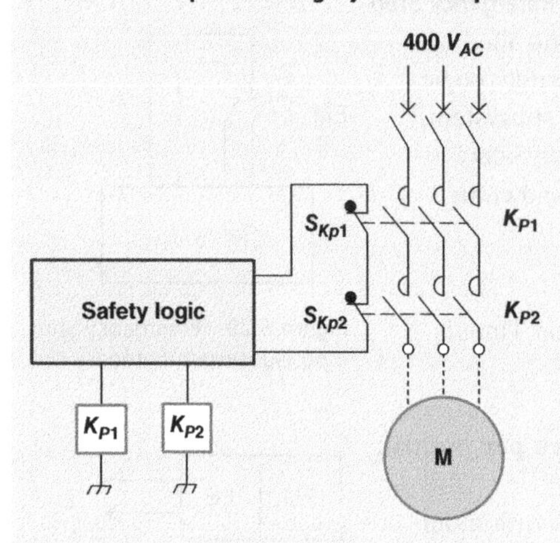

Figure 6.31 Category 4 output subsystem.

In this example, a motor is controlled by two contactors. In case a dangerous situation is detected, the Safety module de-energizes the contactor coils and the motor stops. We focus on the output subsystem.

For the electrical circuit shown in Figure 6.31, the Safety-related **Block Diagram** is shown in Figure 6.32.

- **Safety data:**
  K_{P1} and K_{P2} have $B_{10} = 10^6$ with 73% as percentage of dangerous failures (RDF according to § 1.12.1). Looking at the manufacturer datasheet, the Mission Time of contactors is 20 years.
- **Usage frequency:**
  Contactors are supposed to open **once every week**.
- **Avoidance of Systematic Failures**

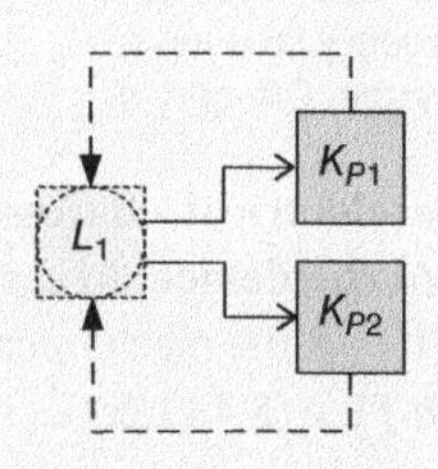

Figure 6.32 RBD of a Category 4 output subsystem.

Since we can claim Category 4 (double channels with monitoring), **basic and well-tried safety principles are applied**. We also verified that enough measures have been applied to prevent **CCF** (score $CCF > 65$).

Contactors have a **direct monitoring** of the status of their mirror contacts (§ 4.12.2.1).

Let's now focus on the calculation of **probability of random failures**:

1. **The first step** is to calculate the **MTTF$_D$** of the contactors. We assume the machine is working 48 weeks in a year.

$$n_{op} = 1 \cdot 48 = 48 \frac{1}{\text{year}}$$

$$B10_D = \frac{10^6}{0.73} = 1.37 \cdot 10^6$$

$$MTTF_{D_K} = \frac{B_{10D}}{0.1 \cdot n_{op}} = \frac{1.37 \cdot 10^6}{0.1 \cdot 48} = 285\,416 \text{ years}$$

2. As a second step, we have to estimate the **DC**. Each channel has a direct monitoring, so we can assume ***DC* = 99%.** Please refer to table E.1 – Output Device – *direct monitoring (e.g. electrical position monitoring of control valves, monitoring of electromechanical devices by mechanically linked contact elements).*

   The total DC average of the two channels is the following

$$DC_{avg} = \frac{(DC_{KP1}/MTTF_{D-KP1}) + (DC_{KP2}/MTTF_{D-KP2})}{(1/MTTF_{D-KP1}) + (1/MTTF_{D-KP2})} = \frac{(99\%/285\,416) + (99\%/285\,416)}{(1/285\,416) + (1/285\,416)} = \mathbf{99\%}$$

3. The last step is to refer to table K.1 of ISO 13849-1 (same as Table 6.2) where, for Category 4, $DC = 99\%$ and $MTTF_D = 2\,500$ years, the $\boldsymbol{PFH_D = 9.06 \cdot 10^{-10}\,h^{-1}}$, which corresponds to PL e.

- **Useful Lifetime (operating lifetime) verification**
  We need to verify if the contactors have to be replaced before their mission time expires. As shown in the formula hereafter, they can be used for 20 years without problems.

$$T_{10D} = \frac{B_{10D}}{n_{op}} = \frac{1.37 \cdot 10^6}{48} = 28\,542 \text{ years}$$

6.3 Simplified Approach for Estimating the Performance Level

Any reliability block diagram, representing a safety function, can be analyzed with a specific Markov modeling, as done in § 6.2.5. Solving the Markov model allows an exact calculation of the PFH_D of the system. However, that is done very rarely.

In ISO 13849-1, Markov models have been developed for five common structures: Category B to Category 4. The different PL levels that are reachable with each category are compiled in the form of a bar chart shown in Figure 6.34. A numerical representation is available in table K.1 of ISO 13849-1 or Table 6.3 of this book.

Many safety functions in Machinery are represented with Reliability Block Diagrams that can be associated with these Categories and that allows a relatively simple way of calculating the Performance Level. That is what is called the **Simplified approach,** and it is the most commonly used method in ISO 13849-1. That is the method used to make all the examples of § 6.2.

6.3.1 Conditions for the Simplified Approach

The simplified approach is based upon the following conditions:

- The component Mission Time (T_M) is at least 20 years.
- Within the Mission Time, the component failure rate is constant.
- Sufficient measures have been applied to prevent CCF: a beta factor of 2% is reached, in case the conditions stated in § 3.6 are fulfilled.

Under those conditions, the Performance Level (PL) **of each subsystem** depends upon three elements:

- The Category,
- The Mean Time to Dangerous Failure ($MTTF_D$) of each channel, and
- The average Diagnostic Coverage or DC_{avg} of each channel.

With the new edition of ISO 13849-1, there is a clear position that the Categories are applicable to subsystems only and each subsystem, within the SRP/CS, can have its own Category. In case the SRP/CS consists of one subsystem only, the designated architecture will be the same as for the entire SRP/CS. **In case the SRP/CS consists of multiple subsystems, each subsystem can have a specific category, and therefore a single SRP/CS can comprise multiple categories.**

Again, when using this simplified approach, the Category of each subsystem, as well as the DC_{avg} and the $MTTF_D$ of each channel, shall be determined. The vertical bands in Figure 6.34 show the range of performance that can be expected from each combination of $MTTF_D$, Category and DC_{avg}. Finding the appropriate range for each of these variables in the bands and then reading across to the vertical axis will indicate the PL that can be achieved with this combination.

6.3.2 How to Calculate $MTTF_D$ of a Subsystem

For the estimation of $MTTF_D$ of a component, the order of priorities is:

1. **Use of manufacturer's data.** This is the recommended option. Today, all standard components like interlocking devices and contactors should be provided with manufacturer's data regarding their Reliability. Please consider that when $MTTF_D$ data of components are provided by the manufacturer, the number of operations indicated by the manufacturer is considered so that the number is lower than the use in the application.
2. **Reference to annex C and annex D** of ISO 13849-1, where conservative Reliability data are provided.
3. **Field data** from specific application failure rates. In Machinery this is probably not easy since failure rate field data should come from identical component applications in similar environments collected over a significant period of time and where the collection and analysis method results in a reasonable level of confidence in the data. Please refer to IEC 61508-7 § B.5.4 for further details.
4. In case of no information at all, a conservative value of **10 years** can be assumed as $MTTF_D$.

Once the $MTTF_D$ is calculated, for each subsystem**, the maximum value for each channel is limited to 100 years**. An $MTTF_D$ value of each channel greater than 100 years is not acceptable because subsystems for high risks should not depend on the Reliability of the components alone; to reinforce the subsystems against systematic and random failure, additional means such as redundancy and testing are necessary. The limitation of $MTTF_D$ of each channel to a maximum of 100 years refers to the single channel of the subsystem which carries out the safety function. Higher $MTTF_D$ values can be used for single components; that means the limitation to 100 years is not applicable while calculating the Reliability of components in a subsystem channel.

For Category 4 subsystems, the maximum value of $MTTF_D$, for each channel, is limited to **2.500 years**; the higher value is justified because, in Category 4, the other quantifiable aspects, structure and DC, are at their maximum point and this allows the series combination of more than 3 subsystems with Category 4 and still achieve PL e.

The value of the $MTTF_D$ of each channel is split in three levels, like shown in Table 6.2 and it should be taken into account for each channel (e.g. single channel, each channel of a redundant system) individually.

Table 6.2 Mean time to dangerous failure of each channel ($MTTF_D$).

	$MTTF_D$
Indication	**Range for each channel**
Low	3 years $\leq MTTF_D <$ 10 years
Medium	10 years $\leq MTTF_D <$ 30 years
High	30 years $\leq MTTF_D \leq$ 100 years (or 2.500 years for Category 4)

6.3.3 Estimation of the Performance Level

If the conditions for the simplified approach are verified, for each subsystem, the user has to estimate:

- The Category,
- The Mean Time To Failure ($MTTF_D$), and
- Diagnostic Coverage.

Let's consider the Output Subsystem shown in Figure 6.33. Let's suppose the $MTTF_D$ of both subsystem elements is the same and = 88 years. The DCs are different and they are shown in the figure. The Average DC is (§ 3.3) 86% as shown in the following calculation.

$$DC_{avg} = \frac{(73\%/88) + (99\%/88)}{(1/88) + (1/88)} = 86\%$$

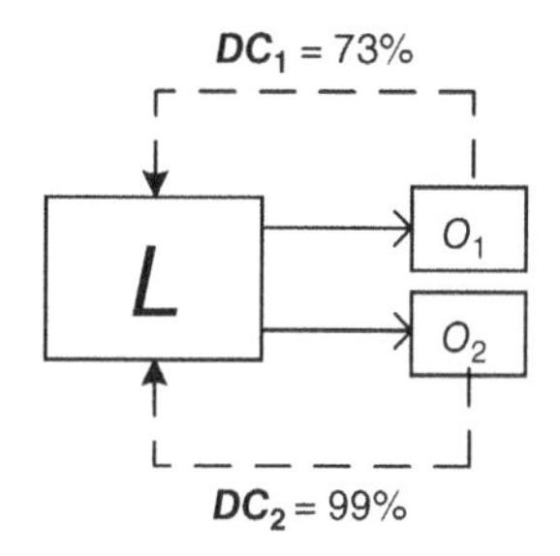

Figure 6.33 Category 4 output subsystem.

6.3.3.1 The Simplified Graph

ISO 13849-1 provides the graph shown in Figure 6.34 as a quick way of estimating the PFH_D of the subsystem.

The Value of PFH_D is shown on the vertical axis of the bar chart, for the different Categories. Category 2 and 3 are subdivided further according to the DC_{avg}. The columns are created by variations of the $MTTF_D$ of the functional channel.

As detailed in Chapter 3, each **Designated Architecture** has been modeled with Markov, in order to be able to calculate the PFH_D values, based upon the input parameters $MTTF_D$ and DC.

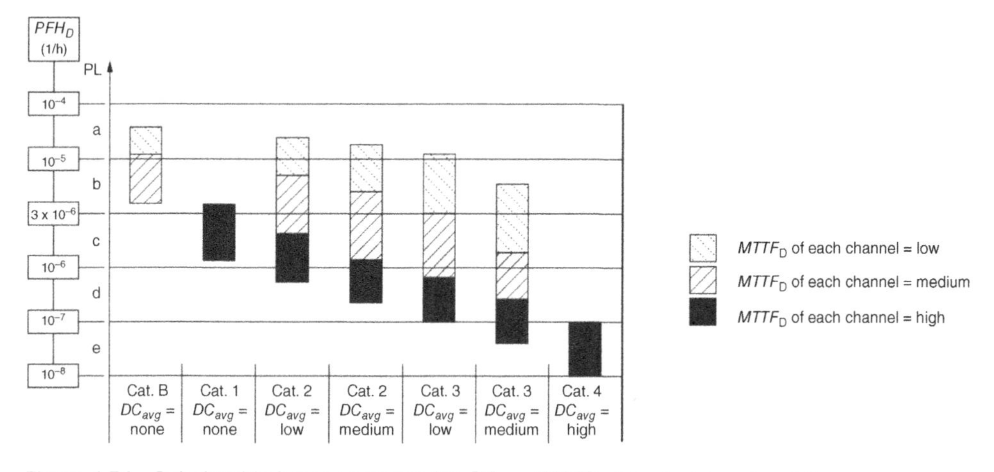

Figure 6.34 Relationship between categories, DC_{avg}, MTTFD of each channel and PL.

The PFH_D intervals were assigned Performance Levels from a to e. Please notice how the PFH_D vertical axis scale is logarithmic. The PFH_D interval from 10^{-6} to 10^{-5} h^{-1} has a peculiarity: it is mapped to the two adjacent Performance Levels b and c. Division of the logarithmic scale in the middle, places the boundary between Performance Levels b and c at the geometric mean of 10^{-6} and 10^{-5} per hour, specifically at $\sqrt{10} \cdot 10^{-6}\ h^{-1} \approx 3{\cdot}10^{-6}\ h^{-1}$.

Going back to the initial example in § 6.3.2, with $DC = 86\%$ (low) and $MTTF_D = 88$ year (high), we are in a Category 3 and we reach **PL d**.

6.3.3.2 Table K.1 in Annex K

Table 6.3 has the same content of table K.1 of the ISO 13849-1 and it contains the numerical values behind the columns of Figure 6.34. It can be used to determine both the PL and the PFH_D more precisely than it is possible using the figure.

Going back to our initial example in § 6.3.2, with $DC = 86\%$ (low) and $MTTF_D = 88$ year (high), we are in a Category 3 and we reach **PL d** and we can determine the value of $\boldsymbol{PFH_D = 1.35\ 10^{-7}\ h^{-1}}$.

Table 6.3 Numerical representation of Figure 6.33.

$MTTF_d$ per ogni canale (Anni)	Cat. B	PL	Cat. 1	PL	Cat. 2	PL	Cat. 2	PL	Cat. 3	PL	Cat. 3	PL	Cat. 4	PL	Cat. 4	PL	$MTTF_d$ > 100 anni per ogni canale (Anni)
		DC_{avg} = none (DC < 60%)		DC_{avg} = none (DC < 60%)		DC_{avg} = low (60% ≤ DC < 90%)		DC_{avg} = med. (90% ≤ DC < 99%)		DC_{avg} = low (60% ≤ DC < 90%)		DC_{avg} = med. (90% ≤ DC < 99%)		DC_{avg} = high (DC ≥ 99%)		DC_{avg} = high (DC ≥ 99%)	
3	3.80×10^{-5}	a			2.58×10^{-5}	a	1.99×10^{-5}	a	1.26×10^{-5}	a	6.09×10^{-6}	b			2.23×10^{-8}	e	110
3.3	3.46×10^{-5}	a			2.33×10^{-5}	a	1.79×10^{-5}	a	1.13×10^{-5}	a	5.41×10^{-6}	b			2.03×10^{-8}	e	120
3.6	3.17×10^{-5}	a			2.13×10^{-5}	a	1.62×10^{-5}	a	1.03×10^{-5}	a	4.86×10^{-6}	b			1.87×10^{-8}	e	130
3.9	2.93×10^{-5}	a			1.95×10^{-5}	a	1.48×10^{-5}	a	9.37×10^{-6}	b	4.40×10^{-6}	b			1.61×10^{-8}	e	150
4.3	2.65×10^{-5}	a			1.76×10^{-5}	a	1.33×10^{-5}	a	8.39×10^{-6}	b	3.89×10^{-6}	b			1.50×10^{-8}	e	160
4.7	2.43×10^{-5}	a			1.60×10^{-5}	a	1.20×10^{-5}	a	7.58×10^{-6}	b	3.48×10^{-6}	b			1.33×10^{-8}	e	180
5.1	2.24×10^{-5}	a			1.47×10^{-5}	a	1.10×10^{-5}	a	6.91×10^{-6}	b	3.15×10^{-6}	b			1.19×10^{-8}	e	200
5.6	2.04×10^{-5}	a			1.33×10^{-5}	a	9.87×10^{-6}	b	6.21×10^{-6}	b	2.80×10^{-6}	c			1.08×10^{-8}	e	220
6.2	1.84×10^{-5}	a			1.19×10^{-5}	a	8.80×10^{-6}	b	5.53×10^{-6}	b	2.47×10^{-6}	c			9.81×10^{-9}	e	240
6.8	1.68×10^{-5}	a			1.08×10^{-5}	a	7.93×10^{-6}	b	4.98×10^{-6}	b	2.20×10^{-6}	c			8.67×10^{-9}	e	270
7.5	1.52×10^{-5}	a			9.75×10^{-6}	b	7.10×10^{-6}	b	4.45×10^{-6}	b	1.95×10^{-6}	c			7.76×10^{-9}	e	300
8.2	1.39×10^{-5}	a			8.87×10^{-6}	b	6.43×10^{-6}	b	4.02×10^{-6}	b	1.74×10^{-6}	c			7.04×10^{-9}	e	330
9.1	1.25×10^{-5}	a			7.94×10^{-6}	b	5.71×10^{-6}	b	3.57×10^{-6}	b	1.53×10^{-6}	c			6.44×10^{-9}	e	360
10	1.14×10^{-5}	a			7.18×10^{-6}	b	5.14×10^{-6}	b	3.21×10^{-6}	b	1.36×10^{-6}	c			5.94×10^{-9}	e	390
11	1.04×10^{-5}	a			6.44×10^{-6}	b	4.53×10^{-6}	b	2.81×10^{-6}	c	1.18×10^{-6}	c			5.38×10^{-9}	e	430
12	9.51×10^{-6}	b			5.84×10^{-6}	b	4.04×10^{-6}	b	2.49×10^{-6}	c	1.04×10^{-6}	c			4.91×10^{-9}	e	470
13	8.78×10^{-6}	b			5.33×10^{-6}	b	3.64×10^{-6}	b	2.23×10^{-6}	c	9.21×10^{-7}	d			4.52×10^{-9}	e	510
15	7.61×10^{-6}	b			4.53×10^{-6}	b	3.01×10^{-6}	b	1.82×10^{-6}	c	7.44×10^{-7}	d			4.11×10^{-9}	e	560
16	7.31×10^{-6}	b			4.21×10^{-6}	b	2.77×10^{-6}	c	1.67×10^{-6}	c	6.76×10^{-7}	d			3.70×10^{-9}	e	620

Probabilità di guasto pericoloso all'ora PFH_d (h^{-1}) e livello di performance level (PL)

18	6.34×10^{-6}	b			3.68×10^{-6}	b	2.37×10^{-6}	c	1.41×10^{-6}	c	5.67×10^{-7}	d			3.37×10^{-9}	e	680
20	5.71×10^{-6}	b			3.26×10^{-6}	b	2.06×10^{-6}	c	1.22×10^{-6}	c	4.85×10^{-7}	d			3.05×10^{-9}	e	750
22	5.19×10^{-6}	b			2.93×10^{-6}	c	1.82×10^{-6}	c	1.07×10^{-6}	c	4.21×10^{-7}	d			2.79×10^{-9}	e	820
24	4.76×10^{-6}	b			2.65×10^{-6}	c	1.62×10^{-6}	c	9.47×10^{-7}	d	3.70×10^{-7}	d			2.51×10^{-9}	e	910
27	4.23×10^{-6}	b			2.32×10^{-6}	c	1.39×10^{-6}	c	8.04×10^{-7}	d	3.10×10^{-7}	d			2.28×10^{-9}	e	1000
30			3.80×10^{-6}	b	2.06×10^{-6}	c	1.21×10^{-6}	c	6.94×10^{-7}	d	2.65×10^{-7}	d	9.54×10^{-8}	e	2.07×10^{-9}	e	1100
33			3.46×10^{-6}	b	1.85×10^{-6}	c	1.06×10^{-6}	c	5.94×10^{-7}	d	2.30×10^{-7}	d	8.57×10^{-8}	e	1.90×10^{-9}	e	1200
36			3.17×10^{-6}	b	1.67×10^{-6}	c	9.39×10^{-7}	d	5.16×10^{-7}	d	2.01×10^{-7}	d	7.77×10^{-8}	e	1.75×10^{-9}	e	1300
39			2.93×10^{-6}	c	1.53×10^{-6}	c	8.40×10^{-7}	d	4.53×10^{-7}	d	1.78×10^{-7}	d	7.11×10^{-8}	e	1.51×10^{-9}	e	1500
43			2.65×10^{-6}	c	1.37×10^{-6}	c	7.34×10^{-7}	d	3.87×10^{-7}	d	1.54×10^{-7}	d	6.37×10^{-8}	e	1.42×10^{-9}	e	1600
47			2.43×10^{-6}	c	1.24×10^{-6}	c	6.49×10^{-7}	d	3.35×10^{-7}	d	1.34×10^{-7}	d	5.76×10^{-8}	e	1.26×10^{-9}	e	1800
51			2.24×10^{-6}	c	1.13×10^{-6}	c	5.80×10^{-7}	d	2.93×10^{-7}	d	1.19×10^{-7}	d	5.26×10^{-8}	e	1.13×10^{-9}	e	2000
56			2.04×10^{-6}	c	1.02×10^{-6}	c	5.10×10^{-7}	d	2.52×10^{-7}	d	1.03×10^{-7}	d	4.73×10^{-8}	e	1.03×10^{-9}	e	2200
62			1.84×10^{-6}	c	9.06×10^{-7}	d	4.43×10^{-7}	d	2.13×10^{-7}	d	8.84×10^{-8}	e	4.22×10^{-8}	e	9.85×10^{-10}	e	2300
68			1.68×10^{-6}	c	8.17×10^{-7}	d	3.90×10^{-7}	d	1.84×10^{-7}	d	7.68×10^{-8}	e	3.80×10^{-8}	e	9.44×10^{-10}	e	2400
75			1.52×10^{-6}	c	7.31×10^{-7}	d	3.40×10^{-7}	d	1.57×10^{-7}	d	6.62×10^{-8}	e	3.41×10^{-8}	e	9.06×10^{-10}	e	2500
82			1.39×10^{-6}	c	6.61×10^{-7}	d	3.01×10^{-7}	d	1.35×10^{-7}	d	5.79×10^{-8}	e	3.08×10^{-8}	e			
91			1.25×10^{-6}	c	5.88×10^{-7}	d	2.61×10^{-7}	d	1.14×10^{-7}	d	4.94×10^{-8}	e	2.74×10^{-8}	e			
100			1.14×10^{-6}	c	5.28×10^{-7}	d	2.29×10^{-7}	d	1.01×10^{-7}	d	4.29×10^{-8}	e	2.47×10^{-8}	e			

(Continued)

Table 6.3 (Continued)

$MTTF_D$ for each channel years	Average probability of a dangerous failure per hour. PFH_D (h^{-1}) and corresponding performance level (PL)													
	Cat. B DC_{avg} = none	PL	Cat. 1 DC_{avg} = none	PL	Cat. 2 DC_{avg} = low	PL	Cat. 2 DC_{avg} = medium	PL	Cat. 3 DC_{avg} = low	PL	Cat. 3 DC_{avg} = medium	PL	Cat. 4 DC_{avg} = high	PL
24	4.76×10^{-6}	b			2.65×10^{-6}	c	1.62×10^{-6}	c	9.47×10^{-7}	d	3.70×10^{-7}	d		
27	4.23×10^{-6}	b			2.32×10^{-6}	c	1.39×10^{-6}	c	8.04×10^{-7}	d	3.10×10^{-7}	d		
30			3.80×10^{-6}	b	2.06×10^{-6}	c	1.21×10^{-6}	c	6.94×10^{-7}	d	2.65×10^{-7}	d	9.54×10^{-8}	e
33			3.46×10^{-6}	b	1.85×10^{-6}	c	1.06×10^{-6}	c	5.94×10^{-7}	d	2.30×10^{-7}	d	8.57×10^{-8}	e
36			3.17×10^{-6}	b	1.67×10^{-6}	c	9.39×10^{-7}	d	5.16×10^{-7}	d	2.01×10^{-7}	d	7.77×10^{-8}	e
39			2.93×10^{-6}	c	1.53×10^{-6}	c	8.40×10^{-7}	d	4.56×10^{-7}	d	1.78×10^{-7}	d	7.11×10^{-8}	e
43			2.65×10^{-6}	c	1.37×10^{-6}	c	7.34×10^{-7}	d	3.87×10^{-7}	d	1.54×10^{-7}	d	6.37×10^{-8}	e
47			2.43×10^{-6}	c	1.24×10^{-6}	c	6.49×10^{-7}	d	3.35×10^{-7}	d	1.34×10^{-7}	d	5.76×10^{-8}	e
51			2.24×10^{-6}	c	1.13×10^{-6}	c	5.80×10^{-7}	d	2.93×10^{-7}	d	1.19×10^{-7}	d	5.26×10^{-8}	e
56			2.04×10^{-6}	c	1.02×10^{-6}	c	5.10×10^{-7}	d	2.52×10^{-7}	d	1.03×10^{-7}	d	4.73×10^{-8}	e
62			1.84×10^{-6}	c	9.06×10^{-7}	d	4.43×10^{-7}	d	2.13×10^{-7}	d	8.84×10^{-8}	e	4.22×10^{-8}	e
68			1.68×10^{-6}	c	8.17×10^{-7}	d	3.90×10^{-7}	d	1.84×10^{-7}	d	7.68×10^{-8}	e	3.80×10^{-8}	e
75			1.52×10^{-6}	c	7.31×10^{-7}	d	3.40×10^{-7}	d	1.54×10^{-7}	d	6.62×10^{-8}	e	3.41×10^{-8}	e
82			1.39×10^{-6}	c	6.61×10^{-7}	d	3.01×10^{-7}	d	1.35×10^{-7}	d	5.79×10^{-8}	e	3.08×10^{-8}	e
91			1.25×10^{-6}	c	5.88×10^{-7}	d	2.61×10^{-7}	d	1.14×10^{-7}	d	4.94×10^{-8}	e	2.74×10^{-8}	e
100			1.14×10^{-6}	c	5.28×10^{-7}	d	2.29×10^{-7}	d	1.01×10^{-7}	d	4.29×10^{-8}	e	2.47×10^{-8}	e
110													2.23×10^{-8}	e
120													2.03×10^{-8}	e
130													1.87×10^{-8}	e
150													1.61×10^{-8}	e
160													1.50×10^{-8}	e
180													1.33×10^{-8}	e
200													1.19×10^{-8}	e
220													1.08×10^{-8}	e
240													9.81×10^{-9}	e
270													8.67×10^{-9}	e
300													7.76×10^{-9}	e
330													7.04×10^{-9}	e

360	6.44×10^{-9}	e
390	5.94×10^{-9}	e
430	5.38×10^{-9}	e
470	4.91×10^{-9}	e
510	4.52×10^{-9}	e
560	4.11×10^{-9}	e
620	3.70×10^{-9}	e
680	3.37×10^{-9}	e
750	3.05×10^{-9}	e
820	2.79×10^{-9}	e
910	2.51×10^{-9}	e
1000	2.28×10^{-9}	e
1100	2.07×10^{-9}	e
1200	1.90×10^{-9}	e
1300	1.75×10^{-9}	e
1500	1.51×10^{-9}	e
1600	1.42×10^{-9}	e
1800	1.26×10^{-9}	e
2000	1.13×10^{-9}	e
2200	1.03×10^{-9}	e
2300	9.85×10^{-10}	e
2400	9.44×10^{-10}	e
2500	9.06×10^{-10}	e

Note 1: If for Category 2 the demand rate is less than or equal to 1/25 of the testrate (see § 4.5.4), then the PFH_D values stated in table K.1 for Category 2 multiplied by a factor of 1.1 can be used as a worst-case estimate.

Note 2: The calculation of the PFH_D values was based on the following DC_{avg}:

- DC_{avg} = low, calculated with 60%;
- DC_{avg} = medium, calculated with 90%;
- DC_{avg} = high, calculated with 99%.

6.3.3.3 The Extended Graph

Occasionally, the DC_{avg} value determined for a subsystem may lie only marginally below one of the thresholds "low" (60%), "medium" (90%), or "high" (99%). If the simplified quantification method in ISO 13849-1 is then applied, purely formal constraints require that the next-lower DC_{avg} level, i.e. "none," "low," or "medium," be used. This procedure constitutes a conservative estimation. Owing to the small number of graduations on the DC_{avg} scale, however, a minor change to the system that has the effect of causing the DC_{avg} value to dip just below one of the thresholds may result in a substantially poorer assessment of the system.

In the IFA Report 2/2017e [53], a detailed graph is published and reported in Figure 6.35. It allows the estimation of the PFH_D of a subsystem with higher precision. An even higher precision is available using the latest software SISTEMA [60].

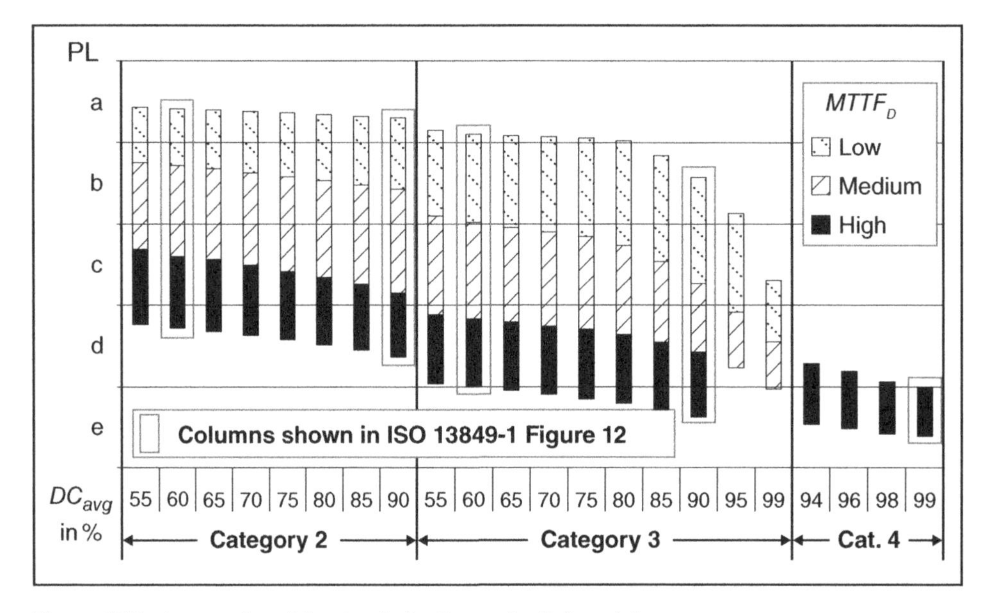

Figure 6.35 Intermediate DC_{avg} levels for Categories 2, 3, and 4.

Referring to our initial example in § 6.3.2, with $DC = 86\%$ (low) and $MTTF_D = 88$ year (high) and using SISTEMA [60], the result is that we are still in Category 3 but we reach $\boldsymbol{PFH_D = 6.2\ 10^{-8}\ h^{-1}}$, which means **PL e.**

6.4 Determination of the Reliability of a Safety Function

An SRP/CS can be made using a combination of subsystems. When combining subsystems with known PFH_D values, the PFH_D value of the SRP/CS is the sum of the PFH_D values of all its subsystems.

Let's take, as an example, a number of subsystems as shown in Figure 6.36. These subsystems operate in a series combination, which, as a whole, performs a safety function. For each subsystem, a PL was evaluated. If the PFH_D value of each subsystem is known, as previously stated, the PFH_D of

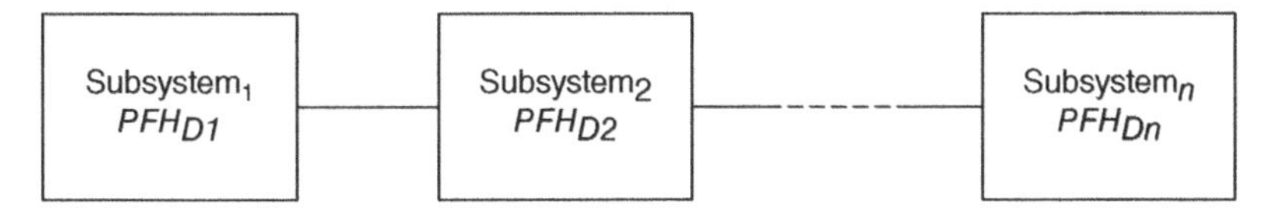

Figure 6.36 Combination of subsystems to achieve overall PL.

the SRP/CS is the sum of all PFH_D values of the *n* individual subsystems. The PL of the SRP/CS is limited by:

- The lowest PL of any individual subsystems involved in performing the safety function and
- The PL corresponding to the PFH_D of the combined SRP/CS according to Table 4.2.

If the PFH_D values of the subsystems are not known, the following methodology is recommended by ISO 13849-1:

a) Identify the lowest PL of all subsystems: this is defined as PL_{low};
b) Identify the number of subsystems with PL_{low}: this number is N_{low}
c) Look up the PL in Table 6.4.

Table 6.4 Calculation of PL for subsystems in series.

PL_{low}	N_{low}	→	PL of the SRP/CS
a	>3	→	None, not allowed
	≤3	→	a
b	>2	→	a
	≤2	→	b
c	>2	→	b
	≤2	→	c
d	>3	→	c
	≤3	→	d
e	>3	→	d
	≤3	→	e

Note: This table is based on the defined PFH_D ranges for each PL forming a kind of logarithmic scale.

7

The Architectures of IEC 62061

7.1 Introduction

7.1.1 The Architectural Constraints

IEC 62061 remains linked to IEC 61508 approach of Route 1_H, described in § 3.4.8.

In low demand mode, components are classified as Type A or Type B, and there are two different tables to be used to decide what is the maximum SIL that a Safety Subsystem can reach. In IEC 62061, one table only is defined for all types of components, and its content is similar to the one used for Type B components.

In the context of **hardware safety integrity**, the highest level that can be claimed by an safety-related control systems or SCS is limited by the hardware fault tolerances (HFT) and safe failure fractions (SFF) of the subsystem that carries out the safety function: the reference to be used is Table 7.1, same as table 6 in IEC 62061.

Table 7.1 Architectural constraints on a subsystem: maximum SIL that can be claimed for an SCS using the subsystem.

Safe failure fraction (*SFF*)	Hardware fault tolerance (*HFT*)		
	0	1	2
$SFF < 60\%$	Not allowed (except if well-tried components are used)	SIL 1	SIL 2
$60\% \leq SFF < 90\%$	SIL 1	SIL 2	SIL 3
$90\% \leq SFF < 99\%$	SIL 2	SIL 3	SIL 3
$SFF \geq 99\%$	SIL 3	SIL 3	SIL 3

Please bear in mind that the SIL limitation does not imply a PFH_D limitation. That means: for each subsystem, due to the architectural constraint, the PFH_D would be smaller than normally indicated for that particular SIL level. Please refer to the examples in this chapter.

In general, the language used and the approaches described in this second edition of the standard are fully understandable and usable. Unfortunately, there still remains language from the 2005 edition that may generate confusion for the approach to be used and that should be finally removed in the next edition.

Table 7.2 compares the Architectural Constraints of IEC 62061 with the limitations given by ISO 13849-1, and it is therefore applicable to all Safety Systems in high-demand mode.

Functional Safety of Machinery: How to Apply ISO 13849-1 and IEC 62061, First Edition. Marco Tacchini.

You can notice that **there is no SIL equivalence to PL a or PL b.** According to ISO 13849-1, it is possible to have a single channel safety system that does not use well-tried components: it is enough to use Category B subsystems with reliability levels equal to PL a or PL b.

Using IEC 62061 that is not possible, since a Basic Subsystem Architecture A (1oo1) can be done only using well-tried components. In other terms, according to ISO 13849-1, it is possible to use a general purpose PLC to implement a Safety System: the maximum PL reachable is PL b. That is not allowed by IEC 62061, neither with a single channel (Architecture A) nor with a redundant channel without diagnostics (Architecture B). Please also refer to § 5.1.4.5.

Table 7.2 Architectural constraints for high-demand mode of operation.

	Hardware fault tolerance – *HFT* (see Note 1)				
	Single channel subsystem *HFT* = 0 (see Note 3)		**Dual channel subsystem *HFT* = 1**		
DC_{avg} (see Note 2)	**Max. PL**	**Category (ISO 13849-1)**	**Max. PL**	**Category (ISO 13849-1)**	**Basic requirements**
SFF	***Max. SIL***	***Basic Subsystem Architecture (IEC 62061)***	***Max. SIL***	***Basic Subsystem Architecture (IEC 62061)***	
"None"	PL a	Category B	—	—	Basic safety principles (see Note 4)
—	—	—	—	—	
"None"	PL b	Category B	—	—	
—	—	—	—	—	
"None"	PL c	Category 1	—	—	Basic safety principles Well-tried safety principles Well-tried Components
<60%	*SIL 1*	*Architecture A (see Note 6)*	*SIL 1*	*Architecture B (see Note 6)*	
"Low"	PL c	Category 2	PL d	Category 3	Basic safety principles Well-tried safety principles CCF is relevant
60 to <90%	*SIL 1*	*Architecture C*	*SIL 2*	*Architecture D*	
"Medium"	PL d	Category 2 (see Note 5)	PL e	Category 3	
90 to <99%	*SIL 2*	*Architecture C*	*SIL 3*	*Architecture D*	
"High"	—	No equivalent category	PL e	Category 4	
$\geq$99%	*SIL 3*	*Architecture C (see Note 3)*	*SIL 3*	*Architecture D*	

Note 1: A hardware fault tolerance of N means that $N + 1$ faults could cause a loss of the safety function.
Note 2: "Low," "medium," and "high" is the denomination used in ISO 13849-1 in the context of quantification and classification of DC_{avg} ranges.
Note 3: For $HFT = 0$ and $SFF \geq 99\%$, the following limitations can be relevant:

- It is highly recommended to limit the maximum to SIL 2, where fault exclusions have been applied to faults that could lead to a dangerous failure (see IEC 62061, 7.3.3.3);
- SIL 3 can only be claimed when there is continuous monitoring of the correct functioning of the element. Typically, electronic technology will be required to achieve this.

Note 4: Where product standards, e.g. IEC 61800-5, IEC 61800-5, IEC 61131-2, ... are used, it can be assumed that basic safety principles can be fulfilled.
Note 5: According to ISO 13849-1, PL d can only be reached when the output (**OTE** or **Fault Reaction** function) initiates a safe state that is maintained until the fault is cleared: it is not sufficient that the output of the test equipment (OTE) only provides a warning.
Note 6: In case of both Basic Subsystem Architecture A and B, we consider here the case of electromechanical components only; please refer to § 3.1.3 and § 3.2.3.

7.1.2 The Simplified Approach

Similarly to the categories of ISO 13849-1, IEC 62061 has four Basic Subsystem Architectures that allow the use of a simplified approach, similar to ISO 13849-1. Instead of a graph or a table showing the PFH_D values, IEC 62061 provides the user with formulas that are, in general, a simplification of Reliability Block Diagrams of the Basic Subsystem Architectures and are intended to provide conservative estimates of PFH_D.

Those formulas are applicable, provided the following two conditions are satisfied:

- $\boldsymbol{\lambda \cdot T_1 \ll 1}$. That means the MTTF is much greater than $\boldsymbol{T_1}$: the minimum between the Proof Test and the Useful Lifetime of the subsystem (please refer to § 3.7.4).

$$\frac{1}{\lambda} = MTTF \gg T_1$$

- During the Useful Lifetime, which is the minimum between the Mission Time and the T_{10D}, **the failure rates are constant.**

7.1.2.1 Differences with ISO 13849-1

Some of the differences between the two standards are the following:

- As explained in § 3.6.2, in IEC 62061 the risk of **common cause failures** is evaluated with a similar table to ISO 13849-1, however there is no minimum value for the scoring.
- In Category 2, ISO 13849-1 requires that the $MTTF_D$ of the Test Channel (TE) is not lower than half the $MTTF_D$ value of the functional channel. The equivalent of Category 2 in IEC 62061 is the Basic Subsystem Architecture C. In this case, however, there is no minimum reliability level of the Fault Handling Function ($\lambda_{D\text{-}FH}$). In case the value is not according to table H.3 of the standard, the simplified formula cannot be used for the calculation of the PFH_D of the subsystem and the general formula for Basic Subsystem Architecture C must be used.
- In ISO 13849-1, the $MTTF_D$ of subsystems is limited to 100 years, except for Category 4; in IEC 62061, there is no limitation of the PFH_D, even when architectural constraint is applied.
- IEC 62061 uses the acronym PFH but that is exactly the same parameter as the PFH_D used in ISO 13849-1 (§ 4.5.1). The reason is to be in line with IEC 61508 series.

7.1.2.2 How to Calculate the PFH_D of a Basic Subsystem Architecture

The following elements have to be taken into consideration, to be able to determine the PFH_D of a subsystem:

- Each subsystem of the safety function has to be associated with one of the four Basic Subsystem Architectures.
- DC and test intervals have to be decided.
- The Common Cause Failure has to be calculated.
- λ_D or $MTTF_D$ of subsystem elements have to be calculated.
- The useful lifetime of components is typically 20 years even if, for components with wear-out characteristics, the useful lifetime is limited to T_{10D}.

7.1.3 The Avoidance of Systematic Failures

Also in the case of IEC 62061, it is important to eliminate the Systematic Failures. The way is also to follow the recommendations listed in Table 7.3.

Table 7.3 Overview of the basic requirements and interrelation to Basic Subsystem Architectures.

	Hardware fault tolerance (*HFT*)			
	0		**1**	
	SFF		***SFF***	
Basic requirements (M = mandatory)	**<60%**	**≥60%**	**<60%**	**≥60%**
Basic safety principles	M	M	M	M
Well-tried safety principles	M	M	M	M
Well-tried components	M	—	M	—
CCF	Not relevant	M	M	M
Type of Basic Subsystem Architecture	**A**	**C**	**B**	**D**

Please refer to ISO 13849-2:2012, annex A–D, for examples of Basic safety principles and well-tried safety principles. Please refer to ISO 13849-2:2012, annexes A and D for examples of well-tried components.

Table 7.3 is the equivalent of table 7 of IEC 62061. In the latter, however, in case of HFT 1 architecture and SFF less than 60%, the use of well-tried components is not required. That is a mistake that will be corrected in the next edition or in an Amendment. That means: **for Basic Subsystem Architecture B (or 1oo2), the use of well-tried components is mandatory**.

7.1.4 Relationship Between λ_D and $MTTF_D$

We remind that both IEC 62061 and ISO 13849-1 assume a constant failure rate of subsystem elements. With that assumption:

$$\lambda_D = \frac{1}{MTTF_D}$$

MTTF and $MTTF_D$ are normally indicated in years. **λ values** are normally given in **FIT** (Failure in Time), whereby 1 FIT means one failure in 10^9 hours.

$$1\,\text{FIT} = 1 \cdot 10^{-9}\,(\text{h}^{-1})$$

One year is approximately **8760** hours. Therefore the MTTF value can be converted into a λ value with the following formula:

$$\lambda_D = \frac{1}{MTTF_D \cdot 8760}$$

Considering that for electromechanical and pneumatic components, the B_{10D} value is given and knowing that:

$$MTTF_D = \frac{B_{10D}}{0.1 \cdot n_{op}}$$

The failure rate λ_D can be calculated from the B_{10D} with the following formula:

$$\lambda_D = \frac{0.1 \cdot n_{op}}{B_{10D} \cdot 8760}$$

Also, in this case, the useful lifetime of the component is limited by its T_{10D}.

7.2 The Four Subsystem Architectures

7.2.1 Repairable vs Non-Repairable Systems

A number of reliability techniques are more or less straightforwardly usable for the analysis of the unreliability of safety-related subsystems, among which **Reliability Block Diagrams** and **Markov Chains**. IEC 62061 first edition used the former and it assumes safety subsystems as non-repairable.

In the new edition of IEC 62061, the formula used for **Basic Subsystem Architecture C** is derived from a Markov State Diagram, that assumes the system as repairable, the remaining formulas are still the ones used in the first edition. That is the reason why, in § 7.2.7, a formula, derived from a reliability block diagram, is given for Basic Subsystem Architecture C.

7.2.2 Basic Subsystem Architecture A: 1oo1

In this Architecture, shown in Figure 7.1 with RBDs, single channel without diagnostic, any dangerous failure of a subsystem element causes a failure of the safety function. This Architecture corresponds to a Hardware Fault Tolerance of 0.

In high or continuous mode of operation, Architecture A shall not rely on a Proof Test interval shorter than lifetime (Mission Time).

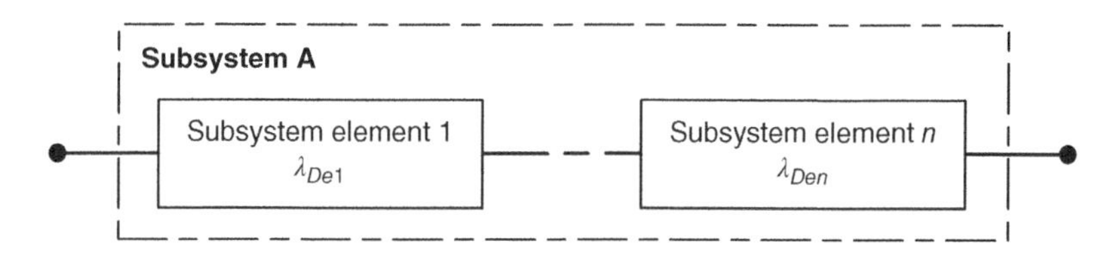

Figure 7.1 Logical representation of Basic Subsystem Architecture A.

The Architecture is equivalent to Category 1 of ISO 13849-1.

This case was analyzed in Chapter 1: the PFH_D of the subsystem is the sum of the dangerous failure rates of all subsystems elements.

$$PFH_D = \sum_{i=1}^{n} \lambda_{De_i} = \lambda_{De_1} + \lambda_{De_2} + \ldots + \lambda_{De_n}$$

7.2.2.1 Implications of the Architectural Constraints in Basic Subsystem Architecture A

Considering Table 7.4, being $HFT = 0$, up to SIL 3 could be reached. That may be valid for electronic components. **However**, for electromechanical components, being usually $SFF = DC = 0$ (§ 3.1.3), **a maximum of SIL 1 can be achieved**, even if the formula gives a lower PFH_D, provided well-tried components are used.

Table 7.4 Architectural constraints for an electromechanical subsystem.

Safe failure fraction (*SFF*)	Hardware fault tolerance (*HFT*)		
	0	1	2
$SFF < 60\%$	SIL 1 if well-tried components are used	SIL 1	SIL 2
$60\% \leq SFF < 90\%$	SIL 1	SIL 2	SIL 3
$90\% \leq SFF < 99\%$	SIL 2	SIL 3	SIL 3
$SFF \geq 99\%$	SIL 3	SIL 3	SIL 3

7.2.2.2 Example of a Basic Subsystem Architecture A

Let's consider the same input subsystem as in Example 6.2.3.1 and shown in Figure 7.2.

The interlocking device has a $B_{10D} = 20{\cdot}10^6$ and it is supposed to open **twice per hour**.

Basic and well-tried safety principles shall be applied. Moreover, well-tried components are used. CCF is not significant.

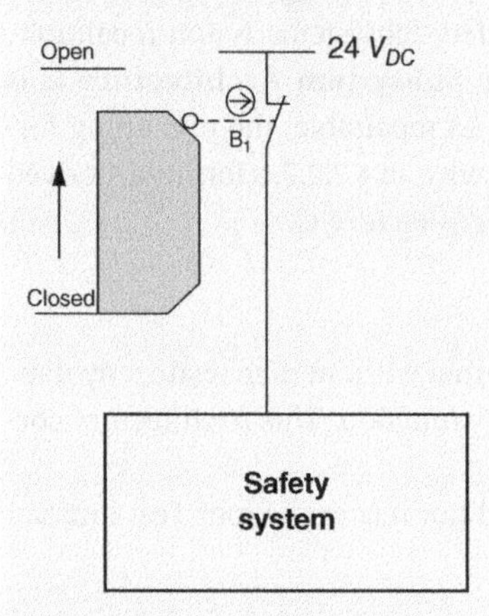

Figure 7.2 Input Subsystem Architecture A.

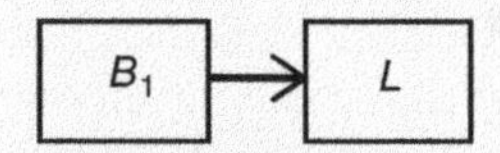

Figure 7.3 Input Subsystem Architecture A represented as an RBD.

Let's now focus on the **probability of random failures.**

- **The first step** is to calculate the λ_D. We assume the machine is working 240 days per year and 8 hours per day.

$$n_{op} = 2 \cdot 8 \cdot 240 = 3840 \frac{1}{\text{year}}$$

$$\lambda_{D_{B1}} = \frac{1}{MTTF_D} = \frac{0.1 \cdot n_{op}}{B10_D} = \frac{0.1 \cdot 3840}{20 \cdot 10^6}$$
$$= 1.92 \cdot 10^{-5} \frac{1}{\text{year}} = 2.19 \cdot 10^{-9}\ \text{h}^{-1} = PFH_D$$

- Being a subsystem Architecture A, there is no diagnostic, which means, we assume $SFF = DC < 60\%$ (§ 3.1.3). Its reliability Block Diagram is shown in Figure 7.3.
- Despite the PFH_D is very low, the fact that $SFF < 60\%$ and $HFT = 0$, the SIL level is limited by the Architectural Constraints to SIL 1, as shown in Table 7.5.

Therefore, the safety subsystem reaches **SIL 1** and has a $\boldsymbol{PFH_D = 2.19{\cdot}10^{-9}\,h^{-1}}$

Table 7.5 Maximum SIL that can be claimed for an SCS based upon its SFF and HFT.

	Hardware fault tolerance (*HFT*)		
Safe failure fraction (*SFF*)	**0**	**1**	**2**
$SFF < 60\%$	SIL 1	SIL 1	SIL 2
$60\% \le SFF < 90\%$	SIL 1	SIL 2	SIL 3
$90\% \le SFF < 99\%$	SIL 2	SIL 3	SIL 3
$SFF \ge 99\%$	SIL 3	SIL 3	SIL 3

7.2.3 Basic Subsystem Architecture B: 1oo2

This Architecture, shown in Figure 7.4 with RBDs, dual channel without a diagnostic, is such that a single fault of any subsystem element does not cause the loss of the safety function. This Architecture corresponds to a Hardware Fault Tolerance of 1.

This Architecture has no equivalent in ISO 13849-1.

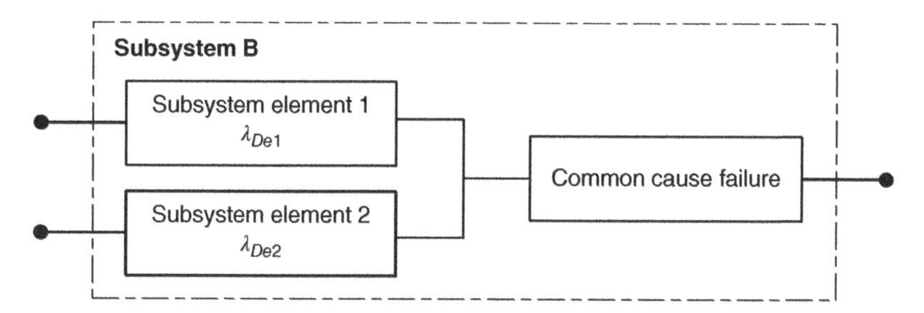

Figure 7.4 Logical representation of Basic Subsystem Architecture B.

For Architecture B, the PFH_D of the subsystem is given by the following formula:

$$PFH_D = (1-\beta)^2 \cdot \lambda_{De_1} \cdot \lambda_{De_2} \cdot T_1 + \beta \cdot \frac{(\lambda_{De_1} + \lambda_{De_2})}{2}$$

where:

- T_1 is the Proof Test interval of the perfect Proof Test or the useful lifetime, whichever is smaller; the **useful lifetime** is the minimum between T_{10D} and the Mission Time of the subsystem
- β is the susceptibility to common cause failures.

7.2.3.1 Implications of Architectural Constraints in Basic Subsystem Architecture B

Considering Table 7.6 for architectural constraints, being $HFT = 1$, up to SIL 3 could be reached. That may be valid for electronic components. **However**, for electromechanical components, being $SFF = DC = 0$ (§ 3.1.3), **a maximum of SIL 1 can be achieved**, even if the formula gives a lower PFH_D.

In case of Basic Subsystem Architecture B and $SFF < 60\%$, only well-tried components can be used.

Table 7.6 Architectural constraints on a subsystem: maximum SIL that can be claimed for an SCS.

Safe failure fraction (*SFF*)	Hardware fault tolerance (*HFT*) 0	1	2
$SFF < 60\%$	SIL 1 if well-tried components are used	**SIL 1**	SIL 2
$60\% \leq SFF < 90\%$	SIL 1	SIL 2	SIL 3
$90\% \leq SFF < 99\%$	SIL 2	SIL 3	SIL 3
$SFF \geq 99\%$	SIL 3	SIL 3	SIL 3

7.2.3.2 Example of a Basic Output Subsystem Architecture B: Electric Motor

For the electrical circuit shown in Figure 7.5, the Safety-related **Block Diagram output subsystem** is shown in Figure 7.6.

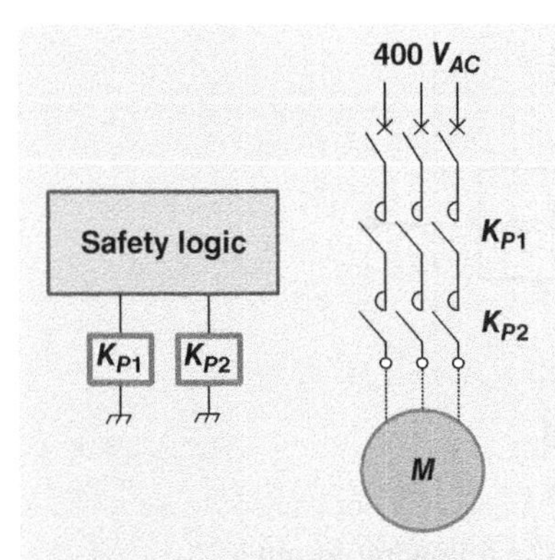

Figure 7.5 Output Subsystem Architecture B.

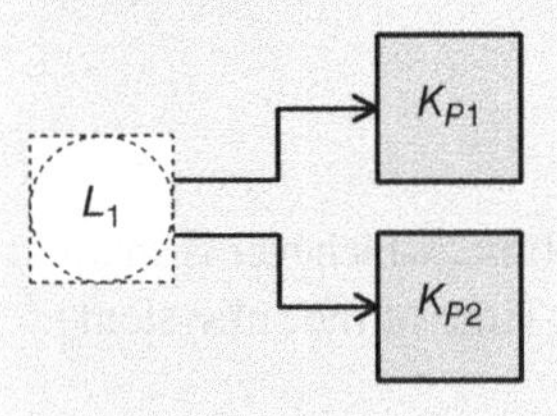

Figure 7.6 RBD of an Output Subsystem Architecture B.

- **Safety data:**
  K_{P1} and K_{P2} have $B_{10} = 10 \cdot 10^6$ with 73% as percentage of dangerous failures (RDF according to § 1.12.1). Looking at the manufacturer datasheet, the Mission Time is 20 years, and we assume a Proof Test of the same length. For the proof test, it is our decision not to have it shorter than the Mission Time.
- **Usage Frequency**
  Contactors are supposed to open **twice every minute**.
- **Avoidance of Systematic Failures**
  Since we want to claim Architecture B, basic and well-tried safety principles are applied. Since $SFF < 60\%$, also well-tried components must be used (in this case two contactors). We also verified that enough measures have been applied to prevent common cause failures: a $\beta = 2\%$ can be assumed (§ 3.6.2).

Let's now focus on the calculation of **probability of random failures**:

1. **The first step** is to calculate λ_D. We assume the machine is working 240 days per year and 8 hours per day.

$$n_{op} = 120 \cdot 8 \cdot 240 = 230\,400\ \frac{1}{\text{year}}$$

$$B_{10D_{KP}} = \frac{10 \cdot 10^6}{0.73} = 1.37 \cdot 10^7$$

$$\lambda_{D_K} = \frac{1}{MTTF_D} = \frac{0.1 \cdot n_{op}}{B10_D \cdot 8760} = \frac{0.1 \cdot 230\,400}{1.37 \cdot 10^7 \cdot 8760} = 1.92 \cdot 10^{-7}\,\text{h}^{-1}$$

2. **The second step** is to estimate T_{10D} and T_1

$$T_{10D_K} = \frac{B_{10D\text{-}K}}{n_{op}} = \frac{1.37 \cdot 10^7}{230\,400} = 59\text{ years} > 20\text{ years}$$

$$T_1 = 20\text{ years} = 175\,200\text{ hours}$$

3. **The last step** is to calculate PFH_D

$$PFH_D = (1-\beta)^2 \cdot \lambda_{De_1} \cdot \lambda_{De_2} \cdot T_1 + \beta \cdot \frac{(\lambda_{De_1} + \lambda_{De_2})}{2}$$

$$PFH_D = (1-0.02)^2 \cdot \left(1.92 \cdot 10^{-7}\right)^2 \cdot 175\,200 + 0.02 \cdot \frac{2 \cdot 1.92 \cdot 10^{-7}}{2}$$

$$PFH_D = 1.004 \cdot 10^{-8}\,\text{h}^{-1}$$

- Being a subsystem Architecture B, there is no diagnostic. That means we assume $SFF = DC < 60\%$ (§ 3.1.3).

- Despite the PFH_D is very low, the fact that $SFF < 60\%$ and $HFT = 1$, the SIL level is limited by the Architectural Constraints to SIL 1, as shown in Table 7.7.
  The Safety Subsystem reaches **SIL 1** and has a $\boldsymbol{PFH_D = 1.01\ 10^{-8}\ h^{-1}}$.

Table 7.7 Maximum SIL that can be claimed for an SCS.

Safe failure fraction (*SFF*)	Hardware fault tolerance (*HFT*)		
	0	1	2
$SFF < 60\%$	SIL 1	**SIL 1**	SIL 2
$60\% \leq SFF < 90\%$	SIL 1	SIL 2	SIL 3
$90\% \leq SFF < 99\%$	SIL 2	SIL 3	SIL 3
$SFF \geq 99\%$	SIL 3	SIL 3	SIL 3

7.2.4 Basic Subsystem Architecture C: 1oo1D

In this Architecture, shown in Figure 7.7 as a RBD, single channel with diagnostic, any undetected dangerous fault of the subsystem element leads to a dangerous failure of the safety function. Where a fault of a subsystem element is detected, the diagnostic function initiates a **fault reaction**. This Architecture corresponds to a Hardware Fault Tolerance of 0.

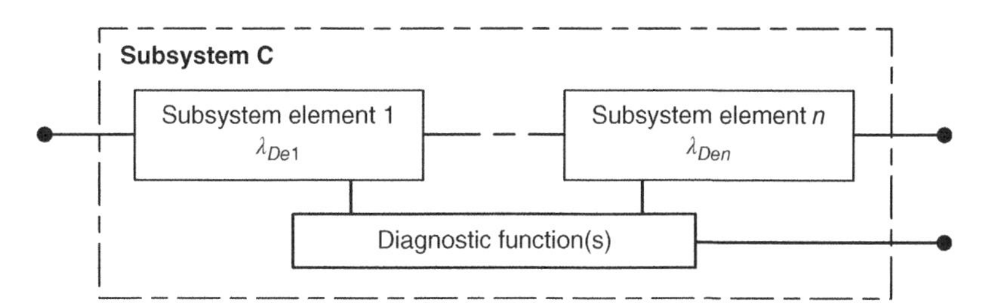

Figure 7.7 Logical representation of Basic Subsystem Architecture "C."

This Architecture corresponds to the Category 2 of ISO 13849-1; in Figure 7.8 there is a more precise representation.

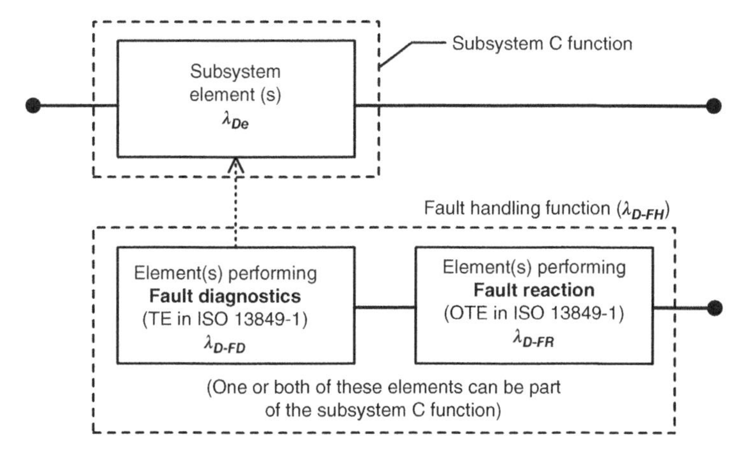

Figure 7.8 Detailed logical representation of Basic Subsystem Architecture "C."

You can notice that the TE (Test Equipment) of Category 2 in ISO 13849-1 is defined as **Fault Diagnostic Function** in IEC 62061. Likewise, the equivalent of the OTE is the **Fault Reaction Function**. The TE + OTE is defined as the **Fault Handling Function** in IEC 62061.

7.2.4.1 Conditions for a Correct Implementation of Basic Subsystem Architecture C

For Architecture C, the calculation of PFH_D assumes a time-optimal fault handling. **Time optimal** fault handling of a subsystem element can be assumed if **one of the following conditions** is satisfied:

- **The diagnostic rate is at least a factor of 100 higher than the demand rate** of the safety function and the time needed for the fault reaction is sufficiently short to bring the system to a safe state, before a hazardous event occurs; **or**
- The fault handling is **performed immediately upon any potential demand** of the safety function and **the time needed** to detect a detectable fault and to bring the system to a safe state is shorter than the process safety time; **or**
- The fault handling is performed continuously, and the time needed to detect a detectable fault and to bring the system to a safe state is shorter than the process safety time; **or**
- The fault handling is performed periodically and the sum of the test interval, the time needed to detect a detectable fault and time needed to bring the system to a safe state is shorter than the process safety time.

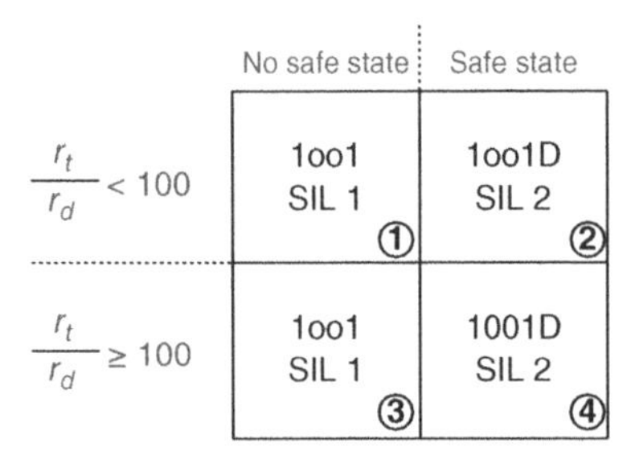

Figure 7.9 Summary of 1oo1D conditions.

Figure 7.9 summarizes all possible conditions related to a correct implementation of a 1oo1D architecture; it is different from Table 6.2, related to ISO 13849-1.

- **Situations 1 and 3.** If no safe state is possible, regardless if there is an optimal testing frequency or not, we cannot claim 1oo1D Architecture. Therefore, **we fall back to a 1oo1** Architecture, where well-tried components must be used.

The reason for the difference with ISO 13849-1, in case of Situation 3, is due to the link of IEC 62061 to IEC 61508.

> ***[IEC 61508-2] 7.4.8 Requirements for system behaviour on detection of a fault***
> ***7.4.8.3*** *The detection of a dangerous fault (by diagnostic tests, proof tests or by any other means) in any subsystem having a hardware fault tolerance of 0 shall, in the case of a subsystem that is implementing any safety function(s) operating in the high demand or the continuous mode,* ***result in a specified action to achieve or maintain a safe state*** *(see Note).*
>
> ***Note:*** *The specified action required to achieve or maintain a safe state will be specified in the E/E/PE system safety requirements (see IEC 61508-1, 7.10). It may consist, for example, of the safe shut-down of the EUC, or that part of the EUC that relies, for functional safety, on the faulty subsystem.*

- **Situation 2.** No Optimal testing is present; however, in case a fault is detected, we can bring the system to a safe state. **We can claim a 1oo1D** Basic Subsystem Architecture and we can reach SIL 2, if all other conditions exist. No need to use well-tried components.
- **Situation 4.** We have both optimal testing and, in case a fault is detected, we can bring the system to a safe state. **We can claim a 1oo1D** Basic Subsystem Architecture and we can reach SIL 2, if all other conditions exist. No need to use well-tried components.

This is a "delicate" Architecture, similarly to the Category 2. The issue is the failure of the Diagnostic Function, called Fault handling Function, while the Functional Channel is still working. The **Fault Handling function** includes both the **Fault Diagnostics** function and the **Fault Reaction** function.

7.2.4.2 Basic Subsystem Architecture C with Fault Handling Done by the SCS

This one is the simplest situation, since the **Diagnostic and Fault Reaction Functions** (equivalent to TE and OTE in ISO 13849-1) are already part of the safety function, as shown in Figure 7.10. The fault handling function is completely performed by **a separate subsystem of the SCS,** which is also involved in performing the safety function, thus contributing to its PFH_D.

In other terms, in this case we can ignore the reliability of the Monitoring channel since that is already taken care while calculating the reliability of the Functional Channel.

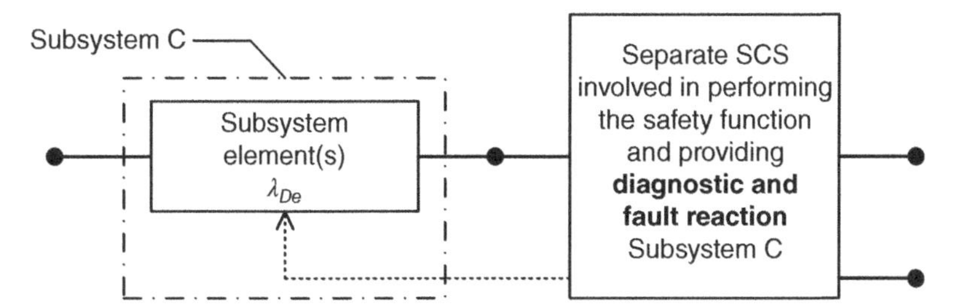

Figure 7.10 Logical representation of Basic Subsystem Architecture C with external Fault handling function.

This could be the case whereby a Safety-related Control System monitors if a motor contactor has opened and, in case of failure, it de-energizes another contactor, that is part of another safety function.

If we consider a Subsystem C, made of n elements numbered from 1 to n, the PFH_D is given by the following formula:

$$PFH_D = (1 - DC_1) \cdot \lambda_{De_1} + ... + (1 - DC_n) \cdot \lambda_{De_n}$$

As you can see, the reliability of the Fault Handling Function (or monitoring function) is not present in the formula. It is the same formula used in the first edition of the standard.

7.2.5 Basic Subsystem Architecture C with Mixed Fault Handling

These are all cases where the reliability of the monitoring channel, or better, the Fault Handling function, is outside of the Safety Function. In order to calculate the PFH_D of this architecture, a new failure rate has to be introduced: it is called the failure rate of the **Fault Handling Function**, or $\lambda_{D\text{-}FH}$.

A Markov model was used to arrive to the different PFH_D formulas.

There are three possible situations:

1. The subsystem C has **external Fault Diagnostics** (TE), but the element performing the fault reaction (it would be the OTE in ISO 13849-1 Category 2) is internal to the subsystem (Figure 7.11).

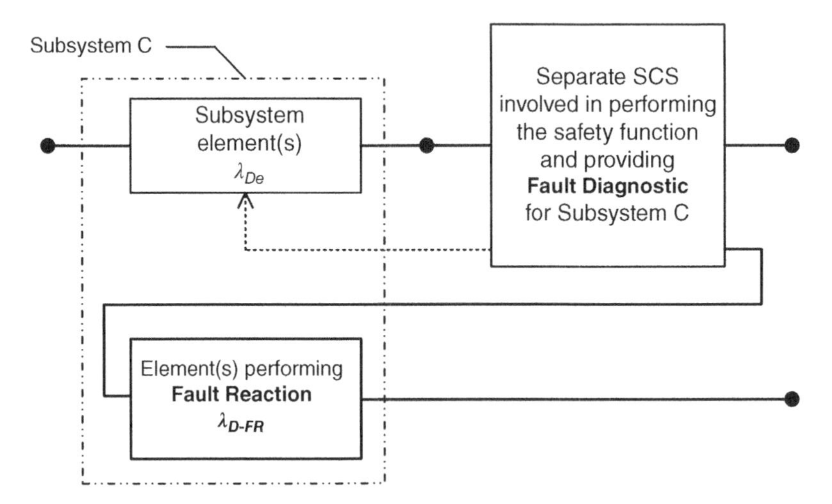

Figure 7.11 Logical representation of Basic Subsystem Architecture C with external Fault diagnostic.

In this case the fault handling failure rate includes the **Fault Reaction** function only.
That means $\lambda_{D\text{-}FH} = \lambda_{D\text{-}FR}$
The Failure Rate of the **Fault Diagnostic** function is not considered, since it is already taken care while calculating the reliability of the Functional Channel.

2. Subsystem C has **external Fault Reaction** (OTE), while the **Fault Diagnostic** (TE) is done Internally to the subsystem, as shown in Figure 7.12.

In this case, the fault handling failure rate includes the **Fault Diagnostic** function only.
That means $\lambda_{D\text{-}FH} = \lambda_{D\text{-}FD}$

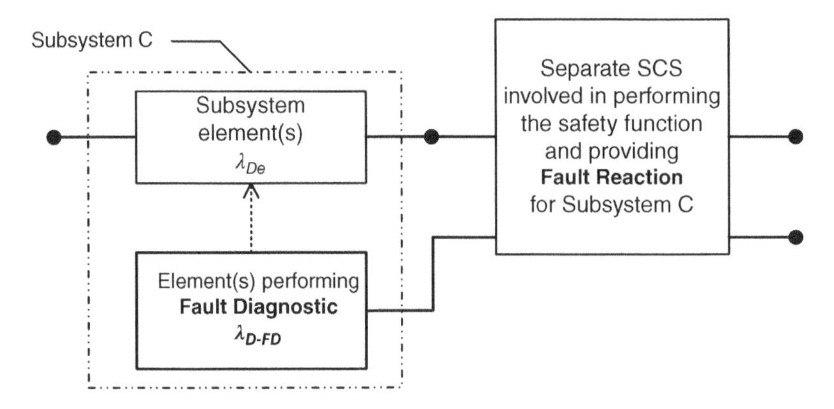

Figure 7.12 Logical representation of Basic Subsystem Architecture C with external Fault reaction.

The Failure Rate of the **Fault Reaction** function is not considered since it is already taken care while calculating the reliability of the Functional Channel.

3. Subsystem C has both **Fault Diagnostics** and **Fault Reaction** internal to the subsystem, as shown in Figure 7.13.

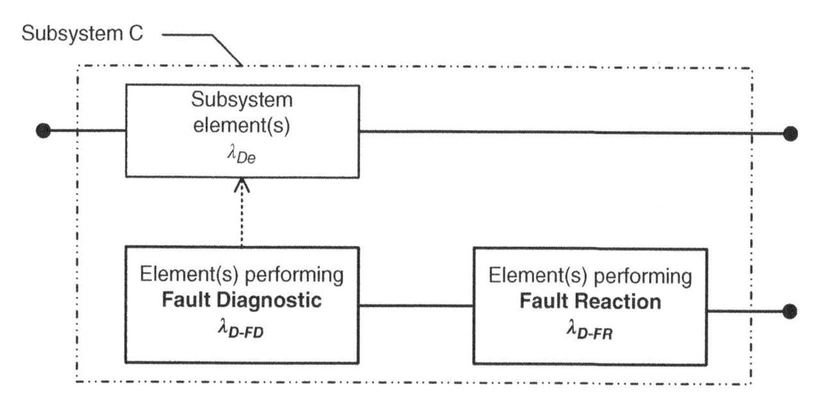

Figure 7.13 Logical representation of Basic Subsystem Architecture C with internal fault diagnostics and internal fault reaction.

In this case, the fault handling failure rate includes both the diagnostic function and the fault reaction function.

That means $\lambda_{D\text{-}FH} = \lambda_{D\text{-}FD} + \lambda_{D\text{-}FR}$

7.2.5.1 PFH$_D$ in Case of Four Conditions Satisfied

In all of the three cases previously described, there is a simple equation that can be used to calculate the PFH$_D$ value, provided **all the following conditions are satisfied**:

1. $\beta \leq 2\%$;
2. $DC \leq 99\%$;
3. $\frac{1}{\lambda_{De}} \leq 1.000$ years;
4. $\frac{1}{\lambda_{D-FH}}$ has at least the minimum value according to Table 7.8;

where $\lambda_{D\text{-}FH}$ is the failure rate of the single element that realizes the fault handling function within the subsystem.

The equation is the following:

$$PFH_D = (1 - DC_1) \cdot \lambda_{De_1} + \dots + (1 - DC_n) \cdot \lambda_{De_n}$$

If the functional channel has more than one element, the related DC is calculated using the following formula.

$$DC = \frac{\sum_{i=1}^{n} (DC \cdot \lambda_{De_i})}{\sum_{i=1}^{n} \lambda_{De_i}}$$

Table 7.8 (Same as table H.3 of IEC 62061).

DC range	Minimum value of $\frac{1}{\lambda_{D-FH}}$ (years)
$60\% \leq DC < 65\%$	44
$65\% \leq DC < 70\%$	59
$70\% \leq DC < 75\%$	100
$75\% \leq DC < 80\%$	170
$80\% \leq DC < 85\%$	300
$85\% \leq DC < 90\%$	550
$90\% \leq DC < 95\%$	1200
$95\% \leq DC \leq 99\%$	5900

7.2.5.2 PFH_D in Case One of the Four Conditions is Not Satisfied

In case **one of the above four conditions** is not satisfied, the simplified formula cannot be used, since it may give a lower value of PFH_D and therefore provide higher reliability than the reality.

If the functional channel is comprised by one element only and the fault handling function, within the subsystem, is realized by another single element, the following equation can be used:

$$PFH_D = \lambda_{De} - DC \cdot [\lambda_{De} - \beta \cdot \min(\lambda_{De}; \lambda_{D-FH})] \cdot \left\{1 - \frac{1}{2} \cdot [\lambda_{D-FH} - \beta \cdot \min(\lambda_{De}; \lambda_{D-FH})] \cdot T_1\right\}$$

where:

- T_1 is the Proof Test interval of the perfect Proof Test or useful lifetime whichever is the smaller;
- λ_{De} is the dangerous failure rate of the single element of the functional channel;
- $\lambda_{D\text{-}FH}$ is the failure rate of the single element that realizes the fault handling function(s) within the subsystem;
- **DC** is the diagnostic coverage for the single element e of the functional channel;
- β is the susceptibility to common cause failures of the functional channel and the channel that realizes the fault handling function(s) within the subsystem.

The same equation can be used even in case all the four above conditions are satisfied: it is the general formula for Architecture C: **it is always valid.**

You may notice that, in case the dangerous failure rate(s) of the fault handling function(s) within the subsystem can be assumed to be zero ($\lambda_{D\text{-}FH} = 0$), the equation becomes the same as the one in § 7.2.6.1. The reason is obvious: the monitoring channel is assumed to have very high reliability.

Please consider that **the general formula can be used even if the reliability of the fault handling function is worse than that required by Table 7.8.**

7.2.5.3 Implications of the Architectural Constraints in Basic Subsystem Architecture C

Architecture C subsystems have an $HFT = 0$. That means, based upon the SFF, a maximum of SIL 3 can be achieved. However, it is important to verify that one of the following conditions is satisfied:

- **the sum of the diagnostic test interval and the time to perform the specified fault reaction** function to achieve or maintain a safe state shall be shorter than the process safety time (e.g. see ISO 13855);

- Or, **the ratio of the diagnostic test rate to the demand rate shall be equal to or greater than 100.** This last condition aligns Architecture 1oo1D with Category 2.

Therefore, a monitored single channel subsystem (1oo1D), where none of the two conditions are met, has to be considered a Basic Subsystem Architecture A (1oo1); please refer to Figure 7.9.

Table 7.9 is applicable. Normally $SFF < 99\%$, since $SFF \geq 99\%$ is only possible when there is continuous monitoring of the correct functioning of the element: typically, only electronic technology can have this.

Table 7.9 Architectural constraints on a subsystem: maximum SIL that can be claimed for an SCS.

	Hardware fault tolerance (*HFT*)		
Safe failure fraction (*SFF*)	**0**	**1**	**2**
$SFF < 60\ \%$	SIL 1 if well-tried components are used	SIL 1	SIL 2
$60\% \leq SFF < 90\%$	**SIL 1**	SIL 2	SIL 3
$90\% \leq SFF < 99\%$	**SIL 2**	SIL 3	SIL 3
$SFF \geq 99\%$	SIL 3	SIL 3	SIL 3

7.2.6 Example of a Basic Subsystem Architecture C

Let's consider the same output subsystem as in Example 6.2.7.3 and shown in Figure 7.14 as a simplified electrical diagram and in Figure 7.15 as a RBD. K_P has $B_{10} = 10^6$ with 73% as percentage of dangerous failures. $U<$ has a $B_{10D} = 4 \cdot 10^5$. The Mission Time of both the contactor and the undervoltage coil is 20 years.

K_P is supposed to open **twice per minute** while $U<$ is supposed to open **once per month.** Let's now focus on the calculation of **probability of random failures.**

1. **The first step** is to calculate the λ_D of the contactor K_P. We assume the machine is working 240 days per year and 8 hours per day.

$$n_{op} = 120 \cdot 8 \cdot 240 = 230\,400\ \frac{1}{\text{year}}$$

$$B_{10D\ KP} = \frac{10^6}{0.73} = 1.37 \cdot 10^6$$

$$\lambda_{D\ KP} = \frac{1}{MTTF_{D_{KP}}} = \frac{0.1 \cdot n_{op}}{B_{10D} \cdot 8760} = \frac{0.1 \cdot 230\,400}{1.37 \cdot 10^6 \cdot 8760} = 1.92 \cdot 10^{-6}\ \text{h}^{-1}$$

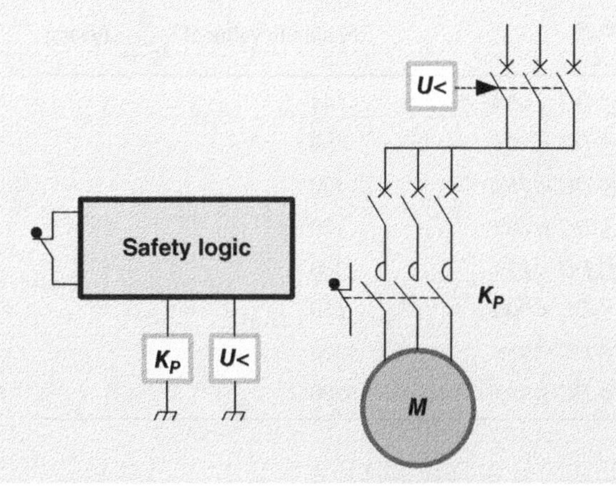

Figure 7.14 Output Basic Subsystem Architecture C.

2. **The second step** is to calculate T_1 and β.
   $T_1 = \min$ (Useful Life; Proof Test);
   Useful Life = min (Mission Time; T_{10D});
   Mission Time = 20 years;

$$T10_{D_{KP}} = \frac{B10_{D_{KP}}}{n_{op}} = \frac{1.37 \cdot 10^6}{230\,400} = 5.9\,\text{year} < 20\,\text{years}$$

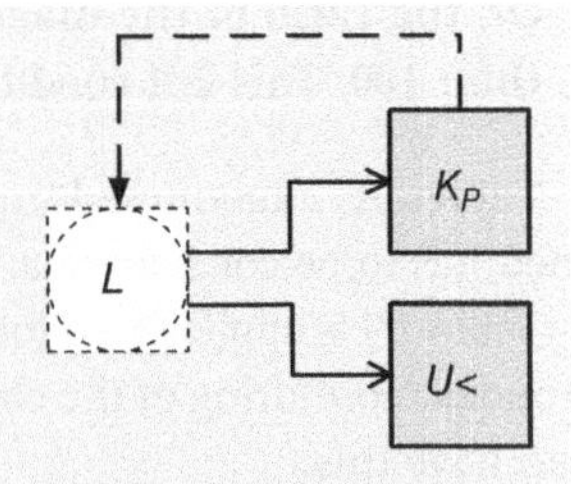

Figure 7.15 RBD of an output Basic Subsystem Architecture C.

   Useful Life = min (20 years; 5.9 years) ➔ Useful Life = 5.9 years;
   Proof Test > 20 years
   $T_1 = \min$ (5.9 years; >20 years) ➔ $\boldsymbol{T_1}$ **= 5.9 years = 51 684 hours;**
   Based upon the CCF criteria listed in annex E of the standard (Table 3.17 of the book), we can assume a $\beta = 0.02$).
3. **The third step** is the determination of the **Diagnostic Coverage**. Being a single channel with a testing element, which directly monitors the status of K_P, we assume $DC = SFF =$ 90% (§ 3.1.3).
4. Now, we have to check that the $\boldsymbol{\lambda_{D\text{-}FH}}$ **value of the test channel is not lower than specified in Table 7.8**. In this case, functional channel and testing channel have the Safety PLC in common, in other terms, we are in the situation of Figure 7.11: **External Fault Diagnostic.** The Fault Reaction (the undervoltage coil) is considered part of the output subsystem while the Fault Handling is not, since its reliability is already taken care in the reliability calculation of the Functional Channel, that include the Safety Logic. For that reason, only the failure rate of the undervoltage coil is important for the $\lambda_{D\text{-}FH}$.
   U< is used only when a fault is detected, once per month:

$$n_{op} = 12\ \frac{1}{\text{year}}$$

$$\frac{1}{\lambda_{D_{U<}}} = \frac{B_{10D}}{0.1 \cdot n_{op}} = \frac{4 \cdot 10^5}{0.1 \cdot 12} = 333\,333\ \text{years}$$

   Since we claim a DC of 90%, based upon Table 7.10, the minimum value of $\frac{\mathbf{1}}{\boldsymbol{\lambda_{D-FH}}}$ is 1200.

   In this case $\frac{1}{\lambda_{D_{U<}}}$ is much higher than required in Table 7.10 the condition is fulfilled;

Table 7.10 (Same as table H.3 of IEC 62061).

DC range	Minimum value of $\frac{1}{\lambda_{D-FH}}$ (years)
$60\% \le DC < 65\%$	44
$65\% \le DC < 70\%$	59
$70\% \le DC < 75\%$	100
$75\% \le DC < 80\%$	170
$80\% \le DC < 85\%$	300
$85\% \le DC < 90\%$	550
$90\% \le DC < 95\%$	1200
$95\% \le DC \le 99\%$	5900

5. **The fourth step** is the determination of the **Diagnostic Coverage**. Being a single channel with a testing element, which directly monitors the status of K_P, we assume $DC = SFF = 90\%$ (§ 3.1.3).
6. **The last step** is to calculate PFH_D
   All conditions listed in § 7.2.4.1 are satisfied:

- $\beta \leq 2\%$;
- $DC \leq 99\%$;
- $1/\lambda_{De} = 1/\lambda_{DKP} = \frac{1}{1.92 \cdot 10^{-6} \cdot 8760} = 59.5 < 1.000$ years;
- $1/\lambda_{D\text{-}FH} = 1/\lambda_{DFR} = 1/\lambda_{DU<} = 333\,333 > 1200$ years;

Therefore the simplified equation can be used.

$$PFH_D = (1 - DC_1) \cdot \lambda_{DK_P} = (1 - 0.90) \cdot 1.92 \cdot 10^{-6} = 1.92 \cdot 10^{-7}\,\text{h}^{-1}$$

Table 7.11 Maximum SIL that can be claimed for an SCS.

	Hardware fault tolerance (*HFT*)		
Safe failure fraction (*SFF*)	**0**	**1**	**2**
$SFF < 60\%$	SIL 1	SIL 1	SIL 2
$60\% \leq SFF < 90\%$	SIL 1	SIL 2	SIL 3
$90\% \leq SFF < 99\%$	**SIL 2**	SIL 3	SIL 3
$SFF \geq 99\%$	SIL 3	SIL 3	SIL 3

With an SFF = 90% and $HFT = 0$, the SIL level is limited by the Architectural Constraints to SIL 2, as shown in Table 7.11.

- The Safety Subsystem reaches **SIL 2** and has a $\boldsymbol{PFH_D = 1.92 \cdot 10^{-7}\,h^{-1}}$.

7.2.7 Alternative Formula for the Basic Subsystem Architecture C

The Technical Standard IEC TS 63394 [80] indicates a different formula to calculate the PFH_D of a Basic Subsystem Architecture C. The formula comes from a Reliability Block Diagram simplification, and therefore it assumes the safety subsystem to be non-repairable. It can be used instead of the one indicated in § 7.2.5.2.

$$PFH_D = (1-\beta) \cdot (1-DC) \cdot \lambda_{De} + (1-\beta) \cdot DC \cdot \lambda_{De} \cdot (\lambda_{D-FH} - \beta \cdot \lambda_{De}) \cdot \frac{(T_1 + T_2)}{2} + \beta \cdot \lambda_{De}$$

where:

- T_1 is the Proof Test interval of the perfect Proof Test or useful lifetime, whichever is the smaller;
- T_2 is the diagnostic test interval;

- λ_{De} is the dangerous failure rate of the single element of the functional channel;
- $\lambda_{D\text{-}FH}$ is the failure rate of the single element that realizes the fault handling function(s) within the subsystem;
- **DC** is the diagnostic coverage for the single element e of the functional channel;
- β is the susceptibility to common cause failures of the functional channel and the channel that realizes the fault handling function(s) within the subsystem.

7.2.8 Basic Subsystem Architecture D: 1oo2D

This Architecture, shown in Figure 7.16, dual channel with a diagnostic, is such that a single fault of any subsystem element does not cause the loss of the safety function. Where a fault of a subsystem element is detected, the diagnostic function(s) initiates a fault reaction function. This Architecture corresponds to a Hardware Fault Tolerance of 1.

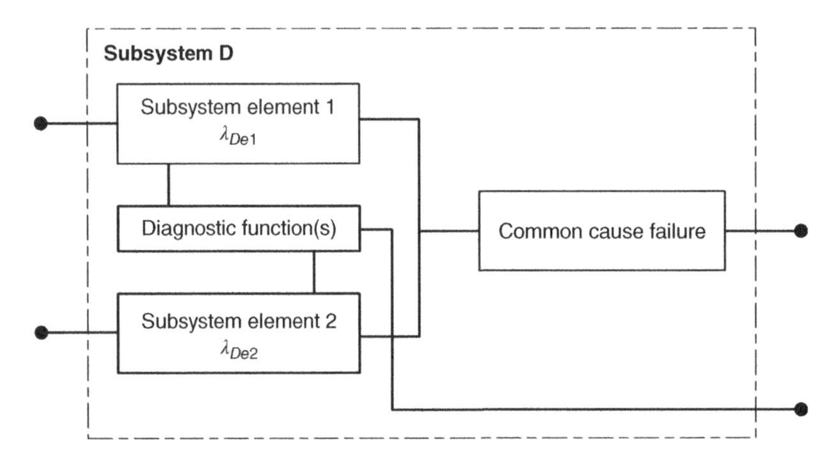

Figure 7.16 Logical representation of Basic Subsystem Architecture "D."

This Architecture corresponds to Category 3 or 4 of ISO 13849-1.
The PFH_D of the subsystem is:

$$PFH_D = (1-\beta)^2 \cdot \left[\lambda_{De_1} \cdot \lambda_{De_2} \cdot (DC_1 + DC_2) \cdot \frac{T_2}{2} + \lambda_{De_1} \cdot \lambda_{De_2} \cdot (2 - DC_1 - DC_2) \cdot \frac{T_1}{2}\right] + \beta \cdot \frac{(\lambda_{De_1} \cdot \lambda_{De_2})}{2}$$

where

- T_1 is the Proof Test interval of the perfect Proof Test or useful lifetime, whichever is the smaller;
- T_2 is the diagnostic test interval;
- β is the susceptibility to common cause failures;
- λ_{De1} is the dangerous failure rate of subsystem element e_1;
- λ_{De2} is the dangerous failure rate of subsystem element e_2;
- DC_1 is the diagnostic coverage for subsystem element e_1;
- DC_2 is the diagnostic coverage for subsystem element e_2.

In case the two subsystem elements are identical, the formula becomes:

$$PFH_D = (1-\beta)^2 \cdot [DC \cdot T_2 + (1-DC) \cdot T_1] \cdot \lambda_{De}^2 + \beta \cdot \lambda_{De}$$

7.2.8.1 Implications of the Architectural Constraints in Basic Subsystem Architecture D

Basic Subsystem Architectures D have an $HFT = 1$. That means a maximum achievable SIL is indicated in Table 7.12, even if the formula gives a lower PFH_D.

Table 7.12 Architectural constraints on a subsystem: maximum SIL that can be claimed for an SCS.

Safe failure fraction (*SFF*)	Hardware fault tolerance (*HFT*)		
	0	1	2
$SFF < 60\%$	SIL 1 if well-tried components are used	SIL 1	SIL 2
$60\% \leq SFF < 90\%$	SIL 1	**SIL 2**	SIL 3
$90\% \leq SFF < 99\%$	SIL 2	**SIL 3**	SIL 3
$SFF \geq 99\%$	SIL 3	**SIL 3**	SIL 3

7.2.8.2 Example of Input Basic Subsystem Architecture D: Emergency Stop

Let's consider the same input subsystem as in § 6.2.9.2, shown in Figure 7.17 as a simplified electrical diagram and in Figure 7.18 as a RBD.

The Emergency Stop has a $B_{10D} = 10^5$ and is supposed to be used **once per month**. The Mission Time is 20 years. Applicable Basic and well-tried safety principles are used.

Let's now focus on the calculation of **probability of random failures**:

1. **The first step** is to calculate the $\lambda_{D\ EM}$ of the Emergency Stop:

$$n_{op} = 12 \frac{1}{\text{year}}$$

$$\lambda_{D\ EM} = \frac{0.1 \cdot n_{op}}{B_{10D} \cdot 8760} = \frac{0.1 \cdot 12}{10^5 \cdot 8760} = 1.37 \cdot 10^{-9}\,\text{h}^{-1}$$

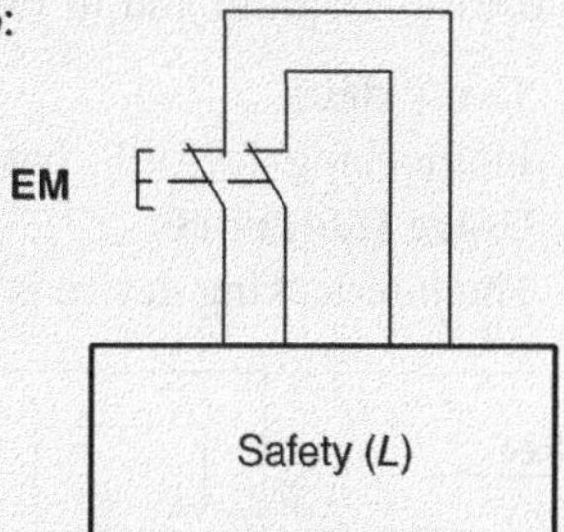

Figure 7.17 Emergency stop Input Subsystem Architecture D.

2. **The second step** is to calculate T_1, T_2, and β.

   T_1 = min (Useful Life; Proof Test);

   Useful Life = min (Mission Time; T_{10D});

   Mission Time = 20 years;

$$T10_{D_{EM}} = \frac{B10_{D\ EM}}{n_{op}} = \frac{10^5}{12} = 8333 > 20\,\text{years}$$

   Useful Life = min (20 years; 8333 years) ➔ Useful time = 20 years;

   Proof Test > 20 years

   T_1 = min (20 years; > 20 years) ➔ $\boldsymbol{T_1}$ **= 20 years = 175 200 hours**;

$$T_2\ (\text{Diagnostic Test Interval}) = \frac{Working\ hours/year}{n_{op}} = \frac{8760}{12} = 730\ \text{hours}$$

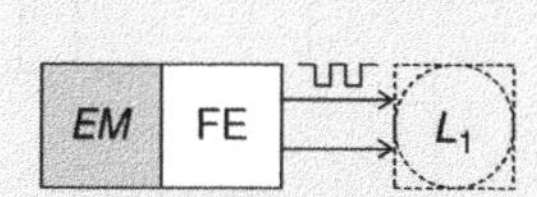

Figure 7.18 RBD of an emergency stop Input Subsystem Architecture D.

We also verified that enough measures have been applied to prevent common cause failures: a $\beta = 2\%$ can be assumed (§ 3.6.2).

3. As a **third step**, we estimate the $\boldsymbol{SFF = DC = 99\%}$
4. **The last step** is to calculate the PFH_D. Using IEC 62061 formula, we have the following results:

$$PFH_D = (1-\beta)^2 \cdot [DC \cdot T_2 + (1-DC) \cdot T_1] \cdot \lambda_{D\,EM}^2 + \beta \cdot \lambda_{D\,EM}$$

$$PFH_D = (1-0.02)^2 \cdot [0.99 \cdot 730 + (1-0.99) \cdot 175\,200] \cdot \lambda_{DEM}^2 + 0.02 \cdot \lambda_{D\,EM} = 2.74 \cdot 10^{-11}\ \mathrm{h}^{-1}$$

Despite the PFH_D is very low, with an $SFF = 99\%$ and $HFT = 1$, the SIL level is limited by the Architectural Constraints to SIL 3, as shown in Table 7.13.

The Safety Input Subsystem reaches **SIL 3** and has a $\boldsymbol{PFH_D = 2.74 \cdot 10^{-11}\ h^{-1}}$.

Table 7.13 Maximum SIL that can be claimed for an SCS.

Safe failure fraction (*SFF*)	Hardware fault tolerance (*HFT*)		
	0	1	2
$SFF < 60\%$	SIL 1	SIL 1	SIL 2
$60\% \leq SFF < 90\%$	SIL 1	SIL 2	SIL 3
$90\% \leq SFF < 99\%$	SIL 2	SIL 3	SIL 3
$SFF \geq 99\%$	SIL 3	**SIL 3**	SIL 3

7.2.8.3 Example of Input Basic Subsystem Architecture D: Interlocking Device

Let's consider the same input subsystem as in § 6.2.8.2, shown in Figure 7.19 as a simplified electrical diagram and in Figure 7.20 as a RBD.

- **Safety data:**
  Interlocking device B_1 has $B_{10D} = 1 \cdot 10^6$ and a Mission Time of 20 years.
- **Usage frequency:**
  The interlocking device is supposed to open **10 times per hour**.
- **Avoidance of Systematic Failures**
  Since we can claim Category 3 (double channels with monitoring), **basic and well-tried safety principles are applied.**

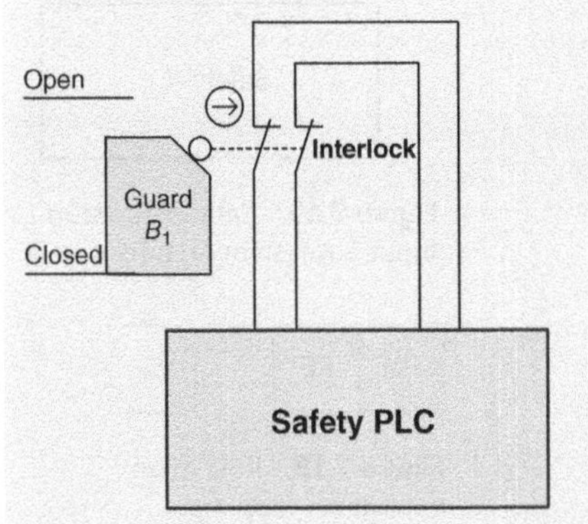

Figure 7.19 Interlocking Device Input Subsystem Architecture D.

1. **The first step** is to calculate the $\lambda_{D\,B_1}$ of the interlocking device. We assume the machine is working 365 days per year and 16 hours per day.

$$n_{op} = 10 \cdot 16 \cdot 365 = 58\,400\ \frac{1}{\text{year}}$$

$$\lambda_{D\,B_1} = \frac{0.1 \cdot n_{op}}{B_{10D} \cdot 8760} = \frac{0.1 \cdot 58\,400}{10^6 \cdot 8760} = 6.67 \cdot 10^{-7}\,\text{h}^{-1}$$

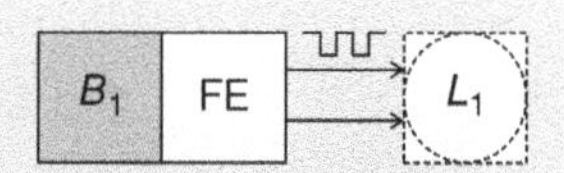

Figure 7.20 RBD of an Interlocking Device Input Subsystem Architecture D.

2. **The second step** is to calculate T_1, T_2, and β.
   T_1 = min (Useful Life; Proof Test);
   Useful Life = min (Mission Time; T_{10D});
   Mission Time = 20 years;

$$T_{10D\text{-}B_1} = \frac{B_{10D\text{-}B_1}}{n_{op}} = \frac{10^6}{58\,400} = 17 < 20\,\text{years}$$

   Useful Life = min (20 years; 17 years) ➔ Useful time = 17 years;
   Proof Test > 20 years
   T_1 = min (17 years; >20 years) ➔ $\boldsymbol{T_1}$ **= 17 years = 148 920 hours**;

$$T_2\,(\text{Diagnostic Test Interval}) = \frac{\text{Working hours/year}}{n_{op}} = \frac{5840}{58\,400} = 0.1\,\text{hours}$$

   We also verified that enough measures have been applied to prevent common cause failures: a $\beta = 2\%$ can be assumed (§ 3.6.2).
3. As a **third step**, we estimate the ***SFF* = *DC* = 99%**
4. **The last step** is to calculate the PFH_D. Using IEC 62061 formula, we have the following results:

$$PFH_D = (1-\beta)^2 \cdot [DC \cdot T_2 + (1-DC) \cdot T_1] \cdot \lambda^2_{D\,B_1} + \beta \cdot \lambda_{D\,B_1}$$

$$PFH_D = (1-0.02)^2 \cdot [0.99 \cdot 0.1 + (1-0.99) \cdot 148\,920] \cdot \lambda^2_{D\,B_1} + 0.02 \cdot \lambda_{D\,B_1} = 1.40 \cdot 10^{-8}\,\text{h}^{-1}$$

Despite the PFH_D is very low, with an *SFF* = 99% and *HFT* = 1, the SIL level is limited by the Architectural Constraints to SIL 3.

The Safety Input Subsystem reaches **SIL 3** and has a $\boldsymbol{PFH_D = 1.40 \cdot 10^{-8}\,h^{-1}}$.

Table 7.14 Maximum SIL that can be claimed for an SCS.

	Hardware fault tolerance (*HFT*)		
Safe failure fraction (*SFF*)	**0**	**1**	**2**
SFF < 60%	SIL 1	SIL 1	SIL 2
60% ≤ *SFF* < 90%	SIL 1	SIL 2	SIL 3
90% ≤ *SFF* < 99%	SIL 2	SIL 3	SIL 3
SFF ≥ 99%	SIL 3	**SIL 3**	SIL 3

7.2.8.4 Example of a Basic Subsystem Architecture D Output

Let's consider the input subsystem shown in Figure 7.21 as a simplified electrical diagram and in Figure 7.22 as a RBD.

K_{P1} and K_{P2} have $B_{10} = 10^6$ with 73% as percentage of dangerous failures and they are supposed open once a week. The machine works 48 weeks per year, 5 days per week, 8 hours per day. The Mission Time is 20 years. Applicable Basic and well-tried safety principles are used.

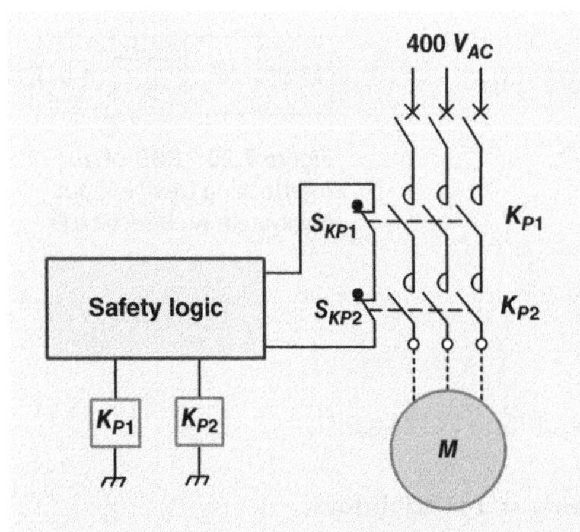

Figure 7.21 Output Subsystem Architecture D.

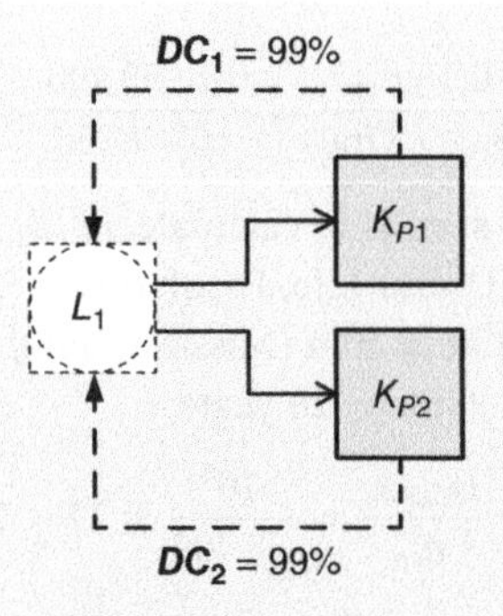

Figure 7.22 RBD of an Output Subsystem Architecture D.

Let's now focus on the calculation of **probability of random failures:**

1. **The first step** is to calculate the $\lambda_{D\ KP}$ of the contactors:

$$n_{op} = 48 \frac{1}{\text{year}}$$

$$B10_D = \frac{10^6}{0.73} = 1.37 \cdot 10^6$$

$$\lambda_{D\ KP} = \frac{0.1 \cdot n_{op}}{B_{10D} \cdot 8760} = \frac{0.1 \cdot 48}{1.37 \cdot 10^6 \cdot 8760} = 4 \cdot 10^{-10}\,\text{h}^{-1}$$

2. **The second step** is to calculate T_1, T_2, and β.
   T_1 = min (Useful Life; Proof Test);
   Useful Life = min (Mission Time; T_{10D});
   Mission Time = 20 years

$$T10_{D_{KP}} = \frac{B10_{D\ KP}}{n_{op}} = \frac{1.37 \cdot 10^6}{48} = 28\,542 > 20\ \text{years}$$

   Useful Life = min (20 years; ≫20 years) → Useful time = 20 years;
   Proof Test > 20 years
   T_1 = min (20 years; >20 years) → T_1 = 20 years = 175 200 hours;

$$T_2\ (\text{Diagnostic Test Interval}) = \frac{\text{Working hours/year}}{n_{op}} = \frac{1920}{48} = 40\ \text{hours}$$

   We also verified that enough measures have been applied to prevent common cause failures: a $\beta = 2\%$ can be assumed (§ 3.6.2).

3. As a **third step**, we estimate the ***SFF* = *DC* = 99%**. Please refer to table D.1 – Output Device – "*Redundant shut-off path with monitoring of the actuators by logic and test equipment*"
4. **The last step** is to calculate the PFH_D. Using IEC 62061 formula, we have the following results:

$$PFH_D = (1-\beta)^2 \cdot [DC \cdot T_2 + (1-DC) \cdot T_1] \cdot \lambda_{D\ KP}^2 + \beta \cdot \lambda_{D\ KP}$$

$$PFH_D = (1-0.02)^2 \cdot [0.99 \cdot 40 + (1-0.99) \cdot 175200] \cdot \lambda_{DKP}^2 + 0.02 \cdot \lambda_{DKP} = 8 \cdot 10^{-12}\,\text{h}^{-1}$$

Despite the PFH_D is very low, with an $SFF = 99\%$ and $HFT = 1$, the SIL level is limited by the Architectural Constraints to SIL 3, as shown in Table 7.14.

The Safety Input Subsystem reaches **SIL 3** and has a $\boldsymbol{PFH_D = 8\ 10^{-12}\,h^{-1}}$.

7.3 Determination of the Reliability of a Safety Function

The PFH_D value of a Safety Function is the sum of the PFH_D values of all of its subsystems, moreover the SIL level is limited by the subsystem having the lowest SIL. In other terms, the SIL that can be achieved by a Safety-related Control System is considered separately for each safety subsystem and is determined from the SIL and the PFH_D of each subsystem as follows:

- The SIL that is achieved is equal to or less than the lowest SIL of any of the subsystems, and
- The SIL is limited by the sum PFH_D value of all subsystems, according to Table 4.3.

Figure 7.23 shows an example of an SCS with safety integrity of SIL 2, despite the overall PFH_D value is suitable for a higher SIL.

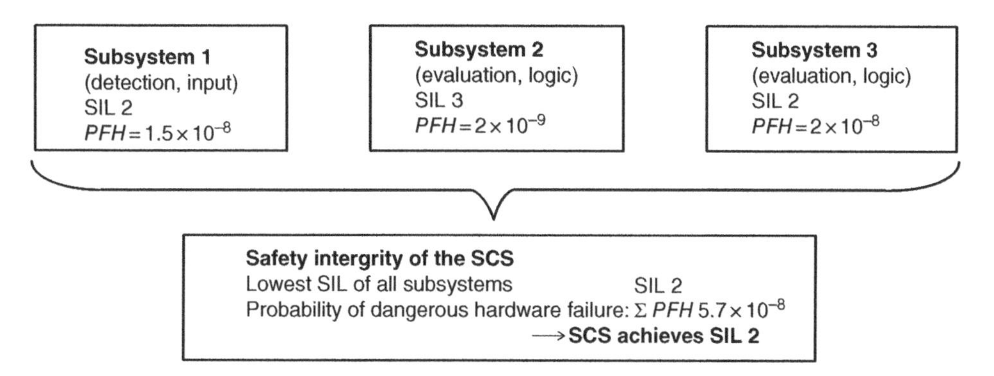

Figure 7.23 Example of SIL calculation of a safety function.

8

Validation

8.1 Introduction

Validation is the activity of demonstrating that the **Safety-related Control System**, SCS or SRP/CS, **meets the Safety Requirements Specification**.

I remind that the whole "Functional Safety House" is based upon

- **Hardware safety Integrity**. The validation goes through the verification that:
  - The components have the required Reliability data
  - The subsystem Architectures or Categories have been properly implemented and
  - The calculations were done correctly.
- **Systematic Safety Integrity, including software**: in other words, the lack of systematic failures (also called Systematic Capability § 3.4.2). This is the most difficult aspect to verify, since its objective is that there were no mistakes in the engineering and construction phase. For example, the cables were correctly sized and protected, the software was correctly configured, or the cabling was properly done. Many of those Systematic Capabilities are listed in the Basic and Well-tried safety principles: annexes A–D of ISO 13849-2.

A loss of the safety function in the absence of a **hardware fault is due to a systematic failure**, which can be caused by errors made during the design and integration stages (a misinterpretation of the safety function characteristics or an error in the logic design or an error in hardware assembly or also an error in typing the software code). Some of these systematic failures will be revealed during the design process, while others should be revealed during the validation process; otherwise, they will remain unnoticed. In addition, it is also possible that an error is made during the validation process: e.g. failure to check a characteristic.

Example

I once analyzed the electrical drawing of a machine and realized a contactor was not properly protected by its overload protection. When I asked to change it, the manufacturer replied that it was not a problem if the contactor would jam, and stay closed, since it was monitored and there were two contactors on the same branch Circuit. He was thinking to implement on the output a Category 4 or Architecture 1oo2D.

However, there was a mistake, a "Systematic mistake" in the design, that would cause a Systematic Failure. The Functional Safety house is built upon the fact that no Systematic Failures or mistakes are present. Provided that is true, you can then reason in terms of (Hardware) Failure rates or B_{10D} or $MTTF_D$, PFH_D, and finally SIL or PL.

Functional Safety of Machinery: How to Apply ISO 13849-1 and IEC 62061, First Edition. Marco Tacchini.

The reason for monitoring a contactor is not to check if it sticks due to a systematic failure: the Diagnostic Coverage assumes no systematic failures are present in the safety circuit and therefore the contactor cannot jam due to an overload. It may have a fault, but it can only be random and not due to a mistake in the engineering phase.

But let's start from the beginning; here is the definition of Validation:

> ***[IEC 62061] 3.2 Terms and definitions***
> ***3.2.65 Validation*** *(of the safety function). Confirmation by examination (e.g. tests, analysis) that the SCS meets the functional safety requirements of the specific application.*
>
> ***[ISO 13849-1] 3.1 Terms and definitions***
> ***3.1.54 Validation****. Confirmation by examination and provision of objective evidence that the particular requirements for a specific intended use are fulfilled.*

The validation includes inspection (for example, by analysis) and testing of the SCS or SRP/CS to ensure that it **achieves the requirements stated in the safety requirements specification**; that means:

- The specified functional requirements of the safety functions set out in the safety requirements specification, or SRS, have been correctly implemented.
- The requirements of the specified PLr or SILr, have been met; which means:
  1. the requirements of the specified category or architecture are met;
  2. the measures for control and avoidance of systematic failures (systematic integrity) are present;
  3. if applicable, the requirements of the software have been met; and
  4. the ability to perform a safety function under expected environmental conditions is present.

8.1.1 Level of Independence of People Doing the Validation

The basic problem is that people make mistakes! They make mistakes in designing the safety function, in configuring the software as well as during the machinery installation.

The idea of the validation is that **the work of a technician is checked by a colleague.** What the standard makes clear is that it is not acceptable that the same person defines the Safety Functions, configures the software, and does the validation: at least a second person is needed, that is the essence of the concept of "level of Independence."

The validation process should be carried out by person(s) who is/are independent from the design of the SRP/CS. **An independent person** is a person not involved in the design of the SRP/CS, and it does not necessarily mean that a third party is required. That is the ISO 13849-1 approach.

IEC 62061 has a bit of a more restricted definition of an **Independent person:** he is a person who may be involved in the same project but shall not be involved in the design activities and shall not have responsibility for project management and shall not have a superior role.

Table 8.1 shows visually the approach of both standards.

Table 8.1 Minimum levels of independence for validation activities.

	Required SIL for the safety function up to		
Minimum level of independence	**SIL 1**	**SIL 2**	**SIL 3**
Same person	Not sufficient	Not sufficient	Not sufficient
Independent person	Sufficient	Sufficient	Sufficient

The analysis can start in parallel with the design process. Problems can then be corrected early while they are still relatively easy to correct.

It's clear that both ISO 13849-1 and IEC 62061 are concerned with the likelihood of mistakes. The manufacturer has to evaluate the risk also considering if he is designing a completely new safety system or simply modifying a software that was already validated and is well known in its organization.

8.1.2 Flow Chart of the Validation Process

The validation process is illustrated in Figure 8.1. Both IEC 62061 and ISO 13849-1 propose a flow chart; we propose the one in ISO 13849-1: the differences are more in the graphics than in the substance. The paragraphs mentioned in the blocks are the ones in ISO 13849-1.

8.2 The Validation Plan

The validation plan shall identify and describe the requirements for carrying out the validation process. The validation plan shall also identify the means to be employed to validate the specified safety functions. It should indicate, where appropriate:

- the operational and environmental conditions during testing;
- the analysis and tests to be done;
- the persons or parties responsible for each step in the validation process; and
- the required equipment.

Subsystems that have previously been validated to the same specification need only reference to that previous validation.

8.2.1 Fault List

Validation involves consideration of the behavior of the SCS for all faults to be considered. A basis for fault consideration is given in the tables of fault lists in ISO 13849-2:2012, annexes A–D, which are based on experience and which contain:

- the components/elements to be included, e.g. conductors/cables;
- the faults to be taken into account, e.g. short circuits between conductors;
- the permitted fault exclusions, taking into account environmental, operating and application aspects; and
- a remarks section giving the reasons for the fault exclusions.

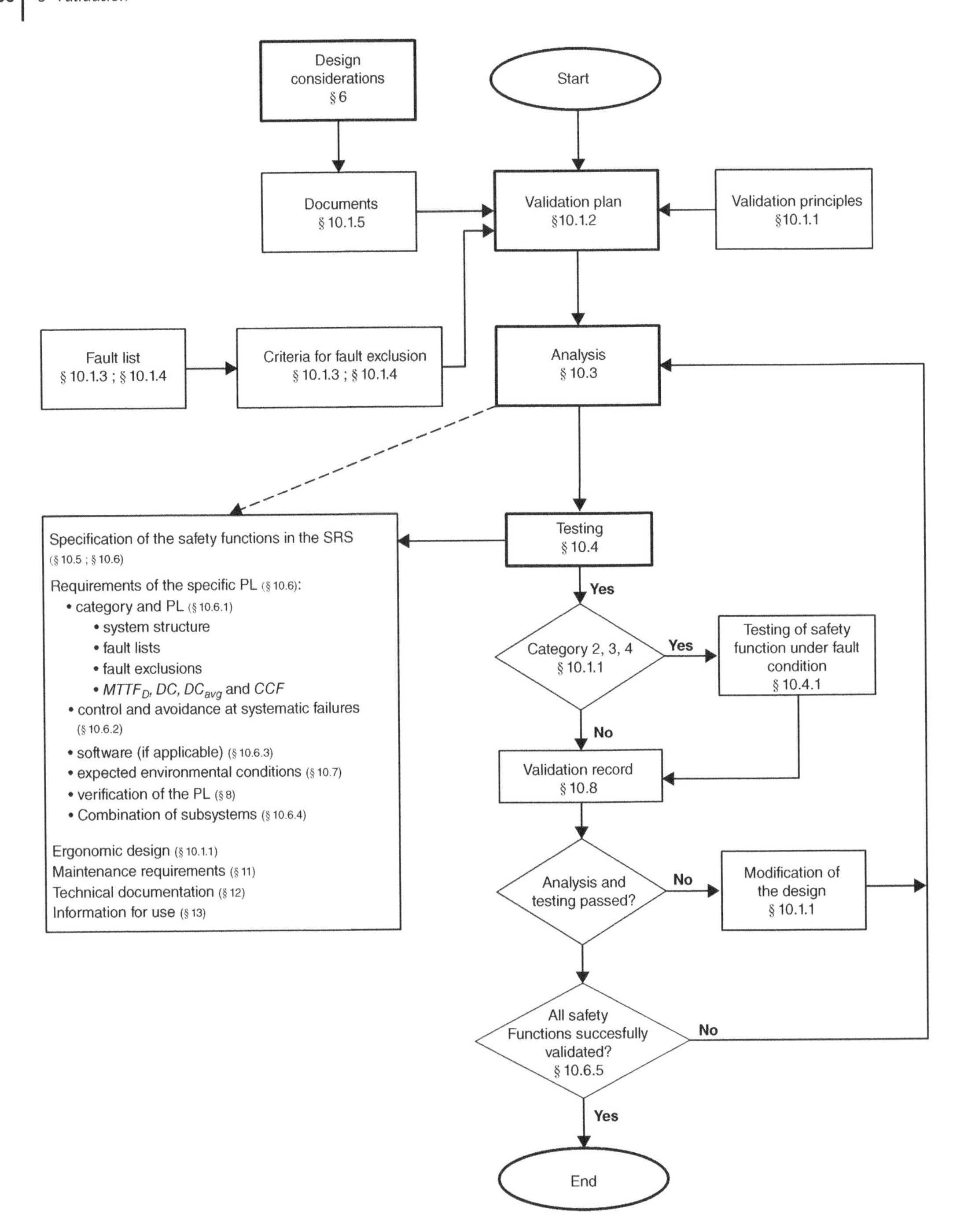

Figure 8.1 The validation process.

8.2.2 Validation Measures Against Systematic Failures

The validation of measures against systematic failures can typically be provided by:

a) inspections of design documents that confirm the application of
   - basic and well-tried safety principles (see ISO 13849-2:2012, annexes A–D);
   - further measures for avoidance of systematic failures; and
   - further measures for the control of systematic failures such as hardware diversity, modification protection, or failure assertion programming;
b) failure analysis (e.g. FMEA);
c) fault injection tests/fault initiation;
d) inspection and testing of data communication, e.g. parameterization, installation;
e) checking that a quality management system avoids the causes of systematic failures in the manufacturing process.

8.2.3 Information Needed for the Validation

Hereafter a possible list:

- safety requirements specification of the required characteristics of each safety function, e.g. Response time, operating mode, PL or SIL;
- drawings and specifications, e.g. for mechanical, hydraulic, and pneumatic parts;
- block diagrams with a functional description of the blocks;
- circuit diagrams, including interfaces/connections;
- functional description of the circuit diagrams, where needed for clarification;
- time sequence diagrams for switching components, signals relevant for safety;
- description of the relevant characteristics of components previously validated;
- report of analysis of all relevant faults, **including the justification of any excluded faults**;
- information for use, e.g. installation and operation manual/instruction handbook.

The information is required on how the SIL or PL and average probability of a dangerous failure per hour is determined. The documentation of the quantifiable aspects shall include:

- The Safety-related block diagram or designated Architectures.
- The determination of λ_D, $MTTF_D$, DC_{avg}, SFF, and CCF.
- The determination of the Category or the Architecture.

Information is required for documentation on systematic aspects of the Safety-related Control System and information is required to describe how the combination of several subsystems achieves the required SIL or PL.

Validation shall be recorded and the records shall demonstrate the validation process for each of the safety requirements

8.2.4 Analysis and Testing

8.2.4.1 Analysis

As already stated, the validation of the SCS or SRP/CS includes an analysis; inputs to the analysis include the following:

- The safety function(s), their characteristics, and the safety integrity or PL defined;
- The system structure (e.g. basic subsystem Architectures);

- The quantifiable aspects (e.g. $MTTF_D$ or λ_D, DC, SFF, and CCF);
- The non-quantifiable, qualitative aspects which affect system behavior (if applicable, software aspects);
- Fault Lists;
- Criteria for Fault Exclusion.

The analysis includes the verification that each Safety Function was designed as required in the Safety Requirements Specification; that the components used were the correct ones (the designer planned to use a certain model of interlocking device: was that model really used?); that the Architectures of the subsystems were properly done (the designer planned to connect the single interlocking device with a monitoring function called "trigger": was the circuit properly done?).

Analysis should also include verification of a correct engineering. Is the common pole of the control transformer properly bonded or, if it not bonded, does it comply with the prescriptions of IEC 60204-1 § 9.4.3? Are the contactors properly sized and protected against overload and short circuit? Is the pneumatic circuit properly protected by a pressure regulator?

The Analysis therefore includes a review of the electrical drawings, and of the software. A check of the hardware components installed in the machine and details of the associated software to confirm their correspondence with the documentation (e.g. manufacturer, type, version) needs to be done.

At a certain moment, there should be the verification, by the independent person or group, that the achieved SIL or PL level is equal to or greater than the SILr or PLr.

8.2.4.2 Testing

Testing shall be carried out to complete the validation. Validation tests shall be planned and implemented in a logical manner. In particular:

a) A test plan shall be produced before testing begins.
b) The activities should be recorded in a Test report with name of the person who did the test, the date, the results, etc.
c) The test records shall be compared with the test plan to ensure that the specified functional and performance targets have been achieved.

Validation of the safety functions by testing shall be carried out by applying input signals, in various combinations, to the SCS. The resultant response at the outputs shall be compared to the appropriate specified outputs.

Functional testing of the safety functions is needed in all required operating modes, as defined in the Safety Requirements Specification, to establish whether they meet the specified characteristics. The functional tests shall ensure that all safety-related outputs are realized over their complete ranges and respond to safety-related input signals in accordance with the specification. The test cases are normally derived from the specifications but could also include some cases derived from analysis of the schematics or software.

From a practical point of view, if the Diagnostic Coverage relies on the trigger function of the logic unit, a short circuit should be realized at the logic unit and a verification that the system detects the fault should be done. Similar simulations can be done by triggering the same two contacts of the interlocking device with the same triggering signal and check that the system was properly configured.

Attention should be paid to the environmental conditions. Generally, for the specification of the $MTTF_D$ or λ_D, an ambient temperature of +40 °C is taken as a basis. During validation, it is important to ensure that, for $MTTF_D$ or λ_D values, the environmental and functional conditions (in particular temperature) taken as a basis are met. Where a device, or component, is operated significantly above (e.g. more than 15 °C) the specified temperature of +40 °C, it will be necessary to use $MTTF_D$ or λ_D values for the increased ambient temperature.

In case of category B or 1, tests under fault conditions are not required and analysis with functional testing may be sufficient.

8.2.4.3 Validation of the Safety Integrity of Subsystems

The safety integrity of each subsystem of the SCS or SRP/CS shall be validated by confirming the requirements of Table 8.2 according to the Category used. The table is present only in ISO 13849-1, and it can be used for IEC 62061 with some considerations:

- Category 3 or Category 4 are equivalent to Architecture D.
- Category 2 is very similar to Architecture C.
- Category 1 is equivalent to Architecture A.
- There is no equivalent of Architecture B in ISO 13849-1.

The validation of $MTTF_D$, DC_{avg}, and CCF is typically performed by analysis and visual inspection.

Table 8.2 Basic requirements for categories.

Requirements	Category				
	B	**1**	**2**	**3**	**4**
Basic safety principles	x	x	x	x	x
Expected operating stress	x	x	x	x	x
Influences of processed material	x	x	x	x	x
Performance during other relevant external influences	x	x	x	x	x
Well-tried components	—	x	—	—	—
Well-tried components for the case of determination of the PL without $MTTF_D$	—	x	x	x	x
Well-tried safety principles	—	x	x	x	x
Mean time to dangerous failure ($MTTF_D$) of each channel	x	x	x	x	x
The check procedure of the safety function(s)	—	—	x	—	—
The recognizable faults and the associated diagnostic measures, including fault reaction	—	—	x	x	x
Checking intervals, when specified	—	—	x	x	x
Diagnostic coverage (DC_{avg})	—	—	x	x	x
Common-cause failures (*CCF*) identified and how to prevent them	—	—	x	x	x
Justification for fault exclusion	x	x	x	x	x
How the safety function is maintained in the case of each of the fault	—	—	—	x	x
How the safety function is maintained for each of the combinations of faults	—	—	—	—	x
Measures against systematic failures	x	x	x	x	x
Measures against software faults	x	—	x	x	x

x, Required; —, not required.

The $MTTF_D$ values for components (including B_{10D}, T_{10D}, and n_{op} values) shall be checked for plausibility. For example, the value given on the supplier datasheet is to be compared with annex C of ISO 13849-1.

Where we have applied a fault exclusion, it means that the particular components or aspects of that component do not contribute to the channel $MTTF_D$. We remind that a fault exclusion implies infinite $MTTF_D$; therefore, fault excluded failures of the component or part of it, will not contribute to the calculation of channel $MTTF_D$. An example can be the mechanical part of an Emergency push button, or the mechanical part of an interlocking device.

The $MTTF_D$ of each channel of the subsystem, shall be checked for correct calculation. In case of ISO 13849-1, the $MTTF_D$ of individual channels shall be restricted to no greater than 100 years (2500 years for Category 4) before the equation is applied.

The DC values for subsystem elements and/or logic blocks shall be checked for plausibility. The correct implementation (hardware and software) of checks and diagnostics, including appropriate fault reaction, shall be validated by testing under typical environmental conditions of use.

The correct implementation of sufficient measures against common-cause failures shall be validated. Typical validation measures are static hardware analysis and functional testing under environmental conditions.

8.2.4.4 Validation of the Safety-related Software

As a first step, check that there is documentation for the specification and design of the safety-related software. This documentation shall be reviewed for completeness and the absence of erroneous interpretations, omissions, or inconsistencies.

In general, software can be considered a "black box." In case of small programs, an analysis by means of reviews or walk-through of control flow, procedures, etc. using the software documentation (control flow chart, source code of modules or blocks, I/O and variable allocation lists, cross-reference lists) can be sufficient.

Black-box testing aims to check the dynamic behavior under real functional conditions, and to reveal failures to meet functional specification, and to assess utility and robustness. Grey-box testing technique can also be used: it is similar to Black-box testing but additionally monitors relevant test parameter(s) inside the software module.

The software can be considered a "black box." Depending on the PLr or SILr, the tests should include:

- black-box testing of functional behavior and performance (e.g. timing performance);
- additional extended test cases based upon limit value analyses, recommended for SIL 2 (PL d) or SIL 3 (PL e);
- I/O tests to ensure that the safety-related input and output signals are used properly; and
- test cases that simulate faults determined analytically beforehand, together with the expected response, in order to evaluate the adequacy of the software-based measures for control of failures.

8.2.4.5 Software-based Manual Parameterization

Some safety-related subsystems need parameterization. For example, a variable speed drive can be parameterized via a PC-based configuration tool for setting the upper safety speed limit parameter. The detection zone of a laser scanner need to be defined: parameters such as angle and distance can be configured per the manufacturer's safety documentation and the machinery risk assessment.

The objective of the requirements for software-based manual parameterization is **to guarantee that the safety-related parameters specified for a safety function or a sub-function are**

correctly transferred into the hardware performing the safety function or a sub-function. Different methods can be applied to set such parameters, such as dip switch-based parameterization or dedicated parameterization software (commonly called configuration or parameterization tools).

During software-based manual parameterization, mistakes can be made and therefore the following can happen:

- parameters are not updated by the parameterization process, completely or in parts without notice to the person responsible for the parameterization;
- parameters are incorrect, completely or in parts;
- parameters are applied to an incorrect device, such as when transmission of parameters is carried out via a wired or wireless network.

Software-based manual parameterization shall use a dedicated tool provided by the manufacturer or supplier **of the SRP/CS or the related subsystem(s)**. The SCS or the related subsystem(s) and the parameterization tool shall have the capability to prevent unauthorized modification, for example by using a dedicated password. Parameterization while the machine is running shall be permitted only if it does not cause an unsafe situation.

The following verification activities shall be performed to verify the basic functionality of the parameterization tool:

- verification of the correct setting for each safety-related parameter (minimum, maximum, and representative values);
- verification that the safety-related parameters are checked for plausibility, for example by detection of invalid values;
- verification that means are provided to prevent unauthorized modification of safety-related parameters.

The initial parameterization, and subsequent modifications to the parameterization, shall be documented. The documentation shall include:

- the date of initial parameterization or change;
- data or version number of the data set;
- name of the person carrying out the parameterization;
- an indication of the origin of the data used (e.g. pre-defined parameter sets);
- clear identification of safety-related parameters;

9

Some Final Considerations

I hope you liked this journey through *Functional Safety of Machinery*.

The book was born from the idea that, with the new editions of ISO 13849-1 (fourth) and IEC 62061 (second), the two standards are now aligned as never before, despite starting from two different approaches.

ISO 13849-1 is based upon the assumption that Safety Related Parts of the Control System are repairable. Markov Chains are used to model the five categories and to calculate the PFH_D for each one of them, based upon the value of key parameters like $MTTF_D$ and DC_{avg}.

IEC 62061 assumes that Safety-related Control Systems or SCS are non-repairable and it uses the reliability block diagram technique to derive the equations for the PFH of the four architectures. In this case, the key elements are the component failure rates, the Safe Failure Fraction (SFF), and the architectural constraints.

The book starts from the mathematics at the base of Functional Safety, and a bit of history was given for those who like to know where we are coming from and how functional safety of machinery is related to other sectors, like the process industry.

The rest of the book details Functional Safety of Machinery as if it were one common approach. Only in Chapters 6 and 7 we give details of the two standards and explain the few differences between them.

9.1 ISO 13849-1 vs IEC 62061

The second edition of IEC 62061 is quite different from the first one. The main change is probably related to the fact that, in case of Electromechanical components (like Pressure Switches, Contactors, or Solenoid Valves), used in High-Demand Safety Control Systems normally, $SFF = DC$.

The same consideration will probably be adopted by the low-demand Technical Standards. At the moment this book was written, a new edition of the IEC 61508 series is at CD level. Inside the IEC 61508-2, a revised language states that, when estimating the SFF of an element, intended to be used in a subsystem **having a hardware fault tolerance of 0**, operating **in low-demand mode,** high-demand mode, or continuous mode of operation, **credit shall only be taken for the diagnostics if it can bring the system to a safe state**. The standard states that in Low-Demand mode, if we want to claim a certain minimum level of diagnostic coverage, we have to assume that the EUC can operate safely in the presence of dangerous faults that are detected by

Functional Safety of Machinery: How to Apply ISO 13849-1 and IEC 62061, First Edition. Marco Tacchini.

the diagnostic tests. All that means, for electromechanical components, having no internal diagnostics, ***SFF* = *DC* in all modes of operations.**

Another significant difference between the old and the new version of IEC 62061, regarding component reliability, is that, instead of annex D, the second edition has annex C for conservative values of B_{10D} and MTTF_D of the most common components and annex D for indication of the Diagnostic coverage of the most common connections.

One of the objectives of the second edition of IEC 62061 was to indicate **how to deal with low-demand mode safety systems**. Unfortunately, the team did not agree on the approach to be followed and that aspect was not included in the new edition. However, the new Technical Standard IEC TS 63394 [80] indicates how to deal with "mixed" Safety Control Systems that typically have input subsystems in both high and low demand, and **the output subsystems are in high demand** (please refer to § 5.7.6). There is now the idea of including the same approach in the next edition of IEC 62061.

There is a relatively straightforward relationship between the Categories of ISO 13849-1 and the Architectures of IEC 62061, with few differences:

- Categories 3 and 4 are equivalent to the Basic Subsystem Architecture D. However, the MTTF_D for Category 3 is limited to 100 years, the one for Category 4 to 2500 years, while there is no limitation of PFH in IEC 62061 Architecture D.
- Category 1 is equivalent to Architecture A.
- There is no equivalence of Architecture B in ISO 13849-1.
- There is no equivalence of Category B in IEC 62061. The reason is that Category B does not require the use of well-tried Components, while those are mandatory in any safety system without diagnostics in IEC 62061 (Architecture A and B).

9.2 High vs Low-Demand Mode Applications

I hope you now realize the importance and the differences in approach between high and low-demand mode applications.

High-demand mode Safety Standards, meaning ISO 13849-1 and IEC 62061, rely on a high-frequency usage of components, meaning, for example, from once a minute to once a week. Within that range, the mathematical models work very well.

Four considerations:

- **The more a component is used, the lower is its Reliability** since it wears out more quickly (think about the relationship between B_{10D} and MTTF_D). Moreover, it may be necessary to replace it after a few years (think about the meaning of T_{10D}).
- But the more a component is used, think about two pressure switches installed on the same piping, **the better is the diagnostic coverage**.
- **The Diagnostics** of an electromechanical component normally happens when there is a demand upon the safety function.
- An **Electromechanical component**, like a pressure switch, normally have no "intelligence" inside: therefore, the diagnostics depends upon the way it is connected to the control system and how good the control system is to detects faults in the connection.

The accumulation of faults is not really an issue and therefore, in the evaluation of the Reliability of a safety system, we can assume one fault only between two demands upon the safety function.

The theory shows some flaws in case the usage is less than once a month. The mathematical models are then "corrected": here are some examples:

- In ISO 13849-1, except in Category 4, the $MTTF_D$ of a sub-function is limited to 100 years.
- Safety functions in **PL e** must be checked once every month or
- In case of a redundant (digital) input signal (for example, two pressure switches on the same piping), the Diagnostic Coverage decreases jointly with the lower demand and testing of the safety Function.

Therefore, if in your machine there are sensors (interlocking devices or digital sensors) or final elements (contactors) that, according to the way the machine is normally used, trigger less than once a month, you may ask the user to open the gate twice a month, for example. In case of a high-pressure safety function, you may ask for the dismantling of the sensor and its test, for example, once a year. The period is just an indication: it depends upon a few factors, like in which environment they are installed and the level of Reliability you need to achieve.

Once again, **in high-demand** mode, we need to have a frequent switch of components because in that way we can automatically check the presence of a detected failures (λ_{DD}).

Low-demand mode theory is quite different.

- First of all, it is based upon electronic components. Take the example of a 4–20 mA flow transmitter. Even if the high threshold happens once every five years, the logic solver receives a continuous status of the health of the component: that means **the Diagnostics is continuous and not only when there is a demand upon the safety function**. Therefore, an **Electronic component** has ways to detect if it is working correctly and, in case of a fault, it has ways to communicate it to the Control System (§ 3.2.4). All that means the **diagnostic coverage** of an electronic component depends mainly on the component itself and less on the way it is connected to the control system.
- The presence of Undetected failures has a high impact upon the Reliability of the safety function. Think about the concept of SFF (§ 3.2). An electronic component in high-demand mode is usually provided with a PFH_D value and, usually, no information is given about the percentage of undetected failures. In low-demand mode that seldom happens and λ_{DU} is one of the most important parameters since it has a direct connection with the PFD (think about its formula [**Equation 1.10.3**]).
- Undetected failures cannot be "ignored": a regular Proof Test has to be done on the component. The more often the test is done, the higher the Safety Function Reliability is.

The theory shows some flaws in case of Electromechanical components like Sensor switches and contactors. Sensors can be replaced, without interrupting the process, but for contactors that is normally an issue. Moreover, **the latter are not normally provided with a Proof Test procedure**.

9.3 The Importance of Risk Assessment

This book is about Risk Reduction in Machinery, achieved through **Functional Safety;** however, we should not forget the number one aim of all is that **people do not get injured when working with machineries or in a process plant**.

To reach that goal you need a thorough risk assessment (machinery) or HAZOP (processes). Without that as a basis, the time spent on functional safety give a suboptimal solution. This is an abstract from [52].

Based on the analysis of 106 accident reports, it was observed that no accidents occurred simply as a result of the designer choosing the incorrect performance levels (PLs) (ISO 13849) or incorrect safety integrity levels (SILs) (IEC 62061) for the safety control system. ***No serious and fatal accidents were caused because the performance level or safety integrity level was too low****. There were 3 reports where the performance levels of the safety control systems were mentioned. It was reported that the control systems were not designed to the required performance levels, based on the level of risk. However, in each case, modification or bypassing of the existing safety control system was also reported.*

A few suggestions on how to perform a Risk Assessment.

9.3.1 Principles of Safety Integration

"Safe" is the state of being protected from recognized hazards that are likely to cause physical harm. There is no such thing as being absolutely safe, that means, a complete absence of risk. Therefore, there is no machine that is absolutely safe. All machinery contain hazards that should be reduced to an acceptable level. The Machinery Directive has a corner stone in the Principles of Safety Integration.

[2006/42/EC] 1.1.2. Principles of safety integration

a) Machinery must be designed and constructed so that it is fitted for its function and can be operated, adjusted, and maintained ***without putting persons at risk*** *when these operations are carried out under the conditions foreseen but also taking into account any reasonably foreseeable misuse thereof.*
The aim of measures taken must be ***to eliminate any risk throughout the foreseeable lifetime of the machinery,*** *including the phases of transport, assembly, dismantling, disabling, and scrapping.*
b) In selecting the most appropriate methods, the manufacturer or his authorized representative must apply the following principles, in the order given:
- ***eliminate or reduce risks*** *as far as possible (inherently safe machinery design and construction),*
- *take the necessary* ***protective measures*** *in relation to risks that cannot be eliminated,*
- ***inform users of the residual risks*** *due to any shortcomings of the protective measures adopted, indicate whether any particular training is required, and specify any need to provide personal protective equipment.*

The second part is called the **3-step method** of risk reduction, and it is described in both ISO 12100 [41] and in B11.0 [45].

The first priority in Risk Reduction is given to inherently safe design measures, because they are the most effective. Some **examples of inherently safe design measures**:

- eliminating the hazard altogether, for example, replacing flammable hydraulic fluid with a non-flammable type, removing risk of falls by having maintenance points easily accessible at ground level rather than at height;
- ensuring the inherent stability of machinery by its shape and the distribution of masses;
- ensuring that accessible parts of the machinery do not have sharp edges or rough surfaces;
- reducing, where possible, the speed and the power of moving parts or the travel speed of the machinery itself;
- locating hazardous parts of machinery in inaccessible places.

When it is not possible to eliminate hazards or sufficiently reduce risk by inherently safe design measures, **the second priority** is given to technical protective measures that prevent people from being exposed to hazards. Some examples of technical protective measures are fixed or interlocking movable guards, light barriers, laser scans, and safety radars.

Finally, for the risks that cannot be adequately reduced neither by inherently safe design measures nor by technical protective measures (**Engineering Controls** in North American language), information must be given to exposed persons in the form, for example, of warnings, signs, and information on the machinery (**Administrative Controls** in North American language). **Examples are** Awareness Means, Information for Use, Supervision, Control of Hazardous Energy (Lockout Tagout), and Personal Protective Equipment.

9.3.1.1 The Glass Dome

What just described is the correct way of doing a risk assessment: that implies to start when the machine is conceived and not when it is nearly finished. In the latter approach, the simple solution is to install guards, fixed or interlocked, to keep the user away from dangerous movements. The issue is that, when approaching safety at the end of the development process, the user may have difficulty in using the machine, and **he is prone to defeat safeguards**. As manufacturers, we should avoid that situation, for example, **implementing appropriate operating modes** that allow the user to operate the machine, even with some safeguards activated. Appropriate operating modes can be **special modes for setting, tool changing, fault finding, maintenance, or process observation**. They depend highly on the type of machine and its application.

The bottom line is that, by following the 3-step method, you should be able to avoid installing what I call a **Glass Dome** on the machine: a Glass Dome makes the machine safe but unusable! Remember that the **best Safeguard is the one the operator does not realize is there**.

9.3.2 How to Run a Risk Assessment

You may now wonder how to do, in practice, a risk assessment. We already mentioned **the importance of a Multidisciplinary team** (§ 4.6). We also mentioned that the machine has to be imagined as **Naked** (§ 4.1.4). Another important aspect is that **the analysis has to be sincere and genuine**: people must speak freely without being afraid to be judged.

But when can I consider the machinery safe? In GT Engineering, we use the following criteria: "the machinery is safe when you would be comfortable in having your son or your daughter working on it." Do the risk assessment sincerely, thinking that your offspring will work on it. When you are comfortable with that idea, the machine you are designing can be considered safe.

Poncarale (BS), Italy *Marco Tacchini*
11 November 2022

Bibliography

1 http://www.electropedia.org/
2 EN 954-1: 1996. Safety of machinery – safety-related parts of control systems – general principles for design.
3 IEC 60204-1: 2016. Safety of machinery – electrical equipment of machines part 1: general requirements.
4 IEC 61508-0: 2010. Functional safety of electrical/electronic/programmable electronic safety-related systems – Part 0: Functional safety and IEC 61508.
5 IEC 61508-1: 2010. Functional safety of electrical/electronic/programmable electronic safety-related systems – Part 1: General requirements.
6 IEC 61508-2: 2010. Functional safety of electrical/electronic/programmable electronic safety-related systems – Part 2: Requirements for electrical/electronic/programmable electronic safety-related systems
7 IEC 61508-3: 2010. Functional safety of electrical/electronic/programmable electronic safety-related systems – Part 3: Software requirements.
8 IEC 61508-4: 2010. Functional safety of electrical/electronic/programmable electronic safety-related systems – Part 4: Definitions and abbreviations.
9 IEC 61508-5: 2010. Functional safety of electrical/electronic/programmable electronic safety-related systems – Part 5: Examples of methods for the determination of safety integrity levels.
10 IEC 61508-6: 2010. Functional safety of electrical/electronic/programmable electronic safety-related systems – Part 6: Guidelines on the application of IEC 61508-2 and IEC 61508-3.
11 IEC 61508-7: 2010. Functional safety of electrical/electronic/programmable electronic safety-related systems – Part 7: Overview of techniques and measures.
12 IEC 62061: 2021. Safety of machinery – functional safety of safety-related control systems.
13 FDIS ISO 13849-1: 2022. Safety of machinery – safety-related parts of control systems – part 1: general principles for design.
14 ISO 13849-2: 2012. Safety of machinery – safety-related parts of control systems – part 2: validation.
15 IEC 62061: 2005. Safety of machinery – functional safety of safety-related electrical, electronic and programmable electronic control systems.
16 IEC 61511-1: 2016. Functional safety – safety instrumented systems for the process industry sector – part 1: framework, definitions, system, hardware and application programming requirements.
17 IEC 61511-2: 2016. Functional safety – safety instrumented systems for the process industry sector – part 2: guidelines for the application of IEC 61511-1:2016.
18 IEC 61511-3: 2016. Functional safety – safety instrumented systems for the process industry sector – part 3: guidance for the determination of the required safety integrity levels.

Functional Safety of Machinery: How to Apply ISO 13849-1 and IEC 62061, First Edition. Marco Tacchini.

19 IEC 61131-3: 2013. Programmable controllers – part 3: programming languages.
20 EN 764-7: 2019. Pressure equipment – part 7: safety systems for unfired pressure equipment.
21 ISO 19973-1: Pneumatic fluid power – assessment of component Reliability by testing – part 1: general procedures.
22 IEC 61800-5-2: 2017. Adjustable speed electrical power drive systems. Part 5-2: safety requirements – functional.
23 ISO 11161: 2022. CD 2 edition. Safety of machinery – integration of machinery into a system – basic requirements.
24 ISO 12895: 2023. Safety of machinery – Identification of whole body access and prevention of associated risks.
25 ISO 19973-1 and 2: Pneumatic fluid power – assessment of component reliability by testing.
26 IEC 61078: 2016. Reliability block diagrams.
27 IEC 60812: 2018. Failure modes and effects analysis (FMEA and FMECA).
28 IEC 61649: 2008. Weibull analysis.
29 ISO/TR 12489: 2013. Petroleum, petrochemical and natural gas industries – reliability modelling and calculation of safety systems.
30 ISO 14119: 2022. DIS 2 Edition – safety of machinery – interlocking devices associated with guards principles for design and selection.
31 IEC 62046: 2018. Safety of machinery – application of protective equipment to detect the presence of persons.
32 IEC 60947-5-1: 2016/COR2:2020. Corrigendum 2 – low-voltage switchgear and controlgear – part 5-1: control circuit devices and switching elements – electromechanical control circuit devices.
33 IEC 60947-5-3: 2013. Low-voltage switchgear and controlgear – part 5-3: control circuit devices and switching elements – requirements for proximity devices with defined behaviour under fault conditions (PDDB).
34 IEC 60947-4-1: 2018. Low-voltage switchgear and controlgear – part 4-1: contactors and motor-starters – electromechanical contactors and motor-starters.
35 IEC 61810-3: 2015. Electromechanical elementary relays – part 3: relays with forcibly guided (mechanically linked) contacts.
36 IEC 60730-2-5: 2013. Automatic electrical controls – part 2-5: particular requirements for automatic electrical burner control systems.
37 IEC 60947-5-8: 2020. Low-voltage switchgear and controlgear – part 5-8: control circuit devices and switching elements – three-position enabling switches.
38 ISO 13851: 2019. Safety of machinery – two-hand control devices – principles for design and selection.
39 IEC 61131-2: 2017. Industrial-process measurement and control – programmable controllers – part 2: equipment requirements and tests.
40 ISO TR 22100-4: 2018. Safety of machinery – relationship with ISO 12100 – part 4: guidance to machinery manufacturers for consideration of related IT-security (cyber security) aspects.
41 ISO 12100: 2010. Safety of machinery – general principles for design – risk assessment and risk reduction.
42 ISO 13850: 2015. Safety of machinery – emergency stop function – principles for design.
43 IEC 61882: 2016. Hazard and operability studies (HAZOP studies) – application guide.
44 NFPA 79: 2021. Electrical standard for industrial machinery.
45 ANSI B11.0-2020. Safety of machinery.
46 ANSI B11.19-2019. Performance requirements for risk reduction measures: safeguarding and other means of reducing risk.

47 ANSI B11.26-2018. Machines – functional safety for equipment: general principles for the design of safety control systems using ISO 13849-1.
48 Z432-04 (R2014). Safeguarding of machinery.
49 European Commission – The 'Blue Guide' on the implementation of EU product rules 2022.
50 Hauptmanns, U. (2015). *Process and Plant Safety*. Springer.
51 Lazzaroni, M., Cristaldi, L., Peretto, L. et al. (2011). *Reliability Engineering, Basic Concepts and Applications in ICT*. Springer.
52 Chinniah, Y. (2015). Analysis and prevention of serious and fatal accidents related to moving parts of machinery. *Safety Science* 75: 163–173.
53 IFA Report 2/2017e – Functional safety of machine controls – application of EN ISO 13849.
54 IFA Report 4/2018e – Safe drive controls with frequency inverters.
55 Safety of Machinery: Significant Differences in Two Widely Used International Standards for the Design of Safety-related control Systems. Yuvin Chinniah and others.
56 Martorell, S., Soares, C.G., and Barnett, J. (2014). Safety, Reliability and Risk Analysis: Theory, Methods and Applications, 3e, vol. 1. CRC Press.
57 Rausand, M. (2014). *Reliability of Safety-Critical Systems - Theory and Applications*. Wiley.
58 Bell, R. – Safety critical systems – a brief history of the development of guidelines and standards.
59 Wikipedia. The Institute for Occupational Safety and Health of the German Social Accident Insurance (German: Institut für Arbeitsschutz der Deutschen Gesetzlichen Unfallversicherung, IFA) is a German institute located in Sankt Augustin near Bonn and is a main department of the German Social Accident Insurance. Belonging to the Statutory Accident Insurance means that IFA is a non-profit institution.
60 Software Systema: by IFA, Germany.
61 VERTICAL RECOMMENDATION FOR USE SHEETS (RfUs) – Status in November 2018 – https://eurogip.fr/wp-content/uploads/2019/12/vertical-rfu_en-November2018.pdf.
62 Eisinger, S., Oliveira, L.F., Tveit, K. et al. (2015). Safety instrumented systems operated in the intermediate demand mode. University of Oslo, Norway, Oslo, Norway. In: *Safety and Reliability of Complex Engineered Systems* (ed. P. Llini et al.). London: Taylor & Francis Group ISBN 978-1-138-02879-1.
63 Proof Testing... A key performance indicator for designers and end users of safety instrumented systems David Green and Ron Bell 2016.
64 Appendix 3 Partial Proof Testing Example – HSE.
65 Smith, D.J. and Simpson, K.G.L. (2016). *Safety Critical Systems Handbook. A Straightforward Guide to Functional Safety, IEC 61508 (2010 Edition) and Related Standards, Including Process IEC 61511 and Machinery IEC 62061 and ISO 13849*, 4e. Elsevier Ltd.
66 Reducing risks, protecting people. HSE's decision-making process – 2001.
67 Goble, W.M. and Cheddie, H. (2005). *Safety Instrumented Systems Verification: Practical Probabilistic Calculations*. ISA.
68 Lundteigen, M.A. (2008). *Safety Instrumented Systems in the Oil and Gas Industry*. NTNU.
69 SERH (2015). *Safety Equipment Reliability Handbook*, 4e, vol. 1,2,3. Exida.
70 Kirkcaldy, K.J. – Exercises in functional safety.
71 Guidelines for Initiating Events and Independent Protection Layers in Layer of Protection Analysis – CCPS (Center for Chemical Process Safety). ISBN: 978-0-470-34385-2 – February 2015
72 IFA: Institut für Arbeitsschutz der Deutschen Gesetzlichen Unfallversicherung. www.dguv.de.
73 Brissaud, F. (2017). Using field feedback to estimate failure rates of safety-related systems. *Reliability Engineering and System Safety* 159: 206–213.
74 https://www.gt-engineering.it/en/Insights/the-emergency-stop-function

75 The way through the standard - Questions and answers on EN ISO 14119:2013 – Euchner publication (https://assets.euchner.de/uploads/124892_05-08-19_Flyer-EN-ISO-14119.en-us.pdf).

76 Yoshimura, I. and Sato, Y. (2008). Safety achieved by the safe failure fraction (SFF) in IEC 61508. *IEEE Transactions on Reliability* 57 (4).

77 Lundteigen, M.A. and Rausand, M. (2009). Architectural constraints in IEC 61508: do they have the intended effect? *Reliability Engineering and System Safety* 94: 520–525.

78 IEC 60947-1 Annex K: Procedure to determine reliability data for electromechanical devices used in functional safety applications.

79 ANSI/RIA R15.06-1999 (R2009) – Industrial robots and robot systems – safety requirements.

80 IEC TS 63394: 2023. Safety of machinery – guidelines on functional safety of safety-related control systems.

81 HSE (2003) Out of control: why control systems go wrong and how to prevent failure.

Index

Note: Page numbers followed with *f* and *t* refer to figures and tables

Functional Safety of Machinery: How to Apply ISO 13849-1 and IEC 62061, First Edition. Marco Tacchini.

S

t

u

v

Printed and bound by CPI Group (UK) Ltd, Croydon, CR0 4YY

16/04/2025

14658590-0002